Lecture Notes in Computer Science 6821

Commenced Publication in 1973
Founding and Former Series Editors:
Gerhard Goos, Juris Hartmanis, and Jan van Leeuwen

W0235112

Karin Anna Hummel   Helmut Hlavacs
Wilfried Gansterer (Eds.)

Performance Evaluation of Computer and Communication Systems

Milestones and Future Challenges

IFIP WG 6.3/7.3 International Workshop, PERFORM 2010
in Honor of Günter Haring
on the Occasion of His Emeritus Celebration
Vienna, Austria, October 14-16, 2010, Revised Selected Papers

 Springer

Volume Editors

Karin Anna Hummel
Helmut Hlavacs
University of Vienna
Research Group Entertainment Computing
Lenaugasse 2/8, 1080 Vienna, Austria
E-mail: {karin.hummel, helmut.hlavacs}@univie.ac.at

Wilfried Gansterer
University of Vienna
Research Group Theory and Applications of Algorithms
Lenaugasse 2/8, 1080 Vienna, Austria
E-mail: wilfried.gansterer@univie.ac.at

ISSN 0302-9743   e-ISSN 1611-3349
ISBN 978-3-642-25574-8   ISBN 978-3-642-25575-5 (eBook)
DOI 10.1007/978-3-642-25575-5
Springer Heidelberg Dordrecht London New York

Library of Congress Control Number: 2011943015

CR Subject Classification (1998): H.4, C.2, D.2, H.3, I.2, H.5

LNCS Sublibrary: SL 1 – Theoretical Computer Science and General Issues

Typesetting: Camera-ready by author, data conversion by Scientific Publishing Services, Chennai, India

Printed on acid-free paper

Springer is part of Springer Science+Business Media (www.springer.com)

Preface

Professor Emeritus Dr. Günter Haring.

Günter Haring has dedicated most of his scientific professional life to performance evaluation and the design of distributed systems, contributing in particular to the field of *workload characterization*. To honor Günter Haring on the occasion of his emeritus celebration, some of the key researchers in the field of performance evaluation were invited to reflect on *Performance Evaluation of Computer and Communication Systems – Milestones and Future Challenges* at the PERFORM 2010 workshop, held during October 14–16, in Vienna, Austria.

PERFORM 2010 aimed at bringing together renowned experts and world leaders in the field of networked systems and performance evaluation not only to review historical milestones, but also to discuss their impact on current and future developments as well as to identify novel, inspiring, and visionary concepts in performance evaluation and future complex networked systems. The reflection on milestones and impacts is particularly timely when thinking about novel emerging technologies such as the Internet of Things, heterogeneous wireless network infrastructures, and socio-technical distributed systems. The contributions presented at PERFORM 2010 and collected in this book demonstrate the strong history of, but also new research directions in, performance evaluation.

The contributions of Günter Haring himself to the field of performance evaluation and distributed systems are manifold as reflected by over 150 publications. His structured way applied to workload characterization led to well-known approaches on how to hierarchically decompose workload by a multi-layer approach, on how to introduce task level descriptions, and on how to apply Markov models to describe the properties of task sequences. In addition to his own contributions and leadership in international research projects, he is and has been an excellent mentor of young researchers demonstrated by their own brilliant scientific careers. It is most admirable that Günter Haring has not only concentrated on his own research, but has also promoted computer science as a pioneer in Austria by – to mention only a few of his achievements – taking the responsibility of being the president of the Austrian Computer Society (1989–1993), being a founding member of the Austrian Center of Parallel Computing (ACPC), and being the founder and first dean of the Faculty for Computer Science at the University of Vienna (2004–2009).

Upon our invitation to contribute to PERFORM 2010, we were glad to accept 20 papers ranging from visionary to in-depth research articles. To assure high quality, the papers were reviewed by a minimum of two referees of the international Technical Program Committee. Upon the recommendation of the referees we introduced the process of supervised adaptation to four papers (shepherding). The strong technical program of PERFORM 2010 is reflected in the sections of this book.

By focusing on "Milestones and Evolutions," Raymond A. Marie opens the discussion about lessons learned in the past of performance evaluation as seen from an expert who contributed, for example, with methods for general queuing networks in the late 1970s. Connie Smith, co-author of a fundamental book on software performance engineering and well-known expert in the area, focuses together with Catalina M. Lladò on model interoperability. From a historical and evolutionary perspective, Giuseppe Serazzi, an expert in workload characterization, bottleneck detection in very large multi-class models, and tools for analyzing the performance of complex systems, and his co-authors Giuliano Casale and Marco Gribaudo give a summary of performance evaluation tools.

Novel challenges for performance evaluation are introduced by the contributions to the section "Trends: Green ICT and Virtual Machines." Jean-Marc Pierson, one of the pioneering European researchers investigating energy efficiency in distributed systems, highlights the importance of including energy as a new criterion to performance evaluation and proposes ways to approach modeling of energy efficiency. Ramon Puigjaner, who contributed with his expertise in performance evaluation in various application domains such as, for example, by the successful sizing of the central computer and communication network during the Olympic games in Barcelona, 1992, and his colleague Carlos Juiz, an expert in performance analysis of Web-based systems, draw the connection from established methods of performance evaluation to "green ICT" demands. Another trend of networked systems, virtualization, is targeted by the predictive scheduling approach of Robert Geist, an expert in performance evaluation of disk scheduling and perception-based measures, and his co-authors Zachary H. Jones, and James Westall.

As "Modeling" is a key aspect in performance evaluation, five profound contributions focus on modeling details. Hermann de Meer, co-author of one of the fundamental books on queuing theory, and his colleagues Patrick Wüchner and Jànos Sztrik introduce finite-source retrial queues for modeling Wireless Sensor Networks. Markov chains and spectral clustering is the topic of the contribution of William J. Stewart, an expert in numerical solution of Markov chains and author of two books on Markov chains, and his co-author Ning Liu. Demetres D. Kouvatsos, who contributed to the field of performance evaluation with results in, for example, entropy maximization, queuing network models, and performance engineering applications, and his co-author Salam A. Assis focus on the analysis of heavy-tailed queues. Hidden Markov models and their use in performance evaluation are discussed by Edmundo de Souza e Silva, Rosa M.M. Leão, and Richard R. Muntz. In this article, the expertise of Edmundo de Souza e Silva, who developed fundamental solution techniques for Markov models, and Rosa M.M. Leão is brought together with the expertise of Richard R. Muntz, who developed pioneering contributions to the theory of queuing networks. Ioannis Stavrakakis, a well-known expert in network analysis research who contributed to various domains of computer networks including recently delay-tolerant networks, proposes the exploitation of linear properties of infinite dimensional linear equations for network protocol performance analysis.

"Mobility and Mobile Networks" are topics of ever-increasing interest as many novel networked services are intended for mobile use. Marco Conti, Andrea Passarella, and Chiara Boldrini, experts in the new field of research on social networks and opportunistic computing, introduce a novel modeling approach for socially aware forwarding schemes. Using mobility information, Vicente Casares-Giner proposes a general formulation of lookahead strategies for location updates in his article which reflects his expertise in applying and using mobility modeling in wireless networking. Concentrating on their expertise in reliability analysis of cellular networks, Fabio Ricciato, Peter Romirer-Maierhofer, and Angelo Coluccia propose a Bayesian estimation of mean failure probabilities in 3G networks. Markus Fiedler, an expert in Quality of Experience and teletraffic modeling, and his co-authors Patrik Arlos, Timothy A. Gonsalves, Anuraag Bhardwaj, and Hans Nottehed detail Web response times in mobile networks.

In the field of general "Communication and Computer Networks," two contributions focus on different aspects and different types of networks. Michal Pióro, who co-authored a widely recognized book on network traffic and network protocol decisions and whose research contributions include traffic modeling, analysis and optimization of communication networks, and his co-authors Walid Ben-Ameur and Pablo Pavon-Marino detail traffic domination in communication networks.

Monique Becker, an expert in evaluating the performance of aggregation techniques in computer networks, and her colleagues Ashish Gupta, Michel Marot, and Harmeet Singh present a summary of their works on clustering in wireless sensor networks.

Finally, "Load Balancing, Analysis, and Management" approaches are presented. Gabriele Kotsis, an expert in workload characterization in parallel and distributed systems, and her colleague Martin Pinzger use analysis insights to manage Web performance in a proactive way. Maria Carla Calzarossa, an expert in workload characterization and benchmarking of complex systems and services, and her co-author Luisa Massari focus on the analysis of Web logs. John C.S. Liu, who contributed to the field of performance analysis by stochastic analysis of computer storage and peer-to-peer systems, and his co-authors Guanlin Lin and Yang Wang present work on a matrix-analytic solution to randomized load balancing.

The technical program of PERFORM 2010 was completed by three additional talks given by Martin Reiser, Alois Ferscha, and Kurt Tutschku. Martin Reiser, inventor of mean value analysis of queuing networks, gave a lively talk entitled "Mean Value Analysis – A Personal Account." Alois Ferscha, who was among the pioneers of proposing a structured way of performance analysis of parallel simulations and, at present, contributes to the field of Pervasive Computing, gave an inspiring talk about "Scenario-Based Modeling for Very Large Scale Simulations." Kurt Tutschku discussed his new concepts and contributions to the challenging field of "Performance Requirements for Future Virtualized and Federated Networks."

Our thanks go to all authors and speakers, and further to the referees for their support in reviewing in spite of busy schedules and in particular to our shepherds for their tireless mentoring support. Many thanks go to Gerry Schneider and his team at the University of Vienna for supporting the event management. We are especially thankful for the support of the organizing team: Andrea Hess for producing the layout and printed content of the workshop program and Shelley Buchinger for organizing the marvelous wine-tasting as one of the social events of the workshop. Many thanks also go to Ewald Hotop and Rudolf Hürner for technical support and producing a nice photo gallery. For precise technical editing support of this book, special thanks go to Andrea Hess. And, finally, we want to thank Günter Haring, not only for always being an encouraging mentor but also for giving us the opportunity to meet his exceptional colleagues and friends at this scientific emeritus celebration.

Thank you and cordial congratulations, Günter Haring!

October 2010

Karin Anna Hummel
Helmut Hlavacs
Wilfried Gansterer

Organization

PERFORM 2010, the Workshop on Performance Evaluation of Computer and Communication Systems – Milestones and Future Challenges, was organized by the Distributed Systems Group, University of Vienna, October 14–16, 2010.

General Chair

Günter Haring   University of Vienna, Austria

Program Chairs

Karin Anna Hummel	University of Vienna, Austria
Helmut Hlavacs	University of Vienna, Austria
Wilfried Gansterer	University of Vienna, Austria

Technical Program Committee

Heinz Beilner	University of Dortmund, Germany
Sem Borst	Bell Labs, Lucent Technologies, USA
Shelley Buchinger	University of Vienna, Austria
Ed G. Coffman	Columbia University, USA
Georges Da Costa	IRIT/Toulouse III, France
Lawrence W. Dowdy	Vanderbilt University, USA
David Erman	Blekinge Institute of Technology, Sweden
Domenico Ferrari	Universita Cattolica del Sacro Cuore, Piacenza, Italy
Markus Fidler	Leibniz University of Hannover, Germany
Claude Girault	Universitè Paris VI, France
Leana Golubchik	University of Southern California, USA
Peter Harrison	Imperial College London, UK
Carlos Juiz	University of the Balearic Islands, Spain
Hisashi Kobayashi	Princeton University, USA
Demetres D. Kouvatsos	University of Bradford, UK
Laurent Lefevre	INRIA/University of Lyon, France
John Lui	Chinese University of Hong Kong, Hong Kong
Richard Muntz	University of California, Los Angeles, USA
Peter Reichl	Forschungszentrum Telekommunikation Wien, Austria
Gerard Reijns	Delft University of Technology, The Netherlands

Hans-Peter Schwefel	Forschungszentrum Telekommunikation Wien, Austria
Don Towsley	University of Massachusetts, USA
Phuoc Tran-Gia	University of Würzburg, Germany
Satish Tripathi	University of Maryland, USA
Hans-Jürgen Zepernick	Blekinge Institute of Technology, Sweden

Local Organizing Committee

Wilfried Gansterer	University of Vienna, Austria
Helmut Hlavacs	University of Vienna, Austria
Karin Anna Hummel	University of Vienna, Austria
Shelley Buchinger	University of Vienna, Austria
Andrea Hess	University of Vienna, Austria

Sponsors and Technical Sponsors

University of Vienna
IFIP WGs 6.3 and 7.3
Euro-NF

Table of Contents

Mobility and Mobile Networks

Communication and Computer Networks

Load Balancing, Analysis, and Management

Disappointments and Delights, Fears and Hopes Induced by a Few Decades in Performance Evaluation

Raymond A. Marie

IRISA, Rennes 1 University,UEB, Campus de Beaulieu,
35042 Rennes Cedex, France
marie@irisa.fr

Abstract. The elements of modelling in general and of performance evaluation of discrete event systems (DES) in particular have undergone a tremendous transformation during these last four decades. The aim of this paper is to look back over all this evolution, trying to retain some particular experiences from the past. I will try to classify these elements according to what I have perceived as their positive or negative potentialities. All the views expressed are my own and entirely subjective. Nothing will be proven since there will be no theorems. We first enumerate a list of events or situations which have occurred during these four decades and which I regard as positive. An opposite set of negative arguments will follow. Then, I will enumerate a list of risks that, from my personal perception, represent the dangers for the domain of modelling and of performance evaluation of systems in the field of computers and telecommunications. Finally, some suggestions to preserve the quality of the expertise of the community will be proposed.

1 Introduction

During these last four decades, the elements of modelling in general, and of performance evaluation of discrete event systems (DES) in particular, have gone through a tremendous transformation. The special event motivating this meeting provides an occasion to look back at this evolution, trying to retain some particular experiences from the past. I will try to classify these elements according to what I have perceived as their positive or negative potentialities, inevitably from a personal and subjective standpoint. Nothing will be proven since there will be no theorems presented. In the following section, I will enumerate a list of events or situations that occurred during these four decades and that I consider to have been positive. On the other hand, over the same period, there have been some developments which I consider negative as discussed in Section 3. I will enumerate in Section 4 a list of risks which represent dangers for the domain of modelling and of performance evaluation of systems in the field of computers and telecommunications. Finally, I will suggest in Section 5 some ideas to preserve the quality of the expertise of the community.

K.A. Hummel et al. (Eds.): PERFORM 2010 (Haring Festschrift), LNCS 6821, pp. 1–9, 2011.
© IFIP International Federation for Information Processing 2011

2   Delights

In the beginning, our scientific ambition was limited by computing power. We were using our imagination to look for approximations in order to reduce a state space to a few hundred states ; or even less if the model was used as a submodel! Often, we were using all the central memory resource of the computer and that required us to be the single user of the main frame ; this type of privilege was only given during night hours (after midnight !).

We were excited by the novelty of the discipline, combining informatics and telecommunications. From time to time, we were lucky enough to get success stories with simple models. For example, the M/M/1 queue with *processor-sharing* discipline was surprisingly good at representing the congestion phenomena on a main frame. In fact, it was realized later that with the processor-sharing discipline the steady state distribution is invariant with respect to the service time distribution; note also the time sharing policy which was used to execute the list of jobs on a main frame (a rare resource at that time !) has the processor-sharing discipline as asymptotic behavior. This was also the time when a small product form queuing network was able to capture the main factors of a computer room covering many hundreds of square-feet in order to predict the response time with reasonable accuracy.

As already mentioned, we were (almost) all young and this situation gave us the opportunity to take on national and international responsibilities before the average age encountered in other scientific communities. We set-up few, but high quality, international conferences which gave us the opportunity to exchange ideas at a time where the Internet did not exist and where postal mail was taking weeks to arrive from the other side of the world. Throughout the following periods, step by step, use of new concepts (e.g., timed Petri nets, neural networks) also stimulated our research activities.

3   Disappointments

On the other hand, we can observe now that huge amounts of available computing resources increase the trend to solve models through simulation and do not encourage researchers to look for tractable analytical solutions.

Sometimes, we see young researchers "reinventing" new methods that were introduced 20 years before just because they often do not look at publications more than 10 years old, especially if they are working on a new application domain such as telecommunications.

The architecture designers do not always take the performance evaluation people seriously. A pessimist could see there the consequence of an eventual competition. The designers are convinced that they know what they are doing ; as a consequence, they sometimes prefer to increase the number of resource units when the expected performances are not attained rather than looking for other solutions.

From an academic point of view, a general trend is that a certain number of courses, very useful for our scientific field, have been withdrawn from the standard curriculum of studies in computer sciences. I am thinking of linear algebra, probabilities, stochastic processes such as Markov processes and linear or non-linear optimization. Note that sometimes you find the theory exposed in the particular context induced by the topic of the course (e.g., dynamic programming introduced in the context of graph theory) and taught this way, the theory loses all its generality. This has a negative impact on the learning process of the student.

Actually, the number of international conferences which are organized per year seems to be continuously increasing. Is it because this is good for Science or is it because each researcher wants to write on his curriculum vitae that he has been general chair of some international conference/symposium/workshop ? Therefore the question becomes : when does a lot become too much ?

4 Fears

Let us now try to present a list of different dangers that are, as I see it, threatening the quality of the work of a modeler of DES.

1. Bad comprehension of the system (this applies to both simulation and analytical approaches) Let us consider the following example ; a processor has to execute two types of job according to a preemptive priority discipline. In order to simplify the example, let us suppose that the service times are exponentially distributed with respective means $1/\mu_1$ and $1/\mu_2$ for class one and class two.

   Let λ_1 and λ_2 denote the respective arrival rates of the two Poisson processes. Let us consider the following numerical values ; $\lambda_1 = 1$, $\mu_1 = 100$, $\lambda_2 = 0.01$ and $\mu_2 = 1$. Then, if the modeling person sees the processor discipline as a preemptive priority repeat different discipline, this person will predict a busy rate of 2 percent for the processor. While, if the real behavior of the system corresponds to a preemptive priority repeat identical discipline, the processor will not be fast enough to allow the treatment of the amount of processing work (since the stability condition, $\lambda_1/\mu_2 < 1$, is not satisfied, the system will blow up). In order to illustrate this example by a figure, let us introduce three complementary notations. Let $\mathbb{E}[T]$ denote the expectation of the total time needed by the server in order to serve a class two customer. Let $\mathbb{E}[B]$ denote the expectation of the service time of a class two customer. Finally let α denote the expectation of the number of preemptions during the service of a class two customer. Figure 1 shows the ratio $\mathbb{E}[T]/\mathbb{E}[B]$ as a function of α. This ratio stays constant if the discipline is a *preemptive priority repeat different* one (because of the memoryless property of the exponential service time distribution). But, if the discipline is a *preemptive priority repeat identical* one, this ratio tends to infinity as soon as α tends to one.

   This second situation may arise if the jobs of the preempted class correspond to executions of different files. Even if a set of execution times corresponding to the different files of class two (each file having a given execution time) can be seen globally as fitting an exponential distribution, once the execution of a particular file is preempted, its successful execution will need a constant time corresponding to the execution of a constant number of bytes. Therefore the modeling person may mix the two cases corresponding to different situations, especially if the service time is exponentially distributed (keeping in mind the memoryless property of the exponential distribution). Additionally we plotted on the figure the ratio for the special case of the constant service time to show a rare case where a randomness behavior (mixed line) looks better than a non-randomness behavior (dashed line). Note that it can be proven that the two priority disciplines give the same result in the special situation where the service time is constant. The scientific derivation of the functions corresponding to these curves can be found in [1].

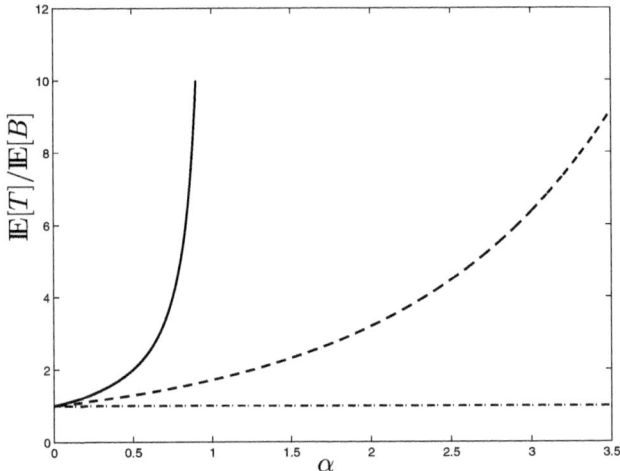

Fig. 1. Ratio $\mathbb{E}[T]/\mathbb{E}[B]$ as a function of α for the *preemptive priority repeat identical* discipline case (solid line), the *preemptive priority repeat different* discipline case (mixed line), and the special case of a constant service time (dashed line)

2. Bad mastery of approximations. There are different categories of approximations ;
    – On the one hand we have approximations at the level of the modeling step.
        For example, we assume that the service time is exponentially distributed while this is not true in reality. This is a classic approximation which is done consciously. The consequence of such an approximation depends generally both on the modeled context and on the performance parameters.
        Another frequent approximation that is sometimes done unconsciously corresponds to the fact of considering that two events are independent

while they are not. This latter approximation is often dangerous because the influence of non independence is generally underestimated. Some examples taken in the context of dependability are convincing. We can exhibit relative errors on the unavailability of several thousand per cent. We can exhibit cases with an unbounded relative error limit on unavailability when time tends to zero.

In order to illustrate this last assertion, let us consider the case of a complex architecture in which there exist multiple copies of a single element type. We are concerned by the probability that, at the end of a mission time T, the system is not available. We assume that the reliability function of each element of the system is exponentially distributed. If we assume that any breakdown of an element is independent of the other breakdowns, the reliability of the global system can be (more or less easily) exactly determined. Things change if, in order to increase the availability of the system, some extra spares are put on the shelf at the start of the mission by the people in charge of the system. In such a situation, the determination of the unavailability of the system at time T becomes more challenging, even if we disregard the exchange time. In order to give numerical data, let us consider the simplest example of a two element redundancy associated to one spare element illustrated on figure 2. If the spare did not exist, the unavailability of this two element system would be given by $(1 - e^{\lambda T})^2$ where λ denotes the failure rate of one element.

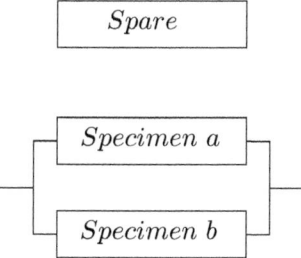

Fig. 2. A minimal system

Using a Markovian model that disregards the exchange time, it is possible to compute exactly the probability $\mathbb{P}(A)$ (resp. $\mathbb{P}(B)$) that the element a (resp. b) is down at time T by lack of spare. We get $\mathbb{P}(A) = \mathbb{P}(B) = (1 - e^{\lambda T})^2$. Having these probabilities, a naive approach would be to consider that the unavailability of this two element system is $(\mathbb{P}(A))^2$. This is the kind of approach that would be easily adopted in the general case where the different elements of the same type could be at quite different locations on the reliability diagram of the global system. This approach assumes independence between the different elements would consider a reliability $(1 - \mathbb{P}(A))$ for each of the different elements of the same type. While, for the two element redundant system, it is easy to find that the exact unavailability at time T equals $1 - e^{-\lambda T}[4 - e^{-\lambda T}(3 + 2\lambda T)]$. Figure 3 shows the ratio of the exact unavailability divided by the naive answer (i.e., $\mathbb{P}(A)^2$) as a function of the product λT. On this figure, we can see first that for $\lambda T = 10^{-3}$, the exact unavailability is approximately 600 times larger than the one obtained when we ignore the dependence between the two elements, this dependence being introduced by the existence of the spare element. Secondly, as suggested by the curve, it can be proven that the ratio tends to infinity when λT tends to zero. This means that the more reliable the system is, the more the unavailability is underestimated.

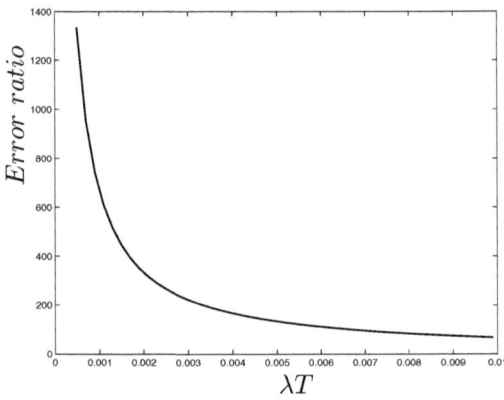

Fig. 3. Unavailability error ratio as a function of λT

- On the other hand, we have approximations at the level of the resolution step. Whatever we are using - a direct or an iterative approximated method, it is important to know if upper and/or lower error bounds have been exhibited. For iterative approximated methods, it is also important to know if the convergence of the method has been proved ; or if nobody has proved the convergence of the method (but nobody so far has obtained a non convergent case study).

3. Use of numerically unstable algorithms. We all know that the main reason for losing accuracy is the execution of a difference of two smaller and smaller positive numbers. Such a situation arises frequently in our domain. For example we encounter it quite often when we look for the original of the generating function of a probability distribution. This is why it is always profitable to try to find an algorithm adding only positive numbers (in addition to the use of the product and division operators). In the special but important case of the use of simulation, let us say that in one way, simulating a process on a finite time interval is making an approximation ; and that too

many, while simulating, forget to give this approximation by means of the confidence intervals...

4. Inadequate use of Markovian models (when is it dangerous to use Markovian models ?). The main grievance done to the use of Markovian models is that in real systems, the time durations of activities are not really exponentially distributed; although it is known that phase-type distributions can reproduce as close as necessary the non exponential distributions of the real system. However, in general, people expressing such grievances do not master the use of fictitious states and it is true that such a procedure has a significant cost, since it can drastically increase the cardinality of the state space. Otherwise there is the possibility of searching for the solution of a semi-Markovian model, but this is not always an easy task. That is why it is important to know when the steady state performance measures do not depend on the type of distributions of the time duration of activities in the real system. We are in such a favorable situation when activities do not execute simultaneously, but sequentially, according to a stochastic routing using probabilities p_{ij} (activity j starts after finishing activity i with probability p_{ij}). In such a (common) situation, the steady state distribution of the semi-Markovian process modeling the real system with non exponential distributions is the same as the one of the Markovian process obtained by taking the rates equal to the inverses of the mean durations of the activities.

   The danger would be to do this simplification while different activities may be executed concurrently. Because of this competition, for equivalent expectations of the execution times, the behavior will depend on the distributions of the random variables. For example, let us consider a simple triple module redundancy with hot repair facility. If the life-time and the repair-time of an element are constant, the redundant system will be always available (supposing the repair-time shorter than the life-time). While if the life-time and the repair-time of an element are exponentially distributed, there is a strictly positive probability to see the redundant system down.

5. Lack of technical background. On the one hand, this risk is highly correlated with the first mentioned danger (bad comprehension of the system) which we will therefore not elaborate futher. On the other hand, it is worthwhile to remark that being technically good is not a sufficient condition for building good models (unfortunately).

6. Lack of scientific background. In my opinion, young people from our domain suffer in particular from a lack of background in applied mathematics and I notice that both simulation and analytical fields are concerned. With respect to the simulation field, is the researcher, involved in a project requiring many months of work, mastering confidence intervals or the special techniques of importance splitting or of importance sampling ? Such techniques are very important for rare event studies, when the probability of an event equals, for example, 10^{-8} (in high speed telecommunication networks, highly

dependable architectures, air-traffic control systems, etc). With respect to the analytical field, does the researcher master Markov regenerative processes ? or stochastic fluid models ? or process algebra ? or timed Petri nets ? or fluid timed Petri nets ? Again, we can mention the case of rare events encountered, for example, with telecommunication networks. Often we are concerned with the dimensioning of a node such that the lost probability in a buffer is lower than 10^{-8}. A realistic queuing model will not have a product form solution, and this will not be reasonable to estimate the probability of this rare event through simulation. A possible solution might be to use the technique of stochastic fluid models, approaching the behavior of the queuing model with a cost independent of the number of customers in the model.

It should be noted that this vision may be biased because of the fact that one enlarges one's expertise in the different mathematical tracks when one spends year after year in the field of modeling.

5   Hopes

Thanks to Research and Development, computational power and data storage have increased the possibilities of performance evaluation studies during the last decades. The development of libraries and of graphical interfaces has increased our productivity. However, if we compare with other industries (space, transport, nuclear), or disciplines (physics, chemistry), we should be more ambitious with respect to our evaluation tools.

There is a place for a large evaluation tool built as a set of cooperative agents including simulation agents and analytical/numerical agents. Following the ideas underlying the notion of "Internet of the future", we could think of a nice virtual machine built on a computer cloud and able to realize all the possible evaluations of performance already done once by one group of an international federation. Of course such a project would need a tremendous effort and the setting-up of standard commissions to define the tasks of tool virtual agents, interface protocols and also to standardize software developments and data structures. There would also be a need for the setting-up of independent teams testing each agent, the possibilities of each method, ranking them with respect to the specifications of the application in case of multiple choice (eg, importance sampling versus importance splitting). Progress in manipulation of UML models and in Model Checking should help the efforts in standardization.

My personal feeling is that the community of the global domain of numerical analysis has done better than we have in the structuring of the efforts. It is true that because of its generality, the task is ambitious (how to standardize the Input/Output in a general way ?) but isn't it worth doing ? Of course, such an action needs a significant financial budget but less than a large program in Physics, and its synergy effect would benefit in the long term.

6   Conclusions

Throughout all this evolution we have seen the development of libraries and of graphical interfaces, increasing our productivity, yet somehow I am pleading for more interfaces and more virtual items. But do the students still understand what the tools are doing below the graphical interfaces ? To do so is necessary to save the level of scientific knowledge ! In fact, from my point of observation, I have come to the conclusion that among the dangers listed in Section 4, the most important one is the lack of scientific background. Discussions on this topic with colleagues from different countries have shown a common agreement on the following facts :

- The attraction of scientific studies is decreasing ; why ? is it because learning theories hurts (isn't it easier to read a book on management than a book on probability theory before going to bed ?) or is it because scientific studies do not maximize the chances of getting rich ?
- The student who comes to CS wants more and more practice of work on keyboard rather than paper and pencil !

In reaction, it is our responsibility not to let the theoretical courses move from compulsory to optional positions in university curricula. But It is also our responsibility to try to make these theoretical courses more like detective stories, i.e., more gripping and more entertaining.

Again this is just a state of my personal vision and such an exercise cannot be fully objective. In addition, there are always exceptions to general observations.

At this point of the story ($E\tau\sigma\iota$ $\epsilon\iota\nu\alpha\iota$ η $Z\omega\eta$[1]), it is now the responsability of the new generation to find a way to keep our young discipline as a field in which reseach remains an attractive way of life.

Reference

1. Marie, R.A., Trivedi, K.S.: A Note on the Effect of Preemptive Policies on the Stability of a Priority Queue. Information Processing Letters 24(6), 397–401 (1987)

---

[1] *So ist das Leben.*

Model Interoperability for Performance Engineering: Survey of Milestones and Evolution

Connie U. Smith[1] and Catalina M. Lladó[2]

[1] Performance Engineering Services, PO Box 2640, Santa Fe, New Mexico, 87504-2640 USA
www.spe-ed.com
[2] Universitat de les Illes Balears, Departament de Ciències Matemàtiques i Informàtica, Ctra de Valldemossa, Km. 7.6, 07071 Palma de Mallorca, Spain
cllado@uib.es

Abstract. Next generation Software Performance Engineering tools will exploit a model interoperability paradigm that uses the performance modeling tool best suited to the software/hardware architecture issues and the life cycle stage of the assessment. The paradigm allows the use of existing tools to the extent possible without requiring extensive changes to them. The performance model solution should be transparent to the user. Significant milestones have been accomplished in the evolution of this paradigm. This paper covers key results in the areas of Model Interchange Formats, model transformations, tools, specification of experiments and results, and extensions for real-time and component-based systems. It then offers conclusions on next steps and the future of the model interoperability paradigm.

1 Introduction

Software performance engineering (SPE) is a systematic, quantitative approach to constructing software systems that meet performance requirements [39]. SPE uses performance models to provide data for the quantitative assessment of the performance characteristics of software systems as they are developed. SPE has evolved over more than 30 years and has been demonstrated to be effective during the development of many large systems.

Although sound performance analysis theories and techniques exist, they are not widely used because they are difficult to understand and require heavy modeling effort throughout the development process [41]. Consequently, software engineers usually resort to testing to determine whether the performance requirements have been satisfied. To ensure that these theories and techniques are used, they must be made more accessible–integrated into the software development process and supported with tools.

First generation performance engineering tools were developed for users with modeling expertise. Examples include [7,17,40]. To use them for SPE requires performance specialists to work with developers to obtain software design and performance data for evolving software. The principal problem is the gap between

K.A. Hummel et al. (Eds.): PERFORM 2010 (Haring Festschrift), LNCS 6821, pp. 10–23, 2011.

software developers who need the techniques and the performance specialists who have the skill to conduct studies with first generation modeling tools. This limits the ability of developers to explore the design alternatives.

The ideal next generation SPE tool will provide support for many SPE tasks in addition to the obvious requirements for performance modeling. The ideal SPE tool will use the performance modeling tool best suited to the software/hardware architecture issues and life cycle stage of the assessment. It also requires a cost-effective solution that works with existing tools to the extent possible without requiring extensive changes to them. The performance model solution process should be transparent to the user of the SPE tool.

Model interoperability seeks cooperation among existing tools that perform different tasks. Thus, model interoperability makes it possible to create a software specification in a development tool, then automatically export the model description and specifications for conducting performance assessments, then obtain results for considering architectural and design options.

This paper examines the milestones in the evolution from first generation performance modeling tools for SPE to the model interoperability framework. Note that our focus was initially on tools for SPE, however the model interoperability framework supports general performance engineering tasks for a variety of modeling tools and application domains. The following sections address the significant milestones. We then offer conclusions on next steps and the future of the model interoperability paradigm.

2   Version 1 Model Interchange Formats

A model interchange format (MIF) is a common representation for performance model data that can be used to move models among modeling tools. A user of several tools that support the format can create a model in one tool, and later move the model to other tools for further work without the need to laboriously translate from one tool's model representation to the other. For example, an analyst might create a model of a server platform to conduct several studies, then move the model to a tool better suited to network analysis.

MIFs require minor extensions to tool functions (import and export) or creation of an external translator to convert file formats to/from interchange formats. They enable easy comparison of results from multiple tools, and the use of tools best suited to the task. Without a shared interchange format, two tools would need to develop a custom import and export mechanism. Additional tools would require a custom interface to every other tool resulting in a N(N-1) requirement for customized interfaces. With a shared interchange format, the requirement for customized interfaces is reduced to 2N.

Related, earlier work in this area is limited. Beilner advocated hierarchical and modular descriptions of models and showed how multiple solution techniques can be used with these model descriptions. The HIT environment demonstrated the feasibility of this approach [7]. Likewise, the Esprit Integrated Modelling Support Environment, IMSE, integrated several individual modeling tools and showed the potential of interchanging tools for performance studies [21]. We seek a mechanism for a loose connection among a variety of modeling tools.

MIFs originated in the VLSI community with the Electronic Design Interchange Format (EDIF) for exchanging VLSI design information among design tools. Subsequently the software community adopted this paradigm for the Case Data Interchange Format (CDIF) standard [4]. In the CDIF standard, the information to be transferred between two tools is known as a model. The contents of a model are defined using a meta-model. A meta-model defines the information structure of a small area of CASE (such as data modeling) known as a Subject Area. Each meta-model is, in turn, defined using a meta-meta-model. The original meta-meta-model is based on the Entity-Relationship-Attribute (ERA) approach.

MIFs for performance modeling were first introduced at the 1995 Tools Conference in Heidelberg. The Software Performance Model Interchange Format (S-PMIF) defined the SPE information requirements for CASE tools with a meta-model and a transfer format for interchange among software performance modeling tools [37]. The Performance Model Interchange Format (PMIF) for the interchange of queueing network models (QNM) was proposed in a panel session and subsequently published in [38]. The session had a lively debate about the concept of model interoperability. It was generally agreed that the concept was good, but the debate centered on the content of the proposals and whether they were ready for technology transfer.

The initial work by Smith and Williams was funded by the National Science Foundation. Without additional funding and without tools to facilitate implementation, the work remained dormant for several years.

3   Version 2 PMIF

With the advent of XML (Extensible Markup Language) tools, MIF implementation became feasible. In 2004, Smith and Lladó developed Version 2 of PMIF [35]. Version 2 converted the meta-model to UML and implemented the XML transfer formats. Modifications were required to exchange QNM among unlike tools. The primary difference was the addition of routing probabilities because the number of visits alone could be insufficient information for some QNM topologies. Probabilities are specified with Transit elements for routing among nodes and for workload entry. Number of visits was retained for tools that must specify the service time per visit rather than total service demand. Other minor changes improved convenience and eliminated redundancy.

We implemented a prototype export mechanism from the *SPE·ED* software performance modeling tool [40] into pmif.xml, and a prototype import mechanism from pmif.xml into the Qnap system performance modeling tool [28]. Subsequent work implemented the reverse exchange exporting from Qnap and importing to *SPE·ED* further demonstrating the soundness of PMIF 2 [34].

The CDIF strategy is export everything you know and provide defaults for other required information; import the parts you need and make assumptions if you require data not in the meta-model. Everything you know is not necessarily everything you use. For example, *SPE·ED* uses visits to specify routing, but it knows about probabilities, and it is relatively easy to calculate them.

We used the prototypes to study several examples. Our use of unlike tools helped us find limitations in the meta-model and find a general way to resolve them. Example solutions confirmed that the pmif.xml transfer was successful. The examples provided a set of models that are well documented, with reproducible results, that may be used by others who wish to explore the pmif.xml approach to model interoperability.

4   Flurry of Related Work

After establishing the viability of performance model interchange formats, there was a flurry of activity among researchers building on this work. It is described in the following sections.

4.1   Design Models to Performance Models

Cortellessa, et.al. collaborated on an unified approach using S-PMIF [32], derived from the original SPE meta-model. They demonstrated the viability of a complete path from software design to performance validation, based on existing tools that were not designed to interact with each other.

This overall process is beneficial because no single monolithic tool is good for everything. Early in development one needs to quickly and easily create a simple model to determine whether a particular architecture will meet performance requirements. Precise data is not available at that time, so simple models are appropriate for identifying problem areas. Later in development, when some performance measurements are available, more detailed models such as Queueing Network Models (QNM), Stochastic Petri Nets (SPN), or Process Algebra (PA) models can be used to study intricacies of the performance of the system. At that time, tools that provide features not in the simpler models are desirable. These industrial strength models are seldom appropriate earlier in development because constructing and evaluating them requires additional time and expertise that is seldom justified when performance details are sketchy at best.

Thus S-PMIF is an intermediate representation between design models and system performance models. It is useful for studying software architecture and design options. Other researchers go directly from software models to system performance models [19,10,25,5,12]. These approaches use XML to transfer design specifications into a particular solver.

Lopez-Grao et al. propose a method to translate several UML diagram types to analyzable GSPN models where performance requirements are annotated according to the UML SPT (Schedulability, Performance and Time) profile [23]. As a slightly more general approach, Gu and Petriu present in [18] a process to transform UML models to Performance Models that can be later specialized for different performance modeling formalisms.

4.2   MIF Extensions

Lately, efforts have developed intermediate models between various design models (different UML diagrams, different UML versions, non UML approaches, etc.) and various system performance modeling tools. Woodside et al. [11] propose the Core Scenario Model (CSM), claiming that it captures the essence of performance specification and estimation as expressed in the UML SPT profile. Grassi et al.[15] focus on the construction of software systems according to the component-based development (CBD) approach. They propose KLAPER (Kernel Language for Performance and Reliability analysis) as an intermediate language between software specification models and performance models. KLAPER is defined as a MOF (Meta-Object Facility) metamodel, which can also be transformed to/from KLAPER models.

Cortellessa investigated the possibility of building an unified ontology for software performance modeling [8]. His study originates from the comparison of three existing metamodels, (i.e. CSM, SPT, PMIF), identifies intersecting areas among the metamodels, and at the same time provides a roadmap to work towards merging metamodels into a unifying ontology.

Harrison et al. [20] generalized the PMIF specifications by considering more abstract collections of interacting nodes using concepts compatible with the Reverse Component Agent Theory (RCAT). The interactions are more general in that they synchronize transitions between a pair of nodes rather than describing traffic flows.

4.3   MIF Tools

Lladó and colleagues developed a Web Service for solving QNM using PMIF [30]. This established the feasibility of a plug-and-play approach for model interoperabilty and provided a foundation for other implementations. Cortellessa and colleagues built a Web Service to a collection of QNM tools along with a tool for creating a PMIF specification from a GUI based tool [31].

The PMIF schema supports syntactic validation of models: it confirms that the XML file contains everything it is supposed to, that IDREFS point to a declared ID, etc. Even if the PMIF is syntactically valid, however, the models might be semantically incorrect. Lladó and colleagues developed a specification for semantic validations and a tool to confirm that a PMIF model is correct before attempting to convert it to a tool-specific file [14,34].

Woodside and colleagues developed a tool architecture, PUMA (Performance by Unified Model Analysis) as a framework into which different kinds of software design tools can be plugged as sources, and different kinds of performance tools can be plugged as targets [11].

5   Automation of Experiments and Results

Interchange formats have been defined for queueing network models (PMIF), software performance models (S-PMIF), layered queueing networks (LQN), UML,

Petri Nets and other types of models. In each interchange format, a file specifies a model and a set of parameters for one run. Automating experiments requires 3 steps: specifying the experiment; executing one or more model runs and collecting output from those runs; and converting the output into meaningful results for the experiment. Each is addressed in the following sections. Figure 1 shows these steps. The performance model MIF is in the top-left while the experiment specification (Ex-SE) is shown at the top-right. These files are combined and used as input for one or more performance modeling tools. Each tool generates the performance metric output as specified for each experiment. The last step transforms the output into desired results as specified by the Results-SE.

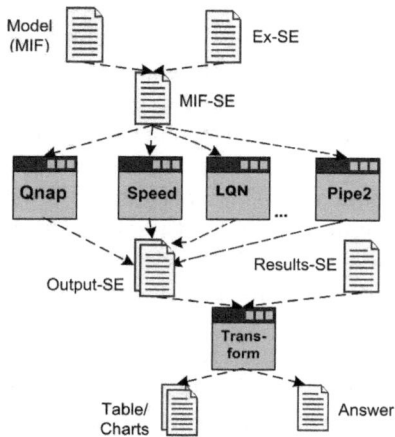

Fig. 1. Model interoperability framework

5.1   Experimentation

Some general approaches to experimentation have been proposed:

- Zeigler [43] has proposed a framework for modeling and simulation that included an experimental frame of the conditions under which the system or its model are to be exercised.
- The IMSE Experimenter tool facilitates performance modeling studies within the Integrated Modeling Support Environment (IMSE) [21] .
- The James II simulation modeling framework [22] allows any experiment definition that has a suitable plug-in for reading and converting the experiment definition (i.e., an instance of IExperimentReader).

Some modeling tools provide an ability to solve models multiple times while varying parameter values. Most current performance modeling tools have a Graphical User Interface (GUI) that leaves it to the user to conduct model experiments. Tools developed before GUIs were prevalent, however, provided experimentation features as part of their user interface: Qnap [28], LQN [3], HIT and Hi-Slang [2].

Tools for other types of modeling have similar capabilities:

- Probabilistic Symbolic Model Checker (PRISM) [24], provides a GUI that finds undefined constants in the model and queries the user for a value.
- The Network Simulator [29] has an object-oriented script language TCL that lets the user set up the model topology, place agents, inject events, etc.
- the Möbius modeling environments [9] supports higher level experimental design with two types of two-level experimental design strategies: Plackett-Burman and two-level factorial design, and two types of response surface designs (central composite and Box Behnken).

Smith, Lladó and colleagues incorporated elements from these approaches to provide a comprehensive model interoperability solution for experimentation. We defined the Experiment Schema Extension (Ex- SE) for specifying a set of model runs and the output desired from them [36,33]. The Ex-SE has the expressive power to specify iterations, alternations, assignments of values, actions based on model results and more. This schema extension provides a means of specifying performance studies that is independent of a given tool paradigm.

We use the term *Schema Extension* because it is used in combination with a host schema and customized to refer to solutions and outputs provided by the host. For example, we have specific instances of Ex-SE for Queuing network models and Petri net models. Figure 2 shows the schema extension instance for PMIF models, PMIF-Ex.

Experiment specifications can be created with an XML editor, with a GUI tool for describing experiments and generating the XML, or by exporting an experiment definition along with the model interchange format from a tool with experimentation functionality. Figure 3 shows a screenshot of the experimenter editor for PIPE2 [1], a Petri net modeling and analysis tool.

5.2 Model Output

An Output schema extension Output-SE is added to the host schema to specify the XML format to be used for output from the experiments. Like the Ex-SE it is customized to the type of output supported by the host schema.

The PMIF Output-SE includes:

- a solution ID for relating the output to the experiment
- the value used for *Ranges* or other variables used in that solution
- the OutputWorkload (overall results by workload)
- OutputNode (overall results by Node)
- OutputNodeWorkload (results by Workload for Nodes).

Tools without experimentation need an Experimenter tool to interpret the experiment, invoke the tool for each <Solve>, and return the output. This can be a general tool that can work with multiple solvers, or tools with experimentation can adapt their Import mechanism to generate the tool specifications for the experiment. This is a more efficient implementation because the tool only needs to be invoked once, and the model does not need to be parsed multiple times.

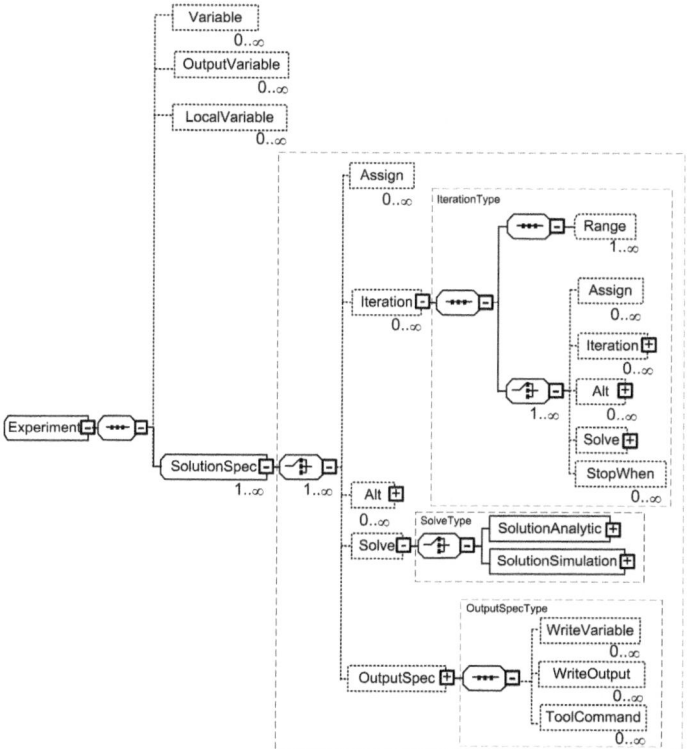

Fig. 2. Experiment Schema Extension, PMIF instance

Lladó and colleagues have developed a tool, EXperiment Output to Spread-Sheet (EXOSS), that takes the XML output from one or more experiments and produces a spreadsheet (xls) file for easily viewing the performance output. The tool works with PMIF queueing network models and Petri net models. It has also been integrated into the PIPE2 Petri net modeling tool.

5.3 Experiment Results

The IMSE Experimenter has an experimental plan with at least one *analysis* specification that describes how experimental *results* are obtained from model outputs. Thus, outputs from multiple runs can be used to obtain overall measures (e.g., mean, standard deviation). The PRISM GUI also provides the ability to specify a graph of the results and to export it in several formats (png, jpg, eps).

We examined Use Cases for building and analyzing performance model results then studied the output metrics and results that are most often desired for the Use Cases [33]. A review of published papers found some examples of common tables of results, such as response times for workloads with varying workload intensities, measured versus model results for response time and utilization for multiple workloads, etc. We also found common charts with line plots, columns

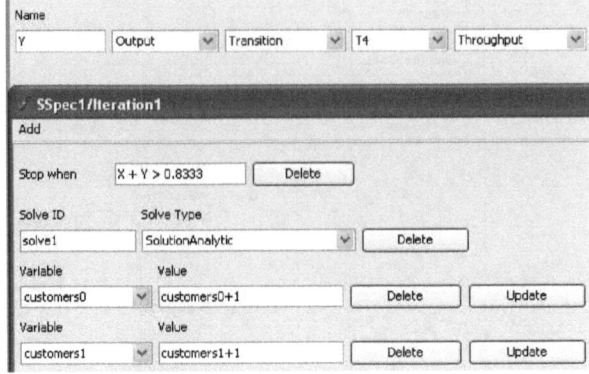

Fig. 3. PIPE2 experiment editor

or bars showing throughput versus response time; response time versus service time; number of users versus response time, etc.

Our conclusion for both QNMs and Petri net models is that the primary results are tables and charts. In the future, more sophisticated techniques for visualization of results may be desired, such as images of multidimensional colored surfaces. We did not address those types of results – we will leave that to the tools that render these types of images. Those tools usually can import the performance values to be displayed from a file and it should be straightforward to transform tool output into such a file format.

The most common format for tables and charts is xls as in spreadsheet tools such as Excel and OpenOffice Calc. However, the most common document preparation system for research publications is LaTeX. Our approach supports transforming the output metrics to tables and charts in xls and LaTeX. The results for an experiment are transformed into one or more tables/charts. Several tools have been implemented for experiment results [33].

5.4   Multiple Domains

In [36], Ex-SE was illustrated with an instance of the extension in which the interchange schema was the PMIF and also LQN. The Ex-SE is compatible with a large number of interchange formats including: PMIF, S-PMIF, LQN, GPMIF, and Petri Net xml specifications. In general, one appends the Experiment type definition into the other XML schema, and changes the Variable type specifications to match new schema if necessary.

Petri nets are different from QNM and other performance model paradigms because they provide additional representation and analysis capabilities in addition to performance analysis. Examples include constraints on tokens in places, invariant analysis, reachability analysis, and so on.

Lladó and colleagues presented a specific, extended instantiation of the Ex-SE for Petri nets (PN-Ex) [26] and analogous extensions to the Output and Results specifications. Both the Output-SE and the Results-SE required specification of an array of values instead of a single value.

The Experiment Schema Extension is not limited to use with models. It could be used with a measurement experimenter. Some of the terminology should be customized, such as changing Solve to Run, and changing the SolutionTypes to something such as ToolSpecification.

6 Real-Time and Component-Based Systems

Some work has addressed the performance analysis of component-based systems. Wu and Woodside use an XML Schema to describe the contents and data types that a Component-Based Modeling language (CBML) document may have [42]. CBML is an extended version of the Layered Queuing Network (LQN) language that adds the capability and flexibility to model software components and component-based systems. Becker et al. address components whose performance behavior depends on the context in which they are used [6]. They address sources of variability such as loop iterations, branch conditions, and parametric resource demand, and then use simulation to predict performance in a particular usage context. Grassi et al. extend the KLAPER MOF meta-model to represent reconfigurable component-based systems in [16]. It is to be used in autonomic systems and enable dynamic reconfiguration to meet QoS goals. These approaches are performance-centric in that they create/adapt a model of component based systems specifically for performance assessment. We prefer to work with generally accepted architecture representations, and use a common interchange format (S-PMIF) that allows the use of a variety of performance modeling tools to provide performance predictions for architecture and design alternatives.

Moreno, Smith and Williams have extended the S-PMIF to include features necessary for evaluating real-time, component-based systems [13]. They subsequently presented a substantially modified S-PMIF, version 2.0, that adapted the meta-model so that it can be expressed in terms of the Ecore meta-meta-model–the core meta-model of the Eclipse Modeling Framework. Several changes were made to better align S-PMIF 2.0 with other performance-related meta-models (PMIF, LQN, and ICM), to clarify terminology, and simplify the M2M transformations [27]. This work demonstrated the use of the model interoperability approach for component-based real-time systems, and implemented a M2M prototype to transform an intermediate constructive model (ICM) to S-PMIF.

We found that preserving the type hierarchy and associations of the S-PMIF meta-model in the schema facilitates the implementation of S-PMIF interchange support by tools using strongly typed modeling technologies to generate the XML such as EMF or ATL. This work opened a door to allow the performance analysis of CCL specifications with other analysis tools without the need for additional integration effort. This means that standard SPE models can easily be used for analysis of systems specified in CCL. We demonstrated the ease with which the S-PMIF can be employed to transform additional design notations into software performance models, thus building on the previous UML-based approaches. In the future, it may be possible to unify the various interchange formats as suggested by [8]. In the meantime, it makes sense to extend the meta-models as necessary to create a superset of the necessary information for performance assessment.

7   Conclusions

This paper introduced the model interoperability approach for Software Performance Engineering. It presented significant milestones and stepping stones in the evolution of the approach: Version 1 Model Interchange Formats; Version 2 PMIF; Flurry of work in model transformations, MIF extensions and MIF tools; Automation of experiments and results; and Extensions for real-time and component-based systems. Our focus was initially on tools for SPE, however the model interoperability paradigm supports general performance engineering tasks for a variety of modeling tools and application domains.

MIFs provide a mechanism for exchanging performance models among tools. The exporting and importing tools can either support the MIF or provide an interface to read/write model specifications from/to a MIF file. The comparison of multiple solution techniques across tools can lead to a wide range of benefits. It is a significant result that MIFs can be used by anyone to exchange information relatively easily between two tools that provide a file input/output capability. It does not require the tool developer to modify code to be of use.

The basic version of PMIF 2 supports queueing network models that may be solved using efficient, exact solution algorithms. It was appropriate to restrict the domain for initial research. Our future work will create an extended PMIF and the supporting tools to cover model features that are supported by most simulation-based tools.

S-PMIF 2 is substantially different from the 2005 version. So, for instance, prototypes developed for the earlier version would have to be modified if they are to support the additional features. Model interchange formats and interfaces, however, must be relatively stable to be viable. S-PMIF was based on concepts embodied in two earlier model interchange formats: EDIF for VLSI designs and the CDIF for software design interchange. Creators of EDIF envisioned the stability problem and addressed it by using a concept of levels that add functionality at each successive level and giving ownership to a standards organization that managed changes. S-PMIF 2 adds a level for analyzing Real-time systems; future levels may add features for additional types of analysis. Tools can continue to support a lower level without change, or may opt to modify interfaces to support additional functionality and/or other changes. Future work should use the newest version, even for the basic level. Using a standard organization to manage the contents of model interchange formats should be considered.

The Experimental Schema Extension (Ex-SE) allows specification of multiple model runs along with the results that are desired from them. The schema extensions provide a means of specifying performance studies that is independent of a given tool paradigm.

Our future work will address a general purpose Experimenter tool with its corresponding editor. We also plan to develop templates for the most frequent results. We will extend the framework to support models that provide simulation solutions and other analysis techniques including real time systems analysis. An interesting extension will include creating rules for specifying threshold values and highlighting results in tables that exceed the threshold. Moreover, we plan to develop a GUI based transformation tool from output to results which will simplify the specification of the most common tables and graphs.

Model interoperability enables the use of the best tool(s) for the particular performance assessments. Early work demonstrated the benefit of comparing results from multiple tools; errors were found in several published case studies. Additional work is needed on model to model transformations among CSM, S-PMIF, PMIF, eDSPN, and others to enable comparison of results from multiple model paradigms. One approach is to unify the various interchange formats as suggested by [8]. Another is to ensure that sufficient modeling power is in each MIF to enable the M2M transformations.

The schemas and other information are available at `www.spe-ed.com/pmif/`. Several prototypes with source code are at `dmi.uib.es/~cllado/mifs`.

Acknowledgements. This work is partially funded by the TIN2009-11711 project of the *Ministerio de Educacion y Ciencia*, Spain. Smith's participation was sponsored by US Air Force Contract FA8750-09-C-0086.

References

1. Platform Independent Petri net Editor 2, `http://pipe2.sourceforge.net/`
2. `http://www4.cs.uni-dortmund.de/{HIT}/`
3. `http://www.sce.carleton.ca/rads/lqn/lqn-documentation/`
4. Electronics Industries Association. CDIF - CASE Data Interchange Format Overview, EIA/IS-106 (1994)
5. Balsamo, S., Marzolla, M.: Performance evaluation of UML software architectures with multiclass queueing network models. In: Proc. 5th International Workshop of Software and Performance, Palma de Mallorca, Spain. ACM (July 2005)
6. Becker, S., Koziolek, H., Reussner, R.: Model-based performance prediction with the palladio component model. In: WOSP 2007. ACM (February 2007)
7. Beilner, H., Mäter, Weißenberg, N.: Towards a performance modeling environment: News on HIT. In: Proc. 4th Int. Conference on Modeling Techniques and Tools for Computer Performance Evaluation. Plenum Publishing (1988)
8. Cortellessa, V.: How far are we from the definition of a common software performance ontology? In: WOSP 2005. ACM (2005)
9. Courtney, T., Gaonkar, S., McQuinn, G., Rozier, E., Sanders, W., Webster, P.: Design of experiments within the möbius modeling environment. In: Proc. of the Fourth International Conference on the Quantitative Evaluation of Systems, Edingurh, UK, September 16-19, pp. 161–162. IEEE Computer Society Press (2007)
10. D'Ambrogio, A.: A model transformation framework for the automated building of performance models from UML models. In: WOSP 2005, pp. 75–86 (July 2005)
11. Woodside, C.M., et al.: Performance by unified model analysis (PUMA). In: WOSP 2005, pp. 1–12 (July 2005)

12. Savino, N., et al.: Extending UML to manage performance models for software architectures: A queuing network approach. In: Proc. 9th Int. Symp. on Modeling, Analysis and Simulation of Computer and Telecommunication Systems, SPECTS (2002)
13. Moreno, G.A., Smith, C.U., Williams, L.G.: Performance analysis of real-time component architectures: A model interchange approach. In: WOSP 2008. ACM Press (June 2008)
14. García, D., Lladó, C.M., Smith, C.U., Puigjaner, R.: Performance model interchange format: Semantic validation. In: International Conference on Software Engineering Advances. INRIA (October 2006)
15. Grassi, V., Mirandola, R., Sabetta, A.: From design to analysis models: A kernel language for performance and reliability analysis of component-based systems. In: Proc. WOSP, pp. 25–36 (July 2005)
16. Grassi, V., Mirandola, R., Sabetta, A.: A model-driven approach to performability analysis of dynamically reconfigurable component-based systems. In: Proc. WOSP. ACM (February 2007)
17. Grummitt, A.: A performance engineer's view of systems development and trials. In: Proceedings Computer Measurement Group, pp. 455–463 (1991)
18. Gu, G., Petriu, D.: From UML to LQN by XML algebra-based model transformations. In: WOSP 2005. ACM (2005)
19. Gu, G., Petriu, D.C.: XSLT transformation from UML models to LQN performance models. In: WOSP 2002, pp. 227–234 (2002)
20. Harrison, P., Lladó, C.M., Puigjaner, R.: A general performance model interchange format. In: Proc. of the First International Conference on Performance Evaluation Methodologies and Tools, Valuetools (2006)
21. Hillston, J.: A tool to enhance model exploitation. Performance Evaluation 22(1), 59–74 (1995)
22. Himmelspach, J., Rhl, M., Uhrmacher, A.M.: Component-based models and simulations for supporting valid multi-agent system simulations. Applied Artificial Intelligence 24(5), 414–442 (2010)
23. López-Grao, J.P., Merseguer, J., Campos, J.: From UML activity diagrams to stochastic Petri nets: Application to software performance engineering. In: WOSP 2004. ACM (2004)
24. Norman, G., Kwiatkowska, M., Parker, D.: Prism: Probabilistic model checking for performance and reliability analysis. ACM SIGMETRICS Performance Evaluation Review 36(4), 40–45 (2009)
25. Marzolla, M., Balsamo, S.: UML-PSI: the UML performance simulator (tool paper). In: Proc. 1st Int. Conf. on Quantitative Evaluation of Systems (QEST). IEEE Computer Society (2004)
26. Melià, M., Lladó, C.M., Smith, C.U., Puigjaner, R.: An experimental framework for PIPEv2.5. In: 5th Int. Conference on Quantitative Evaluation of Systems, St Malo, France, pp. 239–240. IEEE Computer Society Press (September 2008)
27. Moreno, G.A., Smith, C.U.: Performance analysis of real-time component architectures: An enhanced model interchange approach. Performance Evaluation 67, 612–633 (2010)
28. Potier, D., Veran, M.: QNAP2: A portable environment for queueing systems modelling. In: Potier, D. (ed.) First International Conference on Modeling Techniques and Tools for Performance Analysis, pp. 25–63. North Holland (May 1985)
29. The VINT Project. The ns Manual. UC Berkeley, LBL, USC/ISI, and Xerox PARC (2010)

30. Rossello, J., Lladó, C.M., Puigjaner, R., Smith, C.U.: A web service for solving queueing networks models using PMIF. In: Proc. WOSP 2005, Palma de Mallorca, Spain, pp. 187–192. ACM (July 2005)
31. SEAlab Software Quality Group. WEASEL, a web service for analyzing queueing networks with multiple solvers, http://sealabtools.di.univaq.it/Weasel/
32. Smith, C.U., Cortellessa, V., Di Marco, A., Lladó, C.M., Williams, L.G.: From UML models to software performance results: An SPE process based on XML interchange formats. In: Proc. of the Fifth International Workshop of Software and Performance, Palma de Mallorca, Spain, July 12-14, pp. 87–98. ACM (2005)
33. Smith, C.U., Lladó, C.M., Puigjaner, R.: Automatic Generation of Performance Analysis Results: Requirements and Demonstration. In: Bradley, J.T. (ed.) EPEW 2009. LNCS, vol. 5652, pp. 73–78. Springer, Heidelberg (2009)
34. Smith, C.U., Lladó, C.M., Puigjaner, R.: Performance Model Interchange Format (PMIF 2): A comprehensive approach to queueing network model interoperability. Performance Evaluation 67(7), 548–568 (2010)
35. Smith, C.U., Lladó, C.M.: Performance model interchange format (PMIF 2.0): XML definition and implementation. In: Proc. of the First International Conference on the Quantitative Evaluation of Systems, Enschede, The Netherlands, pp. 38–47. IEEE Computer Society Press (September 2004)
36. Smith, C.U., Lladó, C.M., Puigjaner, R., Williams, L.G.: Interchange formats for performance models: Experimentation and output. In: Proc. of the Fourth International Conference on the Quantitative Evaluation of Systems, Edingurh, UK, September 16-19, pp. 91–100. IEEE Computer Society Press (2007)
37. Smith, C.U., Williams, L.G.: Panel presentation: A performance model interchange format. In: Proc. of the International Conference on Modeling Techniques and Tools for Computer Performance Evaluation, Heidelberg, Germany, September 20-22. Springer, Berlin (1995)
38. Smith, C.U., Williams, L.G.: A performance model interchange format. Journal of Systems and Software 49(1), 63–80 (1999)
39. Smith, C.U., Williams, L.G.: Performance Solutions: A Practical Guide to Creating Responsive, Scalable Software. Addison-Wesley (2002)
40. SPE-ED. LS Computer Technology Inc., www.spe-ed.com
41. Woodside, C.M., Franks, G., Petriu, D.C.: The future of software performance engineering. In: International Conference on Software Engineering (ICSE), pp. 171–187. IEEE Computer Society (May 2007)
42. Wu, X., Woodside, C.M.: Performance modeling from software components. In: WOSP 2004, pp. 290–301 (January 2004)
43. Zeigler, B.P., Praehofer, H., Kim, T.G.: Theory of Modeling and Simulation: Integrating Discrete Event and Continuous Complex Dynamic Systems, 2nd edn. Academic Press (2000)

Tools for Performance Evaluation of Computer Systems: Historical Evolution and Perspectives

Giuliano Casale[1], Marco Gribaudo[2], and Giuseppe Serazzi[2]

[1] Imperial College London, London SW7 2AZ
g.casale@imperial.ac.uk
[2] Politecnico di Milano, I-20133 Milan, Italy
{gribaudo,serazzi}@elet.polimi.it

Abstract. The development of software tools for performance evaluation and modeling has been an active research area since the early years of computer science. In this paper, we offer a short overview of historical evolution of the field with an emphasis on popular performance modeling techniques such as queuing networks and Petri nets. A review of recent works that provide new perspectives to software tools for performance modeling is presented, followed by a number of ideas on future research directions for the area.

1 Introduction

Since the early years of computing, software tools have been used to evaluate and improve system performance. This has been soon recognized as fundamental in a number of phases of a computer system's life-cycle, namely design, sizing, procurement, deployment, and tuning. However, due to the inherent complexity of the systems being evaluated and the novelty of the computing field, effective performance evaluation tools took several years to appear on the market. Simulation was the first technique used extensively for evaluating the performance of hardware logic of single components initially, and of entire systems later, see [26] for a review. The introduction of *simulation languages* in the 60s, such as Simscript [23] and GPSS [17], was a milestone since several tools oriented to the simulation of computer systems and networks appeared shortly afterwards on the market. In the early 70s, two simulation packages oriented to computer performance analysis, namely Scert [14,15] and Case, were among the first to reach commercial success. It must be pointed out that, due to its dominant position in the computer market from the 60s to the 80s, almost all tools were developed for modeling systems and network technologies developed by IBM. Features of all generations of IBM systems, such as 360s and MVS, were deeply analyzed through simulation models and with other new analysis techniques that were becoming available. Other types of tools such as *hardware monitors* [4], i.e., electronic devices connected to the system being measured with probes and capable of detecting significant events from which performance indexes can be deduced, were also used in the 70s. These did not reach a great diffusion due to their high costs, the difficulty of use, and the huge effort required to adapt them to different systems and configurations.

In those years, models started to emerge as a new way to evaluate single components and system architectures. Among the various problems approached were the evaluation

K.A. Hummel et al. (Eds.): PERFORM 2010 (Haring Festschrift), LNCS 6821, pp. 24–37, 2011.

of time-sharing supervisors, I/O configurations, swapping, paging, memory sizing, and networks of computers. The commercial interest in simulation modeling tools declined once efficient computational algorithms for *analytical modeling* appeared thanks to the pioneering work of Buzen [3]. Analytical techniques became rapidly popular because of their relatively low cost, general applicability, and easy and flexibility of use with respect to simulation. Such techniques are still popular today and have been the subject of several books and surveys [20,9,12,37]. BEST/1 [8] was the first tool implementing analytical techniques being marketed commercially with great success. Rapidly, tens of tools for analytical modeling appeared on the market. Over the years, as soon as a new analytical technique has been discovered a new tool implementing it has been developed. Thus, we have now performance evaluation tools based on Queuing Networks, Petri Nets, Markov Chains, Fault Trees, Process Algebra, and many other approaches. Hybrid and hierarchical modeling techniques have been introduced in the 70s and 80s to analyze very large and complex systems. Starting from the 90s, due to the increase of the state spaces needed to represent models of modern systems, simulation has become again a fundamental tool for model evaluations. This has been also a consequence of the dramatic increase of computational power in the last two decades, which has made simulation a more effective computational tool than in the past.

Several tools were designed specifically to solve particular class of problems. For example, SPE.ED [35] is a tool focused on the solution of the problems typical of Software Performance Engineering [36]. More recently, in the security domain, the ADVISE method has been introduced to quantitatively evaluate the strength of a system's security [22].

In spite of this long historical evolution, there is a lack of surveys covering the history and current perspectives of the performance tool area. The aim of this paper is to fill this gap and provide an up-to-date review and critique of current software tools for performance modeling. We point to [5] for a special issue on popular open source tools developed in academia in recent years. In this work, we first offer an overview of recent developments, many of which not covered in [5], focusing in particular on Markov chains (Section 2), Queueing Networks (Section 3), Petri Nets (Section 4), Fault Trees (Section 5) and Process Algebras (Section 6). In Section 7 we instead discuss trends and new perspectives in software performance tools architectures. Finally, Section 8 gives final remarks and concludes the paper.

2 Markov Models

Due to limited space, we here give only a brief overview on tools for *Markov modeling* and we focus next on higher-level modeling languages such as *queuing networks* or *Petri nets*.

Markov chains have been extensively used since the beginnings of performance evaluation as the fundamental technique to analyze stochastic models. The power of Markov chains derives from the ease of conditioning probabilities, which depends only on the current active state of the chain. In addition to basic discrete-time Markov chains (DTMCs) and continuous-time Markov chains (CTMCs), the performance evaluation community has intensively investigated the use of absorbing processes, such as phase-type (PH-type) distributions, to represent the statistical properties of measurements and for transient analysis of performance models. Although PH-type distributions and Markov-modulated processes are very active research areas, we here focus only on DTMCs and CTMCs.

Due to their historical importance, many tools exist for the analysis of DTMCs and CTMCs which have been developed both by performance engineers and numerical experts. A comprehensive review of modern numerical techniques for the analysis of Markov chains can be found in [37]. Popular tools include MARCA[1], Mobius[2], SHARPE[3], SMART[4], and PRISM[5]. Such tools include exact and approximate Markov chain solvers, such as the Kronecker-based solution methods proposed in [2]. Advanced techniques for state space generation and storage are also available such as multiway decision diagrams (MDDs), matrix diagrams, and symbolic state-space generation. MDD are an extension of the binary decision diagrams (BDD), a data structure capable of detecting redundancy and similarity in the state space of a model, allowing to reduce significantly the memory requirement to store the states. A discussion on such techniques can be found in [7].

3   Queuing Network Models

Queuing network models (QNMs) have been intensively used for the last three decades to study the effects of resource contention on scalability of computer and communication systems [1,21]. In their basic formulation, a QNM is composed by a set of resources visited by jobs belonging to a set of classes. Each job places a service demand, following some statistical distribution, at each visited resource, and the busy period of a resource depends on the contention placed by other jobs that simultaneously request service. The objective of the study is to compute performance metrics such as server utilizations or job response time distributions. Due to the lack of analytical solutions for general models, a number of approximation methods have been defined in the past, but there is still a lack for widely-applicable analytical approximation tools. In this context, simulation has become important in many practical applications to estimate performance metrics of QNMs, although analytical tools remain fundamental in several contexts, such as optimization studies which require the fast solution of hundreds of thousands models.

Queuing network modeling has a long history and has been addressed by several commercial packages such as BEST/1 [8], RESQ [30], QNAP [28], CSIM [32], and a variety of academic tools such as Tangram-II[6], JINQS[7], SHARPE[8], Java Modelling Tools[9], LQNS[10], and several others[11]. A recent collection of research papers on some

---

[1] http://www4.ncsu.edu/~billy/MARCA/marca.html
[2] http://www.mobius.illinois.edu/
[3] http://people.ee.duke.edu/~kst/
[4] http://www.cs.ucr.edu/~ciardo/SMART/
[5] http://www.prismmodelchecker.org/
[6] http://www.land.ufrj.br/tools/tangram2/tangram2.html
[7] http://www.doc.ic.ac.uk/~ajf/Research/manual.pdf
[8] http://people.ee.duke.edu/~kst/
[9] http://jmt.sourceforge.net
[10] http://www.sce.carleton.ca/rads/lqns/
[11] http://web2.uwindsor.ca/math/hlynka/qsoft.html

of the above academic tools can be found in [5]. It is interesting to point out that, although the networking community has traditionally relied on queueing theory, popular tools such as NS-2[12] have been used quite rarely to simulate QNMs. Indeed, NS-2 and other networking tools are well suited for the description of network components and protocols, but this is usually a level of detail that is excessive for the abstractions used in QNMs. More recently, the OmNet++ framework has tried to invert this trend by publishing several tutorials for QNM analysis[13]. In spite of the large number of tools available, the techniques used for QNM simulation are quite similar: they all implement the classic discrete-event simulation paradigm, where a calendar of events, often based on a priority queue, is maintained in order to process chronologically arrival and departure of jobs from the resources. A variety of papers and books provide help to the developer of such tools to implement the most complex tasks, such as statistical analysis, transient filtering, rare event simulation, and implementation of preemptive disciplines such as processor sharing [10,13,29,27].

More recently, new interesting techniques have been integrated in academic and commercial tools in order to analyse QNMs. We here try to survey for the first time these emerging ideas.

Ψ^2 is a tool[14] for steady-state analysis of QNMs that is based on perfect simulation theory. The fundamental ideas of this new simulation approach is to consider the Markov process underlying the queuing network and first identify a set of representative events such as job arrivals or end of service. A transition function $\Psi(\mathbf{x}, e)$ is then defined to represent the evolution of the current network state $\mathbf{x}$ as a function of each possible event e. The perfect simulation technique applies in its original form to the case where all events e are monotonous, i.e., such that for each pair of states $(\mathbf{x}, \mathbf{x}')$ for which a partial ordering $\mathbf{x} \leq \mathbf{x}'$ exists it is $\Psi(\mathbf{x}, e) \leq \Psi(\mathbf{x}', e)$ for all events e. If such monotonicity condition is satisfied, a case which can be verified for large classes of queuing networks, Ψ^2 can simulate the model efficiency by an adaptation of the coupling-from-the-past (CFTP) algorithm. This algorithm involves an iteration that estimates steady state by randomization of the recent trajectories of the system prior to reaching the steady-state. The computational costs of the techniques grows linearly with the state space size, therefore significantly improving over the cubic or quadratic costs of a direct numerical solution of the infinitesimal generator.

Opedo[15] is a recent tool for the optimization of performance and dependability models. This tool shows a rare case of a complex framework built upon open-source modeling tools such as OmNet++, Java Modelling Tools, APNN[16], and the techniques developed in papers such as [2]. The fundamental idea is to define a black-box interface to describe the output of existing modeling tools and develop a numerical framework for parameter optimization that is based only on black-box descriptions. Opedo uses a number of nonlinear search techniques to estimate a local optimum, such as pattern search and response surface methodologies, or a global optimum, such as evolutionary

---

[12] http://www.isi.edu/nsnam/ns/

[13] http://www.omnetpp.org

[14] http://psi.gforge.inria.fr/

[15] http://www4.cs.uni-dortmund.de/Opedo/

[16] http://www4.cs.uni-dortmund.de/APNN-TOOLBOX/

algorithms and Kriging methods. Integrated frameworks of this type appear promising especially in the context of software performance engineering where the first studies for large automatic software tuning based on performance models have recently appeared [24]. Such frameworks automatically search for a set of design parameters that can ensure desired levels of responsiveness in an application.

Mathworks SimEvents[17] is a commercial extension of the MATLAB/Simulink simulator to support QNMs. Simulink has traditionally focused on simulation of continuous-time dynamical systems based on a number of ODE integrators, therefore the integration of SimEvents inside this framework allows to combine discrete simulation models with continuous-state simulation. Another interesting feature is that the tool description proceeds through the block diagram notations that are popular in control theory, therefore strongly emphasizing the input/output behavior of each component in the simulation. Another advantage of such tool over existing QNM simulators is that it can natively combine finite-state machines and flow charts which are useful for integration with hardware system and complex process models, respectively. Finally, another advantage is the robustness of the Simulink simulator, which is used in real-time critical industrial applications and therefore is affected by very few software bugs due to the high maturity level of the tool.

Another direction explored recently is the idea of considering fast queuing network approximations at the stochastic process level by means of linear programming. An advantage of these approaches over simulation is that linear programming can accurately describe hundreds of thousands or even millions of state probabilities. In the lp-rBm technique in [31], a queuing network can be described as a multidimensional reflected Brownian motion (rBm), which is extremely powerful to represent non-exponential distributions. Linear programming is used to approximate the equilibrium of the rBm which is not available in closed-form. The MAPQN Toolbox[18] applies to closed models with general service time distributions. A number of necessary balance equations between the state of the queue is formulated, leaving equilibrium probabilities as unknown. This returns estimates that are provable bounds on the exact solution.

The wide availability of tools for QNMs suggests that much has been already done in support of the development and application of these models outside pure research. However, a number of additional extensions may be considered that are still lacking in the performance community. First, most tools seem to lack a software regression support in order to validate successive releases on a set of representative models. While these regressions are easy to define, it is harder to find in the literature detailed published solutions for reference models, especially for models with a mixtures of complex features (e.g., non-preemptive multiclass priorities, forking, finite capacity regions). This appears a limitation that the literature should address, since individual groups are currently not sharing their best practices and useful case studies with the rest of the community.

Next, with the exception of few packages such as SMART, Java Modelling Tools, or Opedo, it appears that analytical results have been poorly integrated and exploited in current tools, possibly with the exception of the class of product-form models. While there exist indeed limitations to the accuracy of some approximations, it is a

---

[17] http://www.mathworks.com/products/simevents/
[18] http://www.cs.wm.edu/MAPQN/

contradiction in terms that the largest body of work of the performance modeling community is at all effects marginalized from the software implementation and distribution. Larger research families, such as the linear algebra or parallel computing communities, have addressed these problems by creating public repositories to share standard implementations of important algorithms. Unfortunately, no similar experience has been attempted (at least to the best of the authors' knowledge) in the performance evaluation community. New recent attempts are trying to correct this issue[19], however more cooperation is needed in our community to promote the success of such initiatives.

4 Petri Nets

Petri Nets (PN) are a graph based formalism, capable of visually describing system characterized by parallelism and synchronization. A Petri Net can be seen as a bipartite graph, where nodes are dived into two classes called places and transitions. For an historical review of Petri Nets, the reader can refer here [20]. Applications of Petri Nets to performance evaluation, mainly rely on their stochastic version (SPN - *Stochastic Petri Nets* and its generalization (GSPN - *Generalized Stochastic Petri Nets*). For a tutorial on GSPNs, the interested reader can refer to [19]. A large number of tools are available for GSPNs, e.g., GreatSPN[21], SMART, PIPE2[22].

Petri Nets are usually analyzed in steady state or in transient, either by discrete event simulation or by numerical techniques. In the latter case, the state space of the model is computed and its temporal evolution is mapped to a CTMC. Performance indexes are then obtained from the transient or steady state solution of the obtained CTMC.

Beside steady state and transient analysis, the bipartite graph structure of the model allows several analysis to be performed without explicitly generating the state space. Such analysis allows the determination of invariants, bounding properties, and ability to fire transitions. Petri nets are supported by several tools, each one having its own characteristics for what concerns the analysis techniques and for the capability of verifying different types of structural properties. A reference to the tools supporting PNs analysis can be found here[23]

Throughout the years, several new types of PNs have been devised to simplify the study of computer systems. Each type of PN has its own benefits and it is supported by some specific tools. In the following we will briefly summarize some of the PN families that are currently used to address real-world modelling problems.

The GreatSPN tool supports *Stochastic Well-formed Nets* (SWNs) [6], an important extension to Colored Petri Nets (CPNs) [18]. CPNs improves the concept of marking of place by adding attributes to the tokens. Attributes are called colors, and belong to specific classes called types. Each token has associated a set of types that defines its attributes. When a transition fires, it removes some of the tokens from its input places, and

---

[19] http://www.perflib.net

[20] http://www.informatik.uni-hamburg.de/TGI/PetriNets/history/

[21] http://www.di.unito.it/~greatspn/index.html

[22] http://pipe2.sourceforge.net/

[23] http://www.informatik.uni-hamburg.de/TGI/PetriNets/tools/
quick.html

collects their attributes into variables. At the same time, the firing of a transition inserts tokens into its output places. The attributes of the generated token are computed as functions of the variables collected from the input places. CPNs are important because they allow to use colors to model different types of objects and to model object-dependent behavior in a compact way. SWNs are CPNs where the functions that changes the color of the tokens have special forms. SWN have several interesting properties that allows some analysis to be performed on a reduced symbolic representation of the state space of the model. This allows to significantly reduce the size of the state space, thus increasing the size of the model that can be considered.

As observed in Section 2, the SMART tool has been one of the first tools to encode the state space of the CTMC underlying a GSPN using the MDD and to encode the transition matrix using the Matrix Diagram technique. When applied to PNs, the tool can exploit some of the structural properties of the networks to better organize the MDD levels, and to significantly reduce the time required to compute the state space of the model.

The TimeNET[24] tool has been one of the first tools to support *Non-markovian Stochastic Petri Nets* (NMSPNs) [38]. These type of PN allow transitions to fire following general non-exponential firing time distributions. In this case transitions are characterized by an extra parameter, the memory policy, used to define what happens when a transition, after being disabled, becomes enabled again. Three different policies are possible: *prd* (preemptive repeat different) when a new sample for the distribution is computed every time, *prs* (preemptive resume) when the transition continues its activity by firing after the remaining time, and *pri* (preemptive repeat identical) when, after each time a transition gets enabled, it restarts its activity but maintains the sampled firing time. Non-exponential transition can be solved by approximation as PH-type distributions, or by explicitly considering a memory variable in either the time domain or in the transformed domain.

The Oris[25] and Romeo[26] tools support *Timed Petri Nets* (TPNs). TPNs assigns intervals to timed transitions. Each transition fires after a time that belongs to the associated interval. Nothing is assumed about the distribution of the firing time of a transition, for this reason TPN allows non-determinism, and are particularly suited for Real-time applications. TPN tools transform a TPN model in a set of possible evolution region, each one described by a Difference Bounds Matrices (DBM). Performance indexes are then computed directly from the DBM set.

Fluid Stochastic Petri Nets, Continuous Petri Nets and *Hybrid Petri Nets*, add a new kind of place, the fluid place which contains a continuous marking. The three formalisms are very similar and differs only for small technical details. Even if fluid formalisms have been widely studied in the literature, very few tools actually consider them. One example is the FSPNedit tool [11], which allows for both simulation and numerical analysis of FSPNs. Analytical solution of FSPNs is performed by computing transforming the model into a set of partial derivatives differential equations, and then by computing performance indexes from the solution of the PDEs. Simulation is

---

[24] http://www.tu-ilmenau.de/fakia/TimeNET.timenet.0.html

[25] http://www.stlab.dsi.unifi.it/oris/

[26] http://romeo.rts-software.org/

performed using the time-scale transformation, since dependency on fluid values makes the system non-homogeneous.

Although MDD-based technique have significantly reduced the memory requirements for encoding the CTMC underlying a GSPN, allowing models with billions of states to be stored in few kilobytes and to be generated in fractions of seconds, the probability vector still have to be encoded directly. This actually limits the maximum number of states and thus the complexity of the models that can be addressed. Some research has already been done on techniques to encode the probability vector, but none has provided satisfactory results yet.

For what concerns the use of non-exponential transitions, the current approach tends to increase significantly the state space, limiting thus the number of non-Markovian activities that can be included in a model. Several techniques have been devised to describe the state space of a non-Markovian system using MDD. So a solution to the encoding of the probability vector should also help in allowing the use of an extended number of non-exponential transitions in NMSPNs models.

5   Fault Trees

Fault trees (FTs) is a formalism specifically devised for reliability analysis, and originally created at Bell Labs in the 60s. A fault tree contains a root node called the top event, and several leaves called basic event. Basic events are connected to the top event by arcs that traverse a series of intermediate nodes called gates. Gates usually correspond to boolean operations (the classical and, or and not), but might also contain extended primitives like the "m out of n". Basic events of a FT usually represent the occurrence of a faulty condition (such as the breaking of a component). The top event determines the state of the entire system, which might be compromised whenever one or more of its components fails. Due to their simplicity and their popularity, there exists many tools that can address the solution of FTs. A short list of available tools can be found here[27].

In most cases, the user can assign a probability distribution to the basic event and the tool computes the probability distribution of the top event. If the basic events are independent, the exact distribution can be computed with simple algebraic operations. Difficulties arises when considering correlation among events, repair from faulty states, and cascade of events. In such case FTs are usually analyzed resorting to discrete event simulation, or by mapping them to other formalism such us GSPNs.

One of the most recognized tool is SHARPE, already introduced in Section2, which can perform several different analysis over given FTs. SHARPE also supports other similar formalisms like *Reliability Block Diagrams*, and *Reliability graphs*. The former characterize processes with a block diagram that explicitly shows the introduced redundancy. The latter describe systems with a graph where the failure rates are associated with edges. In this case, the required condition is that there exists a path from one node (called the source) to another node (called the sink). SHARPE solves the proposed models analytically by characterizing the FT with *exponential polynomial distributions*, and then by exploiting the analytical properties of such distributions.

---

[27] http://www.enre.umd.edu/tools/ftap.htm

The tool RADYBAN [25] exploits the analogies between fault trees and another probabilistic formalism: the Bayesian Networks (BAs). BAs are used to represent uncertain knowledge in probabilistic environments, and can be suited to perform reliability analysis. It is possible to prove that BAs can be more powerful than FTs, and that they can be suited to model more advanced features like noisy gates (that is gates that do not perform their and, or, not task deterministically).

6   Process Algebras

Process Algebras are a class of performance evaluation formalisms that describes models using a simple text-based representation. Even if the term was coined in the 80s, studied that lead to the definition of this formalism started in the early 70s. A nice historical introduction to Process Algebra can be found here[28]. In particular the modeling technique split a system into several interacting components. Each component can perform a set of actions, and then evolve to perform other activities. Usually the evolution of each component is represented by a very simple grammar such as:

$$S ::= \alpha.S_1 \mid S_1 + S_2 \mid C_S, \tag{1}$$

where $\alpha.S$ is the prefix operator that tells that component S evolves to component S_1 after performing action α, $S_1 + S_2$ is the *choice* operator that tells that component S can evolve to either S_1 or S_2, and C_S is a constant used to address a sequence of components. Components can then be composed in models, using another very simple grammar such as:

$$P ::= P_1 \bowtie_L P_2 \mid P/L \mid S.$$

Operator $P_1 \bowtie_L P_2$ is the *cooperation* of P_1 and P_2 over the set of actions L. In order to perform one action in set L, the two components P_1 and P_2 have to synchronize, and the action is executed simultaneously. Operator P/L is called *hiding*, and simply prevents the resulting component to synchronize on actions belonging to the set L, by making such actions private (or internal).

In performance evaluation, particular dialects of Process Algebra that associate timing to events are used. Two common timed extensions of Process Algebra are PEPA *(Performance Evaluation Process Algebra)* and EMPA *(Extended Markovian Process Algebra)*. For example, PEPA modifies the grammar presented in Equation 1 to $S ::= (\alpha, r).S_1 \mid S_1 + S_2 \mid C_S$, by adding a rate r to actions (that now are denoted as (α, r)). Each action is executed after an exponential distributed time with rate r.

Process algebras are supported in several tools such as ipc/Hydra[29], PEPA - Workbench[30], Two towers[31], and Mobius. Usually analysis is performed by enumerating the states that can be reached by the model (exploiting symmetries and creating symbolic states to reduce the size of the state space), and by creating a CTMC or a *Generalized Semi-Markov Process* (GSMP) to study the evolution of the model.

---

[28] http://www.win.tue.nl/fm/0402history.pdf
[29] http://www.doc.ic.ac.uk/ipc/
[30] http://www.dcs.ed.ac.uk/pepa/tools/
[31] http://www.sti.uniurb.it/bernardo/twotowers/

The PEPA-Workbench tool, beside offering the possibility to analyze a model using CTMC or discrete event simulation, it allows the use of new approximations based on fluid interpretation and differential equations [16].

Several application-domain specific derivation of Process Algebra have been produced. For example the tool BioPEPA-workbench[32], supports an interesting extension of PEPA called *BioPEPA*, that defines a grammar that is suited for describing the processes that models the chemical reaction happening in biochemical system.

7   Architectures, Trends and Expectations

Several important trends are leading the current researches, such as the conjunction of qualitative (mostly model checking) and quantitative analysis, and the scalability and the parallelization of tools. Due to space constraint, instead of briefly considering several aspects, we focus on a single specific trend: *tool inter-operability*.

Following basic software engineering principles, the internal structure of modern performance tools is often organized around a clear separation of concerns. Separate software modules implement scientific algorithms, user interfaces, managers for performing repeated cycles of experiments, and primitives for generating, storing, and possibly simulate the models. Both in academia and industry, such modularization helps in separating and organizing the activity of scientific programmers (or students) involved in the development of the different parts of the code. On the other hand, this has been hardly combined with software reuse, since most performance groups opted to develop their own libraries instead of creating a public framework for sharing their work with the community.

We believe that such practices does not follow modern trends of software engineering, especially of the open source community that has promoted in recent years the sharing and reuse of software artifacts. In particular, major steps have been done towards software integration by means of standardized programming libraries (e.g., the Java Platform) and data exchange languages (e.g., XML). These technologies create interesting opportunities also to improve the way performance tools are defined.

A proposal for leveraging on these technologies that we describe in this section is to define a new family of *performance meta-tools* that could help the integration of the software artifacts available in the performance community. The general structure of a performance meta-tool, referred to as *p-platform*, is outlined in Fig. 1. Each layer of a p-platform describes a typical concern of a performance modeling study and we propose to organize the interaction between different submodules by means of layers, communicating through standardized meta-languages. In particular we suggest the use of XML as a possible implementation of a meta-language used to describe the interfaces between the layers. The ability of integrate different performance tools into a public framework would substantially improve the robustness and scale of current performance tools. It would also give the ability to users to select the components that best fit the goals of the performance study. The number of components required in the analysis is not fixed and depends on the objectives. Furthermore, the component of a layer can be skipped,

---

[32] http://homepages.inf.ed.ac.uk/stg/software/biopepa/

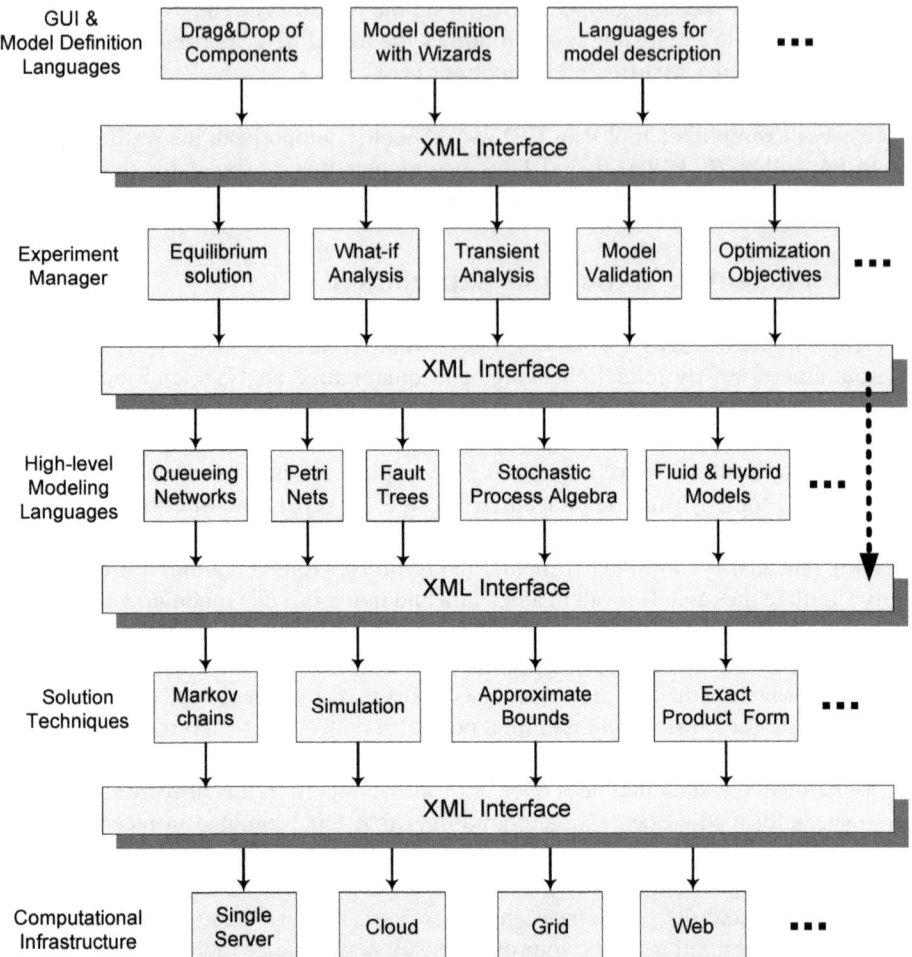

Fig. 1. Main components of the architecture of a performance meta-tool, referred to as p-platform

for example one may evaluate a Markov model without using any high-level modeling language. According to the p-platform description, main steps of a performance study would be:

1. analysis of the intended use of the model based on the study's objectives; identification of the best technique to be used to describe the problem and its characteristics;
2. identification of the solution algorithm required to produce the type of results needed, e.g., equilibrium or transient values, exact or approximate solutions;
3. design of the experiment to be undertaken through the manager module, e.g, what-if, single run, optimization technique;
4. selection of the computing infrastructure to be used to run the numerical or simulation algorithms, single server, cluster, cloud, web, etc.

Although the principles outlined are simple, to the best of the authors' knowledge integration via XML has been poorly adopted by current performance modeling tools. The only notable exceptions are the Java Modelling Suite which coordinates data exchange between modules using XML files, the ongoing performance interchange format project PMIF [33,34], and the Petri Net Markup Language PNML[33] which is supported by a growing number of PN tools. A possible explanation for this is that the majority of the tools have been developed started from the 80s, therefore according to the software engineering principles of the time. We believe that open release of the source code through open platforms such as Sourceforge would represent a first step in the right direction of helping external groups provide ideas, report bugs, and discuss in forums the issues we have outlined in this section.

8 Conclusions

In this paper, we have reviewed past and present efforts towards implementing software tools to support performance evaluation activities. Our analysis has revealed the area to be still very active, with a number of new simulation and analysis techniques still being proposed for classic models such as queuing networks and Petri nets. We have also argued that the recent advent of standardized data exchange languages such as XML opens new opportunities towards integrating existing community efforts into larger performance evaluation frameworks. To support this idea, we have outlined a performance meta-tool architecture, named *p-platform*, that provides high-level intuition on the basic blocks needed to define such frameworks.

Acknowledgement. The authors wishes to thank the anonymous reviewer for the valuable comments that helped to improve the quality of the paper. The work of Giuliano Casale has been supported by the Imperial College Junior Research Fellowship.

References

1. Baskett, F., Chandy, K.M., Muntz, R.R., Palacios, F.G.: Open, closed, and mixed networks of queues with different classes of customers. Journal of the ACM 22(2), 248–260 (1975)
2. Buchholz, P., Ciardo, G., Donatelli, S., Kemper, P.: Complexity of memory-efficient kronecker operations with applications to the solution of markov models. INFORMS Journal on Computing 12(3), 203–222 (2000)
3. Buzen, J.P.: Computational algorithms for closed queueing networks with exponential servers. Comm. of the ACM 16(9), 527–531 (1973)
4. Carlson, G.: A user's view of hardware performance monitors. In: Proc. IFIP Congress, vol. 71, pp. 128–132. North-Holland (1971)
5. Casale, G., Muntz, R.R., Serazzi, G.: Tools for computer performance modeling and reliability analysis. ACM Performance Evaluation Review 36(4) (2009)
6. Chiola, G., Dutheillet, C., Franceschinis, G., Haddad, S.: Stochastic well-formed colored nets and symmetric modeling applications. IEEE Transactions on Computers 42(11), 1343–1360 (1993)

---

[33] http://www.pnml.org/

7. Chung, M.-Y., Ciardo, G., Donatelli, S., He, N., Plateau, B., Stewart, W., Sulaiman, E., Yu, J.: A comparison of structural formalisms for modeling large markov models. In: Proc. of IPDPS, vol. 11, p. 196 (2004)
8. Buzen, J.P., et al.: Best/1 - design of a tool for computer system capacity planning. In: Proc. of the 1978 National Computer Conf., pp. 447–455. AFIPS Press (1978)
9. Ferrari, D., Serazzi, G., Zeigner, A.: Measurement and Tuning of Computer Systems. Prentice-Hall (1983)
10. Fishman, G.S.: Statistical analysis for queueing simulations. Management Science 20(3), 363–369 (1973)
11. Gribaudo, M.: Fspnedit: a fluid stochastic petri net modeling and analysis tool. In: Proc. of Tools of Aachen 2001, pp. 24–28 (2001)
12. Reiser, M., Haring, G., Lindemann, C. (eds.): Dagstuhl Seminar 1997. LNCS, vol. 1769. Springer, Heidelberg (2000)
13. Heidelberger, P., Welch, P.D.: A spectral method for confidence interval generation and run length control in simulations. Comm. of the ACM 24(4), 233–245 (1981)
14. Herman, D.J.: Scert: a computer evaluation tool. Datamation 13(2), 26–28 (1967)
15. Herman, D.J., Ihrer, F.: The use of a computer to evaluate computers. In: Proc. Conf. 1964 SJCC, Washington DC, pp. 383–395. Spartan Books (1964)
16. Hillston, J.: Fluid flow approximation of pepa models. In: QEST 2005, pp. 33–42, 19–22 (2005)
17. IBM. General purpose systems simulator iii user's manual. Technical Report Form H20-0163, IBM (1965)
18. Jensen, K.: Coloured Petri Nets: Basic Concepts, Analysis Methods, and Practical Use, 2nd edn. Springer, Heidelberg (1997)
19. Kartson, D., Balbo, G., Donatelli, S., Franceschinis, G., Conte, G.: Modelling with Generalized Stochastic Petri Nets. John Wiley & Sons, Inc., New York (1994)
20. Kleinrock, L.: Queueing Systems, Theory, vol. 1. John Wiley & Sons, New York (1976)
21. Lavenberg, S.S.: A perspective on queueing models of computer performance. Performance Evaluation 10(1), 53–76 (1989)
22. LeMay, E., Unkenholz, W., Parks, D., Muehrcke, C., Keefe, K., Sanders, W.H.: Adversary-driven state-based system security evaluation. In: MetriSec 2010: Proceedings of the 6th International Workshop on Security Measurements and Metrics, pp. 1–9. ACM, New York (2010)
23. Markowitz, H.M., Hausner, B., Karr, H.W.: Simscript: a simulation programming language. Prentice Hall (1963)
24. Martens, A., Koziolek, H., Becker, S., Reussner, R.: Automatically improve software architecture models for performance, reliability, and cost using evolutionary algorithms. In: WOSP/SIPEW, pp. 105–116 (2010)
25. Montani, S., Portinale, L., Bobbio, A., Codetta-Raiteri, D.: Radyban: A tool for reliability analysis of dynamic fault trees through conversion into dynamic bayesian networks. Reliability Engineering and System Safety 93(7), 922–932 (2008); Bayesian Networks in Dependability
26. Nielsen, N.R.: Computer simulation of computer system performance. In: Proc. of ACM National Meeting, pp. 581–590 (1967)
27. Pawlikowski, K.: Steady-sate simulation of queueing processes: A survey of problems and solutions. ACM Computing Surveys 22(2), 123–168 (1990)
28. Potier, D., Veran, M.: The markovian solver of QNAP2 and examples. In: Hasegawa, T., et al. (eds.) Computer Networking and Perf. Eval., pp. 259–279. North-Holland, Amsterdam (1986)
29. Sauer, C.H., Chandy, K.M.: Computer Systems Performance Modeling. Prentice-Hall (1981)

30. Sauer, C.H., McNair, E.A., Kurose, J.F.: The research queueing (RESQ) package, version 2: Introduction and examples. Technical Report IBM rep. no. RA 138, IBM (1982)
31. Saure, D., Glynn, P., Zeevi, A.: A linear programming algorithm for computing the stationary distribution of semi-martingale reflecting brownian motion (under submission)
32. Schwetman, H.: CSIM Reference Manual (1988)
33. Smith, C., Llado, C.: Performance model interchange format (pmif 2.0): Xml definition and implementation. In: Proc. of QUEST 2004. IEEE Press (2004)
34. Smith, C., Lladó, C., Puigjaner, R.: Performance model interchange format (pmif 2): A comprehensive approach to queueing network model interoperability. Perform. Eval. 67(7), 548–568 (2010)
35. Smith, C.U., Williams, L.G.: Performance Engineering Evaluation of CORBA-Based Distributed Systems with SPE•ED. In: Puigjaner, R., Savino, N.N., Serra, B. (eds.) TOOLS 1998. LNCS, vol. 1469, pp. 321–335. Springer, Heidelberg (1998)
36. Smith, C., Williams, L.: Performance Solutions: A Practical Guide to Creating Responsive, Scalable Software. Addison-Wesley (2001)
37. Stewart, W.J.: Introduction to the Numerical Solution of Markov Chains. Princeton University Press, Princeton (1994)
38. Trivedi, K.S., Bobbio, A., Ciardo, G., German, R., Puliafito, A., Telek, M.: Non-markovian petri nets. In: SIGMETRICS 1995/PERFORMANCE 1995, pp. 263–264. ACM, New York (1995)

Energy: A New Criteria for Performances in Large Scale Distributed Systems

Jean-Marc Pierson

IRIT
University of Toulouse
pierson@irit.fr

Abstract. In the framework of the workshop "Performance Evaluation of Computer and Communication Systems: Milestones and Future Challenges", this paper proposes to address the emerging criteria of energy. Used since a long time in embedded systems where battery operated devices needed a careful handling, the energy metric is taking a large momentum in the last years on large scale systems where thousands of nodes collaborate to serve high end infrastructures like web servers, clouds and grids.

1 Introduction and Motivation

Since the last 5 years, we witness the raise in interest for energy aware infrastructures and computing in large scale systems. What appeared at its beginning as a hype is slowly taking more importance in the everyday life when operating large scale systems. Beside the ecological view coming from the carbon-related global warning concern, attraction is also garnered by several other actors: CEOs and system administrators handling large IT infrastructures caring for their electrical budget or their electricity cap, electricity providers who need to serve optimally a growing demand, and finally computer and mathematical scientists who see an opportunity to explore a new scientific field.

The demand in research in energy-efficiency in large scale systems is supported by several incentives [29,6,33], including financial incentives by government or institutions for energy efficient industries / companies [28]. Indeed, studies like [4] report that the IT consumption accounts between 5 to 10% of the global growing electricity demand, and for a mere 2% of the energy. Data centers hosting web services or cloud computing gather thousands of nodes and every single Watt saved on each machine every second is making a huge difference at the end on the energy bill (either money ROI or ecologically speaking).

While some investigate how to reduce the ecological impact [30], most of the works are driving research for reducing electricity demands in terms of Watts. This paper is focusing on this latter and puts in the perspective the new challenge of energy consumption as a criteria for performances in large scale distributed systems. It is intended to serve as an introduction to the relevant metrics and appropriate methods when one address energy as a criteria to performance evaluation.

This paper is organized as follows. Section 2 reviews some of the standard metrics for performance evaluation in large scale systems. Section 3 is concerned with energy

K.A. Hummel et al. (Eds.): PERFORM 2010 (Haring Festschrift), LNCS 6821, pp. 38–48, 2011.

metrics and energy benchmarks before investigating the integration of energy concerns in section 4. Section 5 concludes the paper.

2 Performance Metrics in Large Scale Systems

Since the early ages of computer science, performance evaluation is related to the number of operations being done during a certain duration. Manufacturers and computer scientists have investigated ways to augment the performances of silicon chips and their usage.

Since the conjecture of Moore in 1965 [26] and refined in 1975 [27] stating that the number of transistors in a chip will double every two years, valid for over 40 years (now we can see its limit), the emphasis is put on the number of operations that can be done during one time interval. While MIPS (Million Instructions per Seconds) was a common measure in first ages to measure the performances of processors it is not anymore that important: Indeed, this measure does not take into account many aspects of a computer not related to CPU, like memory hierarchies or input/output bandwidth, or real application workflows. The preference is now to measure on benchmarks the performances of a computer, from single mono-core node to multicore many-nodes architectures. Specific benchmarks exist to stress the architectures on their CPUs, their memory, their network, their disks. For CPU, we measure performances in terms of number of operations per second (particularly floating point operations per second, or Flops), on well known benchmarks. For instance, SPEC (Standard Performance Evaluation Corporation) [9] and TPC (Transaction Processing Performance Council) [11] both have a collection of benchmarks to evaluate different architectures and applications. Energy concerns are present in these benchmarks as we will detail later in section 3.

The Linpack benchmark [13] is the building brick of the Top500 list [14] that enlists every six months the most powerful operational (super-)computers. Even if not perfect since it does not encompass all the applications characteristics (it is a matrix computation benchmark made for scientific and simulation codes), this benchmark has been accepted by the high performance computing community to compare machines. As for an example, the first in the list is reaching 1.7 Petaflops ($1.7 * 10^{15}$ flops). Another benchmark proposed by J. Dongarra is the HPCC (High Performance Computing Challenge) [24], which encompasses more aspects than just Linpack (but still includes it). Unfortunately, as we will see in section 3 it does not take into account the energy spent for the calculation.

For some time largely ignored by the growing capacities of monocore CPUs, parallelism is coming back on stage due to problems related to physical limits in the chips design and the programmed end of Moore's law. Indeed, smaller integrated chips using CMOS technology induce a heat dispersion problem, due to the existing relations between speed, frequency, voltage and energy (thus heat). Multicore CPUs are now the norm, with trends to build on hundred cores (already produced) or even thousand core CPUs. All the knowledge gained in the late 80s from parallel computing garners a new interest with these new highly integrated architectures.

Many metrics have been built in order to check different aspects of parallel computing and distributed systems.

In the field of parallel programming, the classical metric is the speed-up, which gives the acceleration in terms of time to solution (TTS) of a parallel program using P processors against the same program using 1 processor. It is simply the sequential time over the parallel time. As a first approach, we can already see that the faster a program completes, obviously the less energy it uses.

Another metric is the scale-up. This metric helps to figure out the impact of processing bigger problems with more processors. In the best case, P processors should be able to process a problem of size $P * N$ in the same time than 1 processor is processing a problem of size N. In terms of energy concerns, the scale-up translates to the fact that obviously more processors will consume more energy than one.

Fairness is insuring that the different jobs are processed equally and if there is no starvation. Most of the time a job is processed when allocated all its required resources. Another mean is to allocate only part of the resources. The yield is characterizing how much resources are allocated to a job in comparison with what was asked by this job. Reaching a high minimum yield assures some fairness in the system. We will come back to this notion later in section 3.1 since it allows to trade resources (thus time to solution) for energy. A problem resides in the fact that the resource consumption have to be monitored and may cause overhead when done dynamically.

Other criteria of performances include throughput, latency, user satisfaction, security. Service Level Agreement (SLA) have appeared lately to handle the required level of processing and resources that meets specified needs. In terms of energy awareness, we will see how SLA translates in most common works.

3   Energy Performance of Large Scale Systems

While a major topic since several years in embedded systems with battery operated devices, energy-awareness raised interest since only a few years for large scale systems like super-computers, clusters, grids and clouds. For a long time, energy consumption has simply been ignored in the performance evaluation in parallel architectures, parallel programming, and lately grid computing.

For the sake of comparison, the first UNIVAC I (UNIVersal Automatic Computer) machine in the 1950's was consuming 125 kW for 1905 operations per second[1]. Today's best as mentioned before is reaching 1.7 Petaflops at a cost of 6950 kW.

In 2008, a survey looking back on the TOP500 list elaborates a classification for today's best [25]. The author proposes a new class of Power Efficient Systems with the rise of architectures (and vendors) taking energy concerns seriously.

A common office computer consumes between 120 and 200 Watts while high end servers consumes between 80 to 300 Watts. Several studies [16,23] split the share of the energy consumption in a computer as this: CPU accounts for 37% of energy consumption, memory is 17%, PCI slots are 23%, motherboard is 12%, while disk at 6% and fans with 5% are closing the list. Note here that this does not include the power supply (which is accounting for more or less 20% loss), the networking infrastructure and all the cooling infrastructure. In data-centers for instance the cooling can consume as much electricity than the computers themselves.

---

[1] Source: Wikipedia.

Energy consumption can be considered in mainly two directions: The energy efficiency identifies the performance of the system with respect to the energy costs (Section 3.1 on metrics, Section 3.2 on benchmarks). Another direction is to put either energy savings or performances metrics as optimization goals and others as constraints, or to check for Pareto solutions taking into account both (for instance the power yield in Section 3.1 or in placement and scheduling techniques, see Section 4.3), knowing hardware possibilities (Section 4.1).

3.1   Energy Metrics

The first immediate metric that has been used coming from Flops is Flops/Watt. The idea is to measure the number of flops that can be achieved using one Watt. Simplistic enough, this metric has the merit to be easily understandable and related to its ancestor in Top500. It is used by the Green500 listing (see section 3.2). A problem mentioned with this metric is that it measures the power, but not the energy spent. The power view is instantaneous while energy E relates to power P over a period of time t: $E = P * t$.

Hence, two obvious way can be used to reduce energy consumption: Either by reducing power consumption of the computers, or by reducing time to produce the result. When an infrastructure is always on with the same power consumption factor in average, the time is not an issue. In this situations the Flops/Watt metric makes sense. Less related to number of Flops, a metric considers the number of operations (not only flops) per Watt. Conversely, another approach is to measure the average power the infrastructure needs to achieve a given operation. The Spec-power and TPC-Energy benchmarks (see section 3.2) are using these metrics, respectively.

Still a difficulty appears: Do the maximum power or the average power have to be taken into account? Nowadays, many components have internal or software means to reduce the power consumption (see section 4.1) hence the course of power consumption over time can have big variations. Some infrastructures even rely on unused nodes switch-off to zero power consumption of a set of nodes (see section 4.3).

Another metric used is the energy itself when accounting for the energy of finite applications. The idea is to measure the Time To Solution and the consumed (max, average) power. The result of the multiplication is expressed in Joules (Watts.s) or Watt.hour.

Metrics can be elaborated from other traditional metrics from parallel systems. We can imagine speedup per Watt, scale-up per Watt or any combination of these.

Example of a metric: The power yield. In [5] we extend for instance the yield defined in [35] so that the most energy efficient machines for hosting a set of jobs are chosen. The consumption reduction problem translates to multi-objective optimization problem, where energy adds to CPU, memory (for this work), network bandwidth, and others as constraints of the problem.

Our approach is to rely on the demands a job has on the infrastructure (in terms of CPU and memory bounds), to satisfy at best these demands while minimizing the energy. More formally, [35] defined the yield Y_i of a job i as $Y_i = \sum_{j=1}^{H} (\frac{\alpha_{ij}}{\alpha_i})$ where H is the number of hosts, α_i is the CPU demand of job i and α_{ij} is the allocated part of the CPU of host j for job i. Hence, when a job is allocated what it demands, $Y_i = 1$. Y_i reflects the satisfaction of job i. The memory of a job is considered a rigid need: No allocation can be achieved if a memory need for a job can't be satisfied because of lack of memory.

We extended this definition using a power factor E_{ij}, reflecting the power cost of job i when running on host j. We defined $E_{ij} = \lambda(C_j^{max} - C_j^{min}) \times \alpha_{ij} + (1 - \lambda)\left[A_j(1 - \sum_{i'=1,i'\neq i}^{J}(\alpha_{i'j}))\right]$. The first term $(C_j^{max} - C_j^{min}) \times \alpha_{ij}$ reflects the extra power cost involved by the presence of job i on host j (with C_j^{max} and C_j^{min} being the maximum and minimum power of host j at full charge and when idle). The second term $A_j(1 - \sum_{i'=1,i'\neq i}^{J}(\alpha_{i'j}))$ reflects the attraction of host j for job i at a given time, when other jobs are already placed on it (with A_j an attraction factor that favors the consolidation of the jobs to the (already used) hosts that are consuming less power, for instance we took $A_j = C_j^{min}$). The parameter λ balances the effect of placement (first term, favoring the most energy efficient hosts) and the effect of consolidation (second term, favoring a minimal number of hosts switched-on).

Finally we defined YE_{ij} (the power yield) as $YE_{ij} = \frac{Y_i^{(1-k)}}{E_{ij}^k}$. The k parameter balances between performance (satisfaction) and power E_{ij}. This metric is used in an optimization heuristic to minimize the number of hosts while guaranteeing a given quality of service and/or energy reduction, taking into account additionally the memory used by one job.

As we can see from this example, to capture the effects of the placement of jobs on a set of hosts while still guaranteeing a level of performances is a tedious task: The process must take into account the specificities of the jobs (their demands in CPU and memory), the specificity of the infrastructure (the hosts characteristics, possibly the interconnect), the interactions between jobs (possibly the communications). As for now, an unique metric taking into account all the parameters for task allocation is not existing.

Other metrics. The GreenGrid alliance proposes to use the PUE for data centres infrastructures [34,18]. The PUE (Power Usage Effectiveness) is computed by dividing the amount of power entering a data center by the power used to run the computer infrastructure within it. It encompasses all the surrounding of the infrastructure, including power supplies, chillers, air conditioning. For instance best practices data centres can reach a PUE down to 1.1 while the average data-centers have a PUE of about 1.9.

Finally, one can either optimize the total energy used, or the energy cap. This latter is related to the maximum power consumption over a small period of time. In all infrastructures, the electricity provided by the energy providers is limited, due to physical constraints of the power (electricity) distribution network. A metric is to measure the maximum electricity that can be used in the infrastructure, in case of high workload and extreme situations (for instance when cooling is used at its maximum during hot periods).

3.2   Energy Benchmarks

The Green500 [8] initiative is challenging the most powerful machines in terms of Flops/Watts. In the same manner than the Top500, Linpack is used to compute the performance. The first 3 (numbers from October 2010) in the ranking achieve a 773

MFlops/Watt, with a total consumption of the corresponding machines of 57.54 kW. Notably, 8 out of the 10 first such machines use accelerators (PowerXCell from IBM, GPU from AMD ATI and Nvidia). The Green500 is exploring at the moment a new list, based on the HPCC benchmark.

SpecPower [32] is an industry-standard benchmark that evaluates the power and performance of servers and multi-node computers. The initial benchmark addresses only the performance of server side Java. It exercises the CPUs, caches, memory hierarchy and the scalability of shared memory processors (SMPs) as well as the implementations of the JVM (Java Virtual Machine), JIT (Just-In-Time) compiler, garbage collection, threads and some aspects of the operating system. It computes the overall server-side java operations per Watt, including the idle time on specific workloads. The comparison list includes 172 servers. Among these, a Fujitsu server with 76 quadcores (304 cores) reach a maximum value of 2927 ssj_ops/Watt.

The TPC (Transaction Processing Performance Council) proposes the TPC-Energy benchmark [15] for transactional applications: Web/Application Services, Decision Support, On Line Transaction Processing. TPC measures Watts/operations on the TPC benchmarks (for instance transaction per seconds). Only few servers from HP have now been evaluated. As an example, 5.84 Watts/transaction per seconds is given for an typical online transaction processing workload.

For more details and comparative studies on these energy benchmarks, the reader can refer to [31].

The EEMBC (Embedded Microprocessor Benchmark Consortium) has a similar approach. It is providing a benchmark for energy consumption of processors [2]. It is mainly dedicated for embedded systems and computes number of operations per joule linked to the over performance benchmarks of the consortium, measured on different standard applications for embedded systems.

Manufacturers provide information about the consumption of their components, using average loads. For instance, AMD describes the ACP (Average CPU Power) [1] that characterizes power consumption under average loads (including floating point, integer, java, Web, memory bandwidth, and transactional workloads, subset of TPC and SPEC benchmarks). Interestingly, this work shows for instance that cores can consume between 61% to 80% of the processor power, and processors consume less than 25-35% when idle.

4 Integrating Energy Concerns

Energy concerns have been integrated in many works at hardware, network, middleware, and software levels in large scale distributed systems. This section does not intend to provide an exhaustive view of these works, but rather representative trends followed by researchers worldwide. For further reading, eEnergy Conference proceedings [22] or the COST IC0804 proceedings [21] are providing good insights.

4.1 Hardware Level

Hardware has long been thought as the main (and often only) place where energy savings can be achieved at large. Manufacturers believed (and still believe) that more energy efficient hardware is the key issue, and that quicker processors will achieve the energy saving goals (working more on the t parameter of the energy formula of section3.1). Much developments have been achieved at the processors, memory, motherboards, network card, etc. levels. In [7], the COST Action IC0804 surveys such hardware leverages for energy. As an example, the Performance (P) and CPU (C) states in ACPI compliant components allow to reduce the power consumption in deep sleep or halt modes on processors almost to nothing (down to 2 Watts). These states can be controlled also by software, opening door to Dynamic Voltage Frequency Scaling (DVFS) or efficient consolidation techniques (see section 4.3).

Another aspect at hardware level is the problem of collecting energy usage. Virtually all the researchers are using different power meters: Some use oscilloscopes to measure directly the amperage and voltage at the hardware component level, while the majority in large scale systems are using external power meters. Nowadays, intelligent Power Distribution Units (PDU) allow to collect electricity demands at the plug, distribute it on several servers or nodes and aggregate values to send back to any interested party (like the middleware or software for instance). For instance the GreenNet infrastructure allowed us to measure over a long period of time the pattern of the usage of the Grid5000 platform and to propose appropriate enhancements at the middleware level [12]. Also, the TPC introduced an EMS (Energy Management Systems) to accompany its benchmark. The main advantage of such power meter is that they are not intrusive: They do not change the behavior of the observed nodes. Their drawback is the data acquisition frequency, (often one second), which is far too little when one wants to tune precisely operating systems for instance (where the quantum is order of magnitude smaller).

4.2 Network Level

At the network level, initial works investigated the energy consumption at the hardware level, trying to correlate between network activity and energy spent [20]. Results show that the traffic has little influence on the power consumption compared to the actual switch on of modules or plugs in the switches and routers. The metrics measured the number of bytes per Watt or the achievable bandwidth per Watt. Other works optimize routes so as to prefer less energy demanding technologies (like optical networks) or change dynamically the characteristics of these networks. In Ethernet networks, Adaptive Link Rate [19] is a solution where energy savings can be obtained by quickly changing the speed of network links in response to the amount of data transmitted.

4.3 Middleware Level

At the middleware level, two main complimentary solutions are used: Dynamic hardware adaptation and consolidation. Dynamic hardware adaptation mainly relies on the DVFS capacities. Dynamic Voltage and Frequency Scaling tries to gain energy on inactivity phases. The idea is to find the right clock for the right task. Since $power = voltage^2 * frequency * \alpha$, with α as a hardware and conditions related parameter, reducing frequency and/or voltage allows for a large spare of energy. As we saw in section 4.1, different combinations of frequency/voltage can be tuned by software

(P-C states). The metric is then evaluated against the frequency/voltage combination during the course of a particular application or a lifetime of an infrastructure.

Most of the works are doing consolidation using virtual machines to embed the jobs: In this approach, the main issue is to switch off as much hosts as possible to save energy while still guarantying quality of services. Techniques vary in the choice of the hosts since it depends on each application and on possible links between jobs distributed among the platform. The jobs do not have the same energetic behavior on every node (depending on their CPU, memory, I/O accesses). Moreover if communications occur between jobs, the techniques tend to collocate them on the same hosts. To evaluate the different techniques, the metric is related to the dynamic number and consumption of hosts for the application.

Example with a job allocation formalization and methodology. We proposed in [5] to model the problem of task allocation as a mixed integer linear problem, putting in equation the different constraints of the task allocation process. For instance, using the same notations than in section 3.1, one equation is $\forall i \sum_h \alpha_{ih} \leq \alpha_i$, with α_i is the CPU demand of job i and α_{ih} is the allocated part of the CPU of host h for job i. Afterwards, we defined several problems to be solved, given these constraints: To maximize the minimum yield of the jobs (definition of yield Y as in section 3.1), to minimize the energy (defined as the sum of the dynamic power of the switched-on hosts), and to find a tradeoff between them.

We will not detail here the model itself (due to limitation of space) but we will focus on the methodology and results. Once modeled, the problems can be solved using a mixed integer linear program, finding an optimal: minimal set of efficient hosts to run the jobs so as to guarantee the constraints. The resolution of this NP-hard problem proved to be much too long for real life instances (over 4 hosts and 12 jobs). Therefore we studied different heuristics (greedy-like, binpacking-like, based on the metric described in section 3.1) and compared them with bounds on the optimal. We came to the result that it is possible to find near-optimal placements of jobs in less than 1 second for instances of the problem with 500 hosts and 1500 jobs. ? After these encouraging results, we are now extending this work as to take into account the dynamic of the system, where jobs can change their demands during time and possibly migrate between the hosts of the infrastructure (to enable dynamic optimal consolidation).

4.4   Software Level

Since a long time, offline analysis of codes are performed in embedded systems to evaluate the energetic cost of processors within limited electric constraints. In large scale distributed systems, such offline analysis with apriori consumption models do not exist so far. Techniques like [10] start to dynamically relate a set of observed elements on the system to actual power consumption: Performance counters, load average, memory usage, etc. can be mathematically related to power consumption using linear regression techniques for instance. In these works the more metrics are observed and the more accurate the prediction of energy usage will be. Such techniques are difficult to apply on a very large scale, since a large number of elements have to be observed, possibly intrusively. Moreover, with virtualization mechanisms and communications, such regression techniques based on observations still have to be developed.

Other works include Service Level Agreement (SLA) considerations. In these works, the applications state their performance to achieve or the energy cap not to exceed. For instance, [3] uses machine learning techniques to achieve SLA specifications while [17] use autonomic computing for the same.

5   Discussion and Conclusion

In this paper we propose an overview of performance evaluation under energetic concerns. We describe commonly used metrics and benchmarks, before giving the main trends for energy savings, focusing especially on large scale distributed systems.

We have seen that not a single metric has emerged and that many compete nowadays. Many are useful and complementary and the coming years will tell which ones are used in everyday practices. Finally, as mentioned in the introduction, the ecological case is not under study today. For instance the nature of energy used or the full life-cycle of IT equipment from manufacturing to recycling are most of the time ignored. We believe that the next generation metrics will encompass these ecological parameters as well as the today simple energetic costs.

Acknowledgement. This work was partially supported by the COST (European Cooperation in Science and Technology) framework, under Action IC0804 (www.cost804. org). The author wants to thank particularly D. Borgetto, H. Casanova and G. Da Costa for the common work on the energy yield metric and problem resolution developed in this article.

References

1. AMD-ACP, http://www.amd.com/us/documents/43761c_acp_wp_ee.pdf
2. Embedded Microprocessor Benchmark Consortium Energy benchmark,
   http://www.eembc.org/benchmark/power_sl.php
3. Ll Berral, J., Goiri, I., Nou, R., Julia, F., Guitart, J., Gavalda, R., Torres, J.: Towards energy-aware scheduling in data center using machine learning. In: ACM/IEEE International Conference on Energy-Efficient Computing and Networking (e-Energy), Passau, Germany, April 13-15. ACM (2010)
4. Bertoldi, P., Atanasiu, B.: Electricity consumption and efficiency trends in the enlarged european union (2006), http://re.jrc.ec.europa.eu/energyefficiency/pdf/eneffreport2006.pdf
5. Borgetto, D., Da Costa, G., Pierson, J.-M., Sayah, A.: Energy-Aware Resource Allocation. In: Energy Efficient Grids, Clouds and Clusters Workshop (co-located with Grid 2009) (E2GC2), Banff, October 13-15. IEEE (2009)
6. Cameron, K.W., Pruhs, K., Irani, S., Ranganathan, P., Brooks, D.: Report of the science of power management workshop (April 2009),
   http://scipm.cs.vt.edu/scipm-reporttonsf-web.pdf
7. Careglio, D., Da Costa, G., Kat, R.I., Mendelson, A., Pierson, J.-M., Sazeides, Y.: Hardware leverages for energy reductions in large scale distributed systems. Technical Report IRIT/RT-2010-2-FR, IRIT, University Paul Sabatier, Toulouse (May 2010)

 8. Feng, W.C., Scogland, T.: The green500 list: Year one. In: 23rd IEEE International Parallel and Distributed Processing Symposium (IPDPS) - Workshop on High-Performance, Power-Aware Computing (HP-PAC), Rome, Italy (May 2009)
 9. Standard Performance Evaluation Corporation, http://www.spec.org/
10. Da Costa, G., Hlavacs, H.: Methodology of measurement for energy consumption of applications. Technical Report IRIT/RT-2010-4-FR, IRIT, University Paul Sabatier, Toulouse (July 2010)
11. Transaction Processing Performance Council, http://www.tpc.org/
12. Da Costa, G., De Assuncao, M.D., Gelas, J.-P., Georgiou, Y., Lefèvre, L., Orgerie, A.-C., Pierson, J.-M., Richard, O., Sayah, A.: Multi-Facet Approach to Reduce Energy Consumption in Clouds and Grids: The GREEN-NET Framework. In: ACM/IEEE International Conference on Energy-Efficient Computing and Networking (e-Energy), Passau, Germany, April 13-15, pp. 95–104. ACM (2010)
13. Dongarra, J., Luszczek, P., Petitet, A.: The linpack benchmark: past, present and future. Concurrency and Computation: Practice and Experience 15(9), 803–820 (2003)
14. Dongarra, J.J., Meuer, H.W., Strohmaier, E., Dongarra, J.J., Meuer, H.W., Strohmaier, E.: Top500 supercomputer sites. Technical report, Supercomputer (1997)
15. Transaction Processing Performance Council Energy, http://www.tpc.org/tpc_energy/
16. Fan, X., Weber, W.-D., Barroso, L.A.: Power provisioning for a warehouse-sized computer. In: ISCA 2007: Proceedings of the 34th Annual International Symposium on Computer Architecture, pp. 13–23. ACM, New York (2007)
17. Gadafi, A., Hagimont, D., Broto, L., Pierson, J.-M.: Autonomic Energy Management of Clustered Applications. In: E2GC2: Energy Efficient Grids, Clouds and Clusters Workshop (co-located with Grid 2009), Banff, Canada, October 13-15. IEEE (2009)
18. The Green Grid, http://www.thegreengrid.org/
19. Gunaratne, C., Christensen, K.J.: Ethernet adaptive link rate: System design and performance evaluation. In: LCN, pp. 28–35. IEEE Computer Society (2006)
20. Hlavacs, H., Da Costa, G., Pierson, J.-M.: Energy consumption of residential and professional switches. In: IEEE International Conference on Computational Science and Engineering, vol. 1, pp. 240–246 (2009)
21. Pierson, J-M., Hlavacs, H. (eds.): Proceedings of the COST Action IC804 on Energy Efficiency in Large Scale Distributed Systems - 1st Year. IRIT (July 2010), http://www.cost804.org
22. Katz, R., Hutchison, D. (eds.): ACM/IEEE International Conference on Energy-Efficient Computing and Networking (e-Energy), Passau, Germany, April 13-15. ACM (2010)
23. Lim, K., Ranganathan, P., Chang, J., Patel, C., Mudge, T., Reinhardt, S.: Understanding and designing new server architectures for emerging warehouse-computing environments. SIGARCH Comput. Archit. News 36(3), 315–326 (2008)
24. Luszczek, P., Bailey, D.H., Dongarra, J., Kepner, J., Lucas, R.F., Rabenseifner, R., Takahashi, D.: S12 - the hpc challenge (hpcc) benchmark suite. In: SC, p. 213. ACM Press (2006)
25. Meuer, H.W.: The top500 project: Looking back over 15 years of supercomputing experience. Informatik Spektrum 31(3), 203–222 (2008)
26. Moore, G.E.: Cramming more components onto integrated circuits. Electronics 38(8) (April 1965)
27. Moore, G.E.: Progress in digital integrated electronics. In: IEEE, Technical Digest 1975. International Electron Devices Meeting (1975)
28. Naegel, B.: Energy efficiency: The new sla. The Data Center Journal (December 2008), http://datacenterjournal.com/index.php?option=com_content&task=view&id=2352&itemid=43

29. COST Action IC0804 on Energy Efficiency in Large Scale Distributed Systems, http://www.cost804.org
30. Pierson, J.-M.: Allocating resources greenly: Reducing energy consumption or reducing ecological impact? In: ACM/IEEE International Conference on Energy-Efficient Computing and Networking (e-Energy), Passau, Germany, April 13-15, pp. 127–130. ACM (2010)
31. Poess, M., Nambiar, R.O., Vaid, K., Stephens, J.M., Huppler, K., Haines, E.: Energy Benchmarks: A Detailed Analysis. In: ACM/IEEE International Conference on Energy-Efficient Computing and Networking (e-Energy), Passau, Germany, April 13-15, pp. 131–140. ACM (2010)
32. Standard Performance Evaluation Corporation Power and Performance, http://www.spec.org/power_ssj2008/
33. U.S. Environmental Protection Agency ENERGY STAR Program. Report to congress on server and data center energy efficiency (August 2007), http://www.energystar.gov
34. Rawson, A., Pfleuger, J., Cader, T.: Green grid data center power efficiency metrics: Pue and dcie. In: The Green Grid (December 2008)
35. Stillwell, M., Schanzenbach, D., Vivien, F., Casanova, H.: Resource Allocation using Virtual Clusters. In: Proc. of the 9th IEEE Symp. on Cluster Computing and the Grid (May 2009)

From the Origins of Performance Evaluation to New Green ICT Performance Engineering

Carlos Juiz and Ramon Puigjaner

Universitat de les Illes Balears
Carretera de Valldemossa, km 7.5
07122 PALMA, Spain
cjuiz@uib.es, putxi@uib.cat

Abstract. This paper intends to present an overview of the evolution of performance evaluation since its first steps the Erlang works for modelling telephone networks, based on simple queues until the present current challenges in Green ICT that will require the development of new paradigms and mathematical tools, and rapidly passing across the modelling works of Khintchine and Pollaczeck; Jackson; Baskett, Chandy, Muntz and Palacios; Buzen; Reiser and Lavenberg; and many others, and benchmarking standards that have produced solutions to the problems appearing in these hundred of years. Finally, we analyze some of the challenges of computer performance evaluation appearing today, mainly those related to the energy consumption and sustainability, globally known as Green ICT.

Keywords: Performance evaluation of telephony, computer systems and communication networks, Performance modeling, Queuing theory, Queuing network theory, Simulation, Benchmarking, Green ICT.

1 Introduction

It is well known that you cannot manage what you cannot measure and also that before installing a complex system it is better to have an estimation of what will be its performance behaviour. In the domain of information and telecommunication technologies (ICT), the first case of application of the rules of thumb we mentioned was based on the Erlang works on queuing theory to plan the capacity of the telephone system [1], [2], [3]. Since then a lot of techniques and tools have been developed for predicting the behaviour of ICT systems. These techniques have evolved to be able to tackle the new challenges that have appeared during the last hundred years.

This paper is organized as follows: in the next section the first works of Erlang applied to telephone systems are presented. Section 3 rapidly explores the improvements of the queuing theory experienced until the moment in which computer appears. Section 4 reviews the different models and benchmarks developed to predict the performance of computer systems. In section 5 models and benchmarks developed for communications systems are presented. Section 6 presents the evolution of operational drivers of success for systems and how Performance Engineering is

K.A. Hummel et al. (Eds.): PERFORM 2010 (Haring Festschrift), LNCS 6821, pp. 49–60, 2011.

evolving through these drivers. Section 7 presents the challenges of analyzing the performance of new types of ICT systems or more precisely the challenges of new concepts in performance evaluation. And finally section 8 concludes.

2 Erlang Works

Even if the works of Andrei Andreyevich Markov obtained his results on Markov chains (refined later by Norbert Wiener) before those of Agner Karup Erlang, we prefer to start our history with Erlang because his works are more related to ICT systems. Erlang was a Danish mathematician that worked for long time at the Copenhagen Telephone Company where he had the opportunity to analyze the phenomena related to the new invention that was the telephone. Among his works they are specially important the analysis allowing to represent the telephone communications like a Poisson distribution [1] and the determination of how many circuits were needed to provide an acceptable telephone service (B and C Erlang formulae) and how many telephone operators were needed to handle a given volume of calls [2], [3]. It is necessary to remember that at the beginning of the twentieth century human operators and cord boards were used to switch the telephone calls by means of jack plugs. The main limitation of the Erlang works is the assumption that both the service time and the interarrival time should be exponentially distributed. These results were intensively used for more than sixty years.

However, Erlang was not a pure researcher but a hands-on one. To verify his assumptions he did not refuse to conduct measurements that obliged him to climb into street manholes. He can be considered as the father of the queuing theory and the performance evaluation of ICT systems. He was also an expert in the history and calculation of the numerical tables of mathematical functions, particularly logarithms. He devised new calculation methods for certain forms of tables.

3 Other Advances before the Computers

The next important result in queuing theory came from two mathematicians that simultaneously break one of the limitations of the Erlang assumptions: the service time could have any type of distribution. This result was obtained in 1930 by the Austrian-French Félix Pollaczek and the Russian Aleksandr Yakovlevich Khinchin working independently one of the other. In this case the result was not induced by a technical need but was of academic interest.

Other interesting results, useful in some computer models, were those obtained by Alan Cobham who analyzed the behaviour of a queuing system when there are customers with different priorities. His initial work can be found in [4].

4 Performance Evaluation of Computers

The need of predicting the performance of computers obliged scientists and engineers to develop new techniques and tools adapted to the new challenges.

4.1 Modelling

Although the initial step for tackling the problem of computer system performance modelling was initially thought for modelling a computer system performance, the pioneering works of Jackson [5] and Gordon and Newell [6] with the algorithmic complement of Buzen [7] prepared the arrival of the work of Baskett, Chandy, Muntz and Palacios [8] on product formnetworks. This work established a complete basis for modelling the performance of a computer system allowing an analytical process, even if the possibility of analytical processing was restricted by an important set of conditions. This work was completed by the proposal of convenient algorithms like those of Reiser and Kobayashi [9] and Reiser and Lavenberg [10].

In parallel with these works, Buzen y Denning [11], [12] developed a different approach for tackling the same problem and arrived to similar numerical results; the operational analysis. This approach has a more physical basis without the strong mathematical and statistical conditions of the queuing network approach.

In order to facilitate the use of these results to engineers and researchers, several tools were developed. Among them we can find BEST-1 developed by BGS, the company created by J.P. Buzen, RESQ developed by IBM and QNAP2 developed by INRIA.

In order to free the modelling process of the constraints of the product form networks, a number of approximated methods were developed, like those of Courtois [13], Marie [14], Gelenbe and Mitrani [15] and many others. Some of these approximated algorithms were included in some of the previous tools.

Some of the above mentioned tools also included simulation capabilities to analyze queuing network models not able to be processed by any other technique.

4.2 Benchmarking

However in many cases computer customers need to compare the computing capacity of different systems. Initially this capacity was evaluated simply by the execution time of a typical instruction, habitually the addition instruction. With the increasing complexity of computer systems (clever architectures, sophisticated operating systems, etc.) this simple comparison was no more valid. Computer customers proposed the execution of a sample of their real workload on the different systems they had to compare. Sophisticated techniques were developed to characterise the system workload by a reduced number of programmes extracted from this workload.

The next step was to standardise these sets of programmes depending on the type of workload we were interested in. The definition of these standards was reached by consensus like in the case of LINPACK [16], a collection of Fortran subroutines that analyze and solve linear equations and linear least-squares problems for the scientific computation or established by organizations participated by computer manufacturers and customers, like in the case of SPEC [17].

The first benchmarks proposed by SPEC were devoted to batch environments, then the target was a conversational system and currently cover most of the typical working environments of computer systems. An aspect particularly interesting is the possibility of scaling the benchmarks in order to adapt the size to the particular needs of the customer. Also SPEC organization publishes results of its benchmarks with different sizes running over a variety of systems.

5 Performance Evaluation of Networks

When computers were no more able to work isolated the need of connection initially with terminals and then with other computers appears. The complexity of the systems increased and consequently the complexity of the tools and techniques used to evaluate their performance.

5.1 Modelling

When the computers ran in an isolated way it was quite easy to build analytical models able to be processed with the help of appropriated tools. However, with the increasing of the networked systems complexity (based on wired or wireless LANs, or in ad-hoc networks, or in sensor networks, etc.), frequently it was very hard to find an enough accurate model for representing such systems and able to find the solution analytically. Fortunately, the increasing computer speed allowed building simulation models representing with enough accuracy the behaviour of the corresponding systems and having the possibility of obtaining results with small enough confidence intervals in computing times reasonably short.

Among the different simulation engines there two that are used very frequently: OPNET [18] and ns-2 [19].

5.2 Benchmarking

In terms of standard benchmarks, and restricting our comments to SPEC [17], the variety of benchmarks proposed by this institution has been increased to cover the most typical systems including a network in their structure. To cover these needs benchmarks for analyzing conversational systems, transactional systems, web systems (an e-commerce sales systems), etc. have been developed and proposed to the community of computer systems users.

6 New Performance Challenges of ICT Systems

The classical drivers of operational success at ICT have been always: Performance, Dependability, Quality, Flexibility and Cost [59]. Some of them seem to be clearly non-functional and other functional, but all five are closely related. In fact, the cost has the key of selecting particular solutions at datacenters for the other four operational drivers. Lately, other two drivers have been emerged as important as the first four: Security (some authors included it into Dependability) and Energy (classically included it in Cost). In Sections 2-5 we have overviewed the evolution, just considering Performance, then considering Performance and Quality or Performance and Dependability, and other combinations of operational drivers for Performance Engineering and Performance Evaluation of software, hardware, networks, etc. In last ten years, a lot of research effort has been put on Performance and Security issues, but the issue of this paper is about the new opportunities in the combination of Performance and Energy efficiency, which is part of Green ICT arena. That is, most of old studies in Performance Evaluation were done with a single driver in mind: Performance, but nowadays, there is a shift between this single objective to the consideration of multiple, particularly, Performance and Energy.

Turning to the drivers of success, it is clear that taking only one of them usually produce the detriment of some of the others. For example, maximizing the performance may not guarantee quality of service or even do more inflexible a particular solution, while maximizing security can reduce yields by overexposure to controls, and these situations always increase the cost. Consequently, the challenge of the new performance engineers is to find a balance among these factors. Similarly, one of the challenges is to find the Green ICT energy efficiency acceptable performance, i.e. the energy cost control while maintaining the performance necessary for operation. In the next section, we try to explain the factors that can influence the performance engineering and energy, and the opportunities offered by these factors for research.

7 Green ICT and Performance

The prevailing international scientific opinion on climate change is that human activities resulted in substantial global warming from the mid-20th century, and that continued growth in greenhouse gas concentrations [48] caused by human-induced emissions would generate high risks of dangerous climate change [32]. Even this debate is controversial, what it is clear is that some human activities, as the use of ICT, are producing huge greenhouse gases.

With the overall improvement in the quality of connections and the spread of the wireless Internet, the web applications, easily updatable and often "in the cloud and free", are gaining ground rapidly to traditional software. But not only distributed applications of cloud computing are growing, traditional web applications based on client-server paradigm are also growing, despite the alleged gratuity of some popular web applications.

Data center power consumption continues to grow at an alarming pace [37, 28]. There are five distinct industry segments accounting for most of a data center's energy-efficiency: (1) servers and storage systems, (2) power conditioning equipment, (3) cooling and humidification systems, (4) networking equipment, and (5) lighting/physical security. Therefore, sustainable data centers require the multidisciplinary collaboration of mechanical engineers, electrical engineers, computer scientists, and others. Particularly, the hardware infrastructure of the data center consists of hundreds (or thousands) of servers hosting revenue-generating services, interconnected with each other and the outside world via networking equipment, and relying on storage devices for persistent data. These hardware elements are managed by a datacenter-wide software stack that spans the platforms and virtualization layers [50].

On one hand, data centers that host web applications and user files are huge warehouses with a lot of servers, additional electronic equipment and complex cooling systems. While computers, mobile phones and telecommunications networks are today primarily responsible for energy consumption, energy spending caused by data centers that maintain the quality of internet services in operation is growing faster. But giving more capacity to data centers usually means more servers and consequently all five factors mentioned above consume more energy [51].

On the other hand, Performance engineering is the set of roles, skills, activities, practices, tools, and deliverables applied at every phase of the Systems Development Lifecycle, which ensures that a solution will be designed and implemented to meet performance requirements [44]. The increasing number of Internet users and systems connected to it, especially due to the popularization of applications and Web-based services, produces continuously capacity problems. In these growing web systems, if no action is taken, a loss in quality of service is produced and, ultimately, worsens the performance of any web application. Additionally, this growth of web systems, may be in a slightly irregular shape and unplanned in most cases, due to time-to-market pressures [46]. Variables that could be affected by these changes in system workload are those perceived by the user and that constitute the quality of service (Quality of Service, QoS, among which is singular importance, the response time) as those need to know the server's responsibility to provide quality of service desired (which shall include the use, overhead and the throughput) [56]. The performance improvement of these services can be done at different layers, from server side to applications, from network protocol to routing, etc. [45], [47]. When even the techniques of performance and modelling of web-based computer systems and distributed has not reached a sufficiently scientific mature, a new challenge directly related to this branch of research emerges: the relationship between performance and energy cost is i.e., the energy efficiency of systems; this section is aimed by these combination of two topics and the future challenges for performance engineers due to this combination.

7.1   State of the Art

The solutions to achieve better energy efficiency in sharing distributed resources will address different topics. The first is at the computer architecture level, techniques have been proposed to reduce processor energy usage through multi-core processors [34], designing energy proportional hardware [20], and increasing the number of performance and sleep states [41]. Benchmarking techniques to quantify the energy costs of different computer architectures have been proposed [48].

Even though, the current possibilities of on-the-fly adaptation of the underlying hardware behaviour will be studied: for instance even it is possible to decrease the voltage of the processors to slow them down and save energy, there will be the need to characterize the electrical consumption of the systems, as a function of the workload of the system. This implies first to have a means to measure or estimate the energy consumption volume (by actual power-consumption measurement equipments), and then to model the behaviour of the computations related to the obtained performances. However, the price to be paid is reducing the system components longevity. Actions have to be taken at different levels of decision to improve the energy-efficiency of the distributed web servers:

- At the networking level, the idea would be to decrease the overall usage of the network. For instance, moving large amount of data will operate on routes consuming less energy and the placement of the data (or the replication of the data) will be optimized to minimize the equipment usage; complementarily, approaches to minimizing the time to transfer the data, and to reducing the data to be transmitted (e.g., by using forecasting techniques) will be investigated.

The performance forecasting for several network layers [31], [39] should provide some skills performance and power tools development in order to know how applications are using the network.

- At the server level, in order to maximize energy efficiency, whether in a single system or over an ensemble of systems, users and data center operators need to understand the relationship between resource usage and system-level power consumption. This understanding enables such optimizations as consolidating workloads on as few machines as possible and turning others off [22], [25], [35], since current hardware is highly inefficient at low utilization [20]. While several types of full-system power models have been proposed, often in the context of enabling a particular optimization, they have not been systematically compared over a variety of software workloads and hardware platforms.

  Performance engineering will contribute to these efforts building several power models for servers executing web different workloads in extensive studies including regression models and also simulation. Other techniques include multi-server power coordination [43], hard disk spin-down [40], synchronization aware multithreading [42], and compiler driven optimizations [49].

  Another interesting issue is server virtualization, where hardware elements are managed by a datacenter-wide software stack that spans the platforms and virtualization layers. However, adding a new software layer also adds additional overhead and risk of over utilization of hardware with clear consequences in energy but also savings.

  Performance engineers will also study how performance and energy varies with virtualization of servers in comparison with the non-virtual ones.

- At the application level, context awareness of the applications should be favored: it aims at developing applications that are aware of the power state of the computers and the currently set power policy, such that they are tuned appropriately, and respond to changes dynamically [38]. Developing energy-aware data mining techniques will support energy efficiency at the applications level. For instance, clustering/classifying of data sources, identifying usage patterns and cluster users will help to decrease the energy consumed in the network. This last level is very attractive to performance research for several reasons. First, this layer has the most information on the actual user impact of performance and energy tradeoffs, enabling more aggressive performance sacrifice in unimportant areas compared to lower layer techniques. For instance, an application may throttle processor frequency to the minimum value required to satisfy user perceptible delay behavior. Second, application specific optimizations can be made at this layer such as changing the algorithm used, accuracy of computation (e.g. changing from double precision to single), or quality of service provided [29]. Third, energy usage at the application layer may be made dynamic [27]. For instance, an application hosted in a data center may decide to turn off certain low utility features if the energy budget is being exceeded, and an application on a mobile device may reduce its display quality [30] when battery is low. This is different from system layer techniques that may have to throttle the throughput resulting in users being denied service.

7.2   Some Green ICT Research Challenges for Performance Engineering

Many of the current green performance activities represent isolated optimizations; others attempt to standardize approaches and solutions and to promote a common understanding of the issues and complexities. While recognizing their vital importance, the literature is replete with such solutions and experiences. The new challenges will be how to reduce power consumption in a collaborative manner not only as individual efforts on devices but holding an optimal quality of service.

These new performance challenges will propose realistic energy-efficient alternate models and also study new solutions for benchmarks. While much effort is nowadays put into hardware specific solutions to lower energy consumptions, the need for a complementary approach is necessary from the application viewpoint. Performance Engineering will characterize the energy consumption and energy efficiencies of applications at servers. Then based on the current hardware adaptation possibilities and interconnection networks, performance engineers will characterize the trade-off between energy savings and functional and non-functional parameters, particularly the web system performance. Most of the current effort consists of building performance and power models for servers from standard benchmarking, subsequently deriving new simulation engines to derive performance and energy results and finally propose a new benchmark models for servers taking into account energy issues.

For this reason, we propose an incomplete list of Green ICT performance topics putting together computing technologies, networking, optimization, and their performance evaluation and power issues. Our proposed research opportunities list focuses on several integrated energy and ICT related issues, from simulation to software applications, from networking to data centres, from performance engineering to consumption models, etc. trying to transit from research on only ICT functional issues to green ICT considerations, together.

The current open list of research topics can be started with the first volume of results in COST IC0804 [58], where there are included the results of this European action. However, our position is that there are research issues very appealing for the performance and power research advance. Here there is our open and incomplete list:

- Energy consumption/performance models: models and tools where system administrator concerns about energy-efficiency of ICT usage are considered together with performance.
- New simulation and benchmarking and monitoring engines considering devices, networks and their corresponding energy consumption
- Energy Efficient Computing Centres: this will be addressed at different levels, i.e. at a data centre level, workload and server management, respectively.
- Energy Efficient Networks: at multiple levels, including networking devices, communication and control protocols, network architectures design, and network performance.
- Green Software Design: considering the compiler level on algorithmic and data structures optimizations for grid systems. Also, at application level to investigate what are the key aspects in the applications that specifies the performance requirements/power consumption that user expects from those services

- Design for Energy Awareness: empowering social in their use of products and services by increasing visibility of energy choices during operation.
- ICT Management Green Metrics: indicators of performance/energy from data centre monitoring to ICT governance/management
- Power-aware middleware for data centres: more intelligent power management for virtualised machines, power-aware virtualisation and benchmarking.
- Power-efficient routing: design of packet classifiers for high speed and low power consumption.
- Energy-efficient bandwidth allocation: making minimum consumption of energy resources in network equipment.
- Energy and distributed computing: placement of mobile agents, workload placement, virtual machine migration, P2P resource allocation, etc, providing means to develop energy-aware allocation strategies.
- Eco Data Base Management Systems: managing the energy consumed during query processing and considering query aggregation in workload.
- Energy-Efficient algorithms: Dynamic Speed scaling and power-down client integration but holding resilience of systems.
- Dynamic performance management: programming models for the cloud rely on the use of tasks scheduled across large data centres, taking advantage of the parallelism/performance and power.
- Etc.

For sure that there are many more, in fact, all performance engineering issues are directly or indirectly related to power consumption, so that we may select almost every performance issue to be studied from the energy-efficiency viewpoint. Thus, we may review the performance engineering research from the very beginning but now considering power consumption as a primary factor for the new Systems Development Lifecycle.

8 Conclusions

Performance evaluation concept has evolved along with he changes experienced by the changes in the technology: hardware, software and networks, and in the goals of the end users with respect to the systems they were using.

We have overviewed the evolution, just considering Performance evaluation, then considering other engineering factors as Performance and Quality or Performance and Dependability, and other combinations of operational drivers for Performance Engineering of software, hardware, networks, etc. In last ten years, a lot of research effort has been put on Performance and these new issues, but the topic of this paper is about the new opportunities in the combination of Performance and Energy efficiency, which is part of Green ICT arena. That is, most of old studies in Performance Evaluation were done with a single driver in mind: Performance but nowadays, there is a shift between this single objective to the consideration of multiple, particularly, Performance and Energy. We have seen how the tools and techniques used to evaluate or predict the computer performance have evolved and the new challenges derived from Green ICT will oblige to find new tools and techniques addressed to overcome the current challenges as they are suggested in the last part of this paper.

Acknowledgements. This work has been partially funded by the Spanish Ministry of Science and Innovation and the FEDER funds, both in the framework of the projects TIN2007-60440 and TIN2009-11711.

References

1. Erlang, A.K.: The Theory of Probabilities and Telephone Conversations. Nyt Tidsskrift for Matematik, B 20 (1909)
2. Erlang, A.K.: Solution of some Problems in the Theory of Probabilities of Significance in Automatic Telephone Exchanges. Elektroteknikeren 17 (1917)
3. Erlang, A.K.: Telephone Waiting Times. Matematisk Tidsskrift, B 31 (1920)
4. Cobham, A.: Priority Assignment in Waiting Line Problems. Journal of the Operations Research Society of America (1954)
5. Jackson, J.R.: Jobshop like Queueing Sysems. Management Science 10 (1963)
6. Gordon, W.J., Newell, G.F.: Closed Queueing Systems with Exponential Servers. Operations Research 15(2) (1967)
7. Buzen, J.P.: Computational Algorithms for Closed Queueing Networks with Exponential Servers. CACM 16(9) (1973)
8. Baskett, F.T., Chandy, K.M., Muntz, R.R., Palacios, F.G.: Open, Closed and Mixed Networks with Different Classes of Customers. JACM 22(2) (1975)
9. Reiser, M., Kobayashi, H.: Queueing networks with multiple closed chains: Theory and computational algorithms. IBM J. Res. Develop. 19, 3 (1975)
10. Reiser, M., Lavenberg, S.S.: Mean Value Analysis of Closed Multichain Queueing Networks. JACM 27(2) (1980)
11. Denning, P.J., Buzen, J.P.: The Operational Analysis of Queueing Network Models. ACM Computing Surveys 10(3) (1978)
12. Buzen, J.P., Denning, P.J.: Operational Treatment of Queue Length Distributions and Mean Value Analysis. Computer Performance 1(1) (1980)
13. Courtois, P.J.: Decomposability: Queueing and Computer Systems Applications. Academic Press (1977)
14. Marie, R.: Modélisation par Réseaux de Files d'Attente. Thèse de Docteur ès Sciences Mathematiques. Université de Rennes (1977)
15. Gelenbe, E., Mitrani, I.: Analysis and Synthesis of Computer Systems. Academic Press (1981)
16. LINPACK: http://www.netlib.org/linpack/
17. SPEC: Standard Performance Evaluation Corporation, http://www.spec.org/
18. OPNET: Applications and Network Performance, http://www.opnet.com/
19. ns-2: The Network Simulator, http://www.isi.edu/nsnam/ns/
20. Barroso, L.A., Hözle: The case for energy proportional computing. IEEE Computer 40, 12 (2007)
21. COST IC0804. Energy efficiency in large scale distributed systems, http://www.cost804.org/
22. Chen, G., He, W., Liu, J., Nath, S., Rigas, L., Xiao, L., Zhao, F.: Energy-aware server provisioning and load dispatching for connection-intensive internet services. In: NSDI, San Francisco, CA (2008)
23. Standard Performance Evaluation Corporation (SPEC). SPEC power, http://www.spec.org/

24. Fan, X., Weber, W.D., Barroso, L.A.: Power provisioning for a warehouse-sized computer. In: Proceedings of the International Symposium on Computer Architecture, ISCA (2007)
25. Heath, T., Diniz, B., et al.: Energy conservation in heterogeneous server clusters. In: Proceedings of the Symposium on Principles and Practice of Parallel Programming, PPoPP (2005)
26. Chun, B.G., Iannaccone, G., Katz, R.H., Lee, G., Niccolini, L.: An Energy Case for Hybrid Datacenters. In: Workshop on Power Aware Computing and Systems, HotPower 2009 (2009)
27. Flinn, J., Satyanarayanan, M.: Energy-aware adaptation for mobile applications. In: SOSP 1999: Proceedings of the 17th ACM Symposium on Operating Systems Principles (1999)
28. Mankoff, J., Kravets, R., Blevis, E.: Some computer science issues in creating a sustainable world. IEEE Computer 41(8) (2008)
29. Im, C., Ha, S.: Energy optimization for latency- and quality-constrained video applications. IEEE Design and Test of Computers 21(5) (2004)
30. Iyer, S., Luo, L., Mayo, R., Ranganathan, P.: Energy-adaptive display system designs for future mobile environments. In: ACM MobiSys (2003)
31. Gilly, K., Alcaraz, S., Juiz, C., Puigjaner, R.: Analysis of burstiness monitoring and detection in an adaptive Web system. Computer Networks 53(5) (2009)
32. Executive Summary. Chapter 9: Projections of Future Climate Change. Climate Change 2001: The Scientific Basis (2005),
    http://www.grida.no/climate/ipcc_tar/wg1/339.htm
33. Feng, X., Ge, R., Cameron, K.W.: Power and Energy Profiling of Scientific Applications on Distributed Systems. In: Proceedings of the 19th IEEE International Parallel and Distributed Processing Symposium, IPDPS 2005 (2005)
34. Kumar, R., Farkas, K.I., Jouppi, N.P., Ranganathan, P., Tullsen, D.M.: Single-ISA heterogeneous multi-core architectures: The potential for processor power reduction. In: ACM/IEEE MICRO (2003)
35. Pinheiro, E., Bianchini, R., et al.: Load balancing and unbalancing for power and performance in cluster-based systems. In: Proceedings of the Workshop on Compilers and Operating Systems for Low Power, COLP (2001)
36. Ranganathan, P., Rivoire, S., Moore, J.: Power modeling and measurement. In: Advances in Computers. Elsevier (2009)
37. U.S. EPA. Report to congress on server and data center energy efficiency. Tech. Rep. (2007)
38. Kusic, D., Kephart, J.O., Hanson, J.E., Kandasamy, N., Jiang, G.: Power and performance management of virtualized computing environments via lookahead control. Cluster Computing 12(1) (2009)
39. Alcaraz, S., Gilly, K., Juiz, C., Puigjaner, R.: Handling HTTP flows over a DiffServ framework. In: LANC 2007 (2007)
40. Narayanan, D., Donnelly, A., Rowstron, A.: Write off-loading: Practical power management for enterprise storage. In: FAST (2008)
41. Nedevschi, S., Popa, L., Iannaccone, G., Ratnasamy, S., Wetherall, D.: Reducing network energy consumption via sleeping and rate-adaptation. In: NSDI (2008)
42. Park, S., Jiang, W., Zhou, Y., Adve, S.: Managing energy-performance tradeoffs for multithreaded applications on multiprocessor architectures. In: SIGMETRICS (2007)
43. Raghavendra, R., Ranganathan, P., Talwar, V., Wang, Z., Zhu, X.: No power struggles: A unified multi-level power management architecture for the data center. In: ASPLOS (2008)

44. Menasce, D.A., Almeida, V., Dowdy, L.: Performance by Design: Computer Capacity Planning by Example. Prentice Hall PTR (2004)
45. Standard Performance Evaluation Corporation (SPEC). SPECweb2009, http://www.spec.org/web2009/
46. Menasce, D.A., Almeida, V.: Capacity Planning for Web Services: metrics, models, and methods. Prentice Hall PTR (2001)
47. Killelea, P.: Web Performance Tuning. O'Reilly (2002)
48. National greenhouse gas inventory data for the period 1990-2007. UN FCCC (2009), http://unfccc.int/resource/docs/2009/sbi/eng/12.pdf
49. Xie, F., Martonosi, M., Malik, S.: Compile-time dynamic voltage scaling settings: opportunities and limits. SIGPLAN Not. 38(5) (2003)
50. Banerjee, P., Patel, C.D., Bash, C., Ranganathan, P.: Sustainable Data Centers: Enabled by Supply and Demand Side Management. In: DAC 2009. ACM Press (2009)
51. Katz, R.: Tech Titans Building Boom. IEEE Spectrum (February 2009)
52. Patel, C.D., Ranganathan, P.: Enterprise power and cooling. ASPLOS Tutorial (2006)
53. Laudon, J.: Performance/Watt: The new server focus. SIGARCH Computer Architecture News 33(4) (2005)
54. Sun Microsystems. SWaP (Space, Watts and Performance) metric, http://search.sun.com/main/
55. Rivoire, S., Shah, M.A., Ranganathan, P., Kozyrakis, C.: JouleSort: A Balanced Energy-Efficiency Benchmark, http://www.hpl.hp.com/environment/datacenters.html
56. Molero, X., Juiz, C., Rodeño, M.J.: Evaluación y Modelado del Rendimiento de Sistemas Informáticos. Pearson Prentice Hall (2004)
57. Deloitte España. Barómetro de Empresas (2009).
58. Pierson, J.M., Hlavacs, H. (eds.): Proceedings of the COST Action IC0804 on Large Scale Distributed Systems, 1st Year (2010)
59. Slack, N., Chambers, S., Betts, A., Johnston, R.: Operations and Process Management: Principles and practice for strategic impact. FT Prentice Hall (2005)

Predicting Disk Scheduling Performance with Virtual Machines

Robert Geist, Zachary H. Jones, and James Westall

Clemson University

Abstract. A method for predicting the performance of disk scheduling algorithms on real machines using only their performance on virtual machines is suggested. The method uses a dynamically loaded kernel intercept probe (*iprobe*) to adjust low-level virtual device timing to match that of a simple model derived from the real device. An example is provided in which the performance of a newly proposed disk scheduling algorithm is compared with that of standard Linux algorithms. The advantage of the proposed method is that reasonable performance predictions may be made without dedicated measurement platforms and with only relatively limited knowledge of the performance characteristics of the targeted devices.

1 Introduction

In the last five years, the use of virtual computing systems has grown rapidly, from nearly non-existent to commonplace. Nevertheless, system virtualization has been of interest to the computing community since at least the mid-1960s, when IBM developed the CP/CMS (Control Program/Conversational Monitor System or Cambridge Monitor System) for the IBM 360/67 [1]. In this design, a low-level software system called a *hypervisor* or *virtual machine monitor* sits between the hardware and multiple guest operating systems, each of which runs unmodified. The hypervisor handles scheduling and memory management. *Privileged* instructions, those that trap if executed in user mode, are simulated by the hypervisor's trap handlers when executed by a guest OS.

The specific architecture of the host machine essentially determines the difficulty of constructing such a hypervisor. Popek and Goldberg [2] characterize as *sensitive* those machine instructions that may modify or read resource configuration data. They show that an architecture is most readily virtualized if the sensitive instructions are a subset of the privileged instructions.

The principal roadblock to widespread deployment of virtual systems has been the basic Intel x86 architecture, the de facto standard, in which a relatively large collection of instructions are sensitive but not privileged. Thus a guest OS running at privilege level 3 may execute one of them without generating a trap that would allow the hypervisor to virtualize the effect of the instruction. A detailed analysis of the challenges to virtualization presented by these instructions is given by Robin and Irvine [3].

K.A. Hummel et al. (Eds.): PERFORM 2010 (Haring Festschrift), LNCS 6821, pp. 61–72, 2011.
© IFIP International Federation for Information Processing 2011

The designers of VMWare provided the first solution to this problem by using a binary translation of guest OS code [4]. Xen [5] provided an open-source virtualization of the x86 using *para-virtualization*, in which the hypervisor provides a virtual machine interface that is similar to the hardware interface but avoids the instructions whose virtualization would be problematic. Each guest OS must then be modified to run on the virtual machine interface.

The true catalysts to the widespread development and deployment of virtual machines appeared in 2005-2006 as extensions to the basic x86 architecture, the Intel VT-x and AMD-V. The extensions include a "guest" operating mode, which carries all the privilege levels of the normal operating mode, except that system software can request that certain instructions be trapped. The hardware state switch to/from guest mode includes control registers, segment registers, and instruction pointer. Exit from guest mode includes the cause of the exit [6]. These extensions have allowed the development of a full virtualization Xen, in which the guest operating systems can run unmodified, and the Kernel-based Virtual Machine (KVM), which uses a standard Linux kernel as hypervisor.

Nevertheless, VMWare, Xen, and KVM are principally aimed at facilitating user-level applications, and thus they export only a fairly generic view of the guest operating system(s), in which the devices to be managed are taken from a small, fixed collection of emulated components. This would seem to preclude the use of virtual machines in testing system-level performance, such as in a comparative study of scheduling algorithms. The devices of interest are unlikely to be among those emulated, and, even if they are, the emulated devices may have arbitrary implementation, thus exhibiting performance characteristics that bear little or no resemblance to those of the real devices. For example, an entire virtual disk may be cached in the main memory of a large NAS device supporting the virtual machine, thus providing a disk with constant ($O(1)$) service times.

In [7], we introduced the *iprobe*, an extremely light-weight extension to the Linux kernel that can be dynamically loaded and yet allows the interception and replacement of arbitrary kernel functions. The *iprobe* was seen to allow a straightforward implementation of PCI device emulators that could be dynamically loaded and yet were suitable for full, system-level driver design, development, and testing.

The goal of this effort is to extend the use of the *iprobe* to allow performance prediction for real devices using only virtual machines. We suggested this possibility in [8]. In particular, we will predict the performance of a collection of disk scheduling algorithms, one of which is new, for a targeted file-server workload on a specific SCSI device. We then compare the results with measurements of the same algorithms on the real hardware. We will see that the advantage of our approach is that reasonable performance predictions may be made without dedicated measurement platforms and with relatively limited knowledge of the performance characteristics of the targeted devices.

The remainder of the paper is organized as follows. In the next section we provide background on both kernel probes and disk scheduling. In section 3, we propose a new disk scheduling algorithm that will serve as the focus of our tests.

In section 4, we show how the kernel probes may be used to implement a *virtual performance throttle*, the mechanism that allows performance prediction from virtual platforms. In section 5 we describe our virtual machine, the targeted (real) hardware, and a test workload. Section 6 contains results of algorithm performance prediction from the virtual machine and measurements of the same algorithms on the real hardware. Conclusions follow in section 7.

2   Background

2.1   Kernel Probes

The Linux kernel probe or *kprobe* utility was designed to facilitate kernel debugging [10]. All *kprobes* have the same basic operation. A *kprobe* structure is initialized, usually by a kernel module, i.e., a dynamically loaded kernel extension, to identify a target instruction and specify both pre-handler and post-handler functions. When the *kprobe* is registered, it saves the targeted instruction and replaces it with a breakpoint. When the breakpoint is hit, the pre-handler is executed, then the saved instruction is executed in single step mode, then the post-handler is executed. A return resumes execution after the breakpoint.

A variation on the *kprobe*, also supplied with Linux, is the *jprobe*, or jump probe, which is intended for probing function calls, rather than arbitrary kernel instructions. It is a *kprobe* with a two-stage pre-handler and an empty post-handler. On registration, it copies the first instruction of the registered function and replaces that with the breakpoint. When this breakpoint is hit, the first-stage pre-handler, which is a fixed code sequence, is invoked. It copies both registers and stack, in addition to loading the saved instruction pointer with the address of the supplied, second-stage pre-handler. The second-stage pre-handler then sees the same register values and stack as the original function.

In [7], we introduced the intercept probe or *iprobe*, which is a modified *jprobe*. Our *iprobe* second-stage pre-handler decides whether or not to replace the original function. If it decides to do so, it makes a backup copy of the saved (function entry) instruction and then overwrites the saved instruction with a no-op. As is standard with a *jprobe*, the second-stage pre-handler then executes a *jprobe* return, which traps again to restore the original register values and stack. The saved instruction (which now could be a no-op) is then executed in single step mode. Next the post-handler runs. On a conventional *jprobe*, this is empty, but on the *iprobe*, the post-handler checks to see if replacement was called for by the second-stage pre-handler. If this is the case, the single-stepped instruction was a no-op. The registers and stack necessarily match those of the original function call. We simply load the instruction pointer with the address of the replacement function, restore the saved instruction from the backup copy (overwrite the no-op) and return. With this method, we can intercept and dynamically replace any kernel function of our choice. Note that it is possible to have two calls to the same probed function, one that is to be intercepted and one that is not. A discussion of the handling of potential attendant race conditions in SMP systems may be found in [9].

2.2   Disk Scheduling

Scheduling algorithms that re-order pending requests for disks have been studied
for at least four decades. Such scheduling algorithms represent a particularly
attractive area for investigation in that the algorithms are not constrained to be
work-conserving. Further, it is easy to dismiss naive (but commonly held) beliefs
about such scheduling, in particular, that a greedy or shortest-access-time-first
algorithm will deliver performance that is optimal with respect to any common
performance measure, such as mean service time or mean response time. Consider
a hypothetical system in which requests are identified by their starting blocks
and service time between blocks is equal to distance. Suppose the read/write
head is on block 100 and requests in queue are for blocks 20, 82, 120, and 200.
The greedy schedule and the differing, optimal schedule are shown in Table 1.

Table 1. Sub-optimal performance of the greedy algorithm

algorithm	mean service	mean response
greedy 82, 120, 200, 20	79.0	131.5
optimal 120, 82, 20, 200	75.0	124.5

Disk scheduling algorithms are well-known to be analytically intractable with
respect to estimating response time moments. Early successes in this area, due
to Coffman and Hofri [10] and Coffman and Gilbert [11] were restricted to highly
idealized, *polling* servers, in which the read/write head sweeps back and forth
across all the cylinders, without regard to the extent of the requests that are
actually queued.

Almost all knowledge of the performance of real schedulers is derived from
simulation and measurement studies. Geist and Daniel described UNIX system
measurements of the performance of a collection of "mixture" algorithms that
blended scanning and greedy behavior [12]. Worthington, Ganger, and Patt [13]
showed, in simulation, that scheduling with full knowledge of disk subsystem
timing delays, including rotational delays and cache operations, could offer major
performance improvements. They also concluded that knowledge of the cache
operation was far more important than an accurate mapping of logical block to
physical sector, a point which we will address.

More recently, Pratt and Heger [14] provided a comparison of the four schedulers distributed with Linux 2.6 kernels. Until 2.6, Linux used a uni-directional or
circular scan (CSCAN), in which requests are served in ascending order of logical
block number until none remains, whereupon the read/write head sweeps back
down to the lowest-numbered pending request. With recognition that the best
scheduler is likely workload-dependent, Linux authors changed the 2.6 kernel to
allow single-file, modular, drop-in schedulers that could be dynamically switched.
Four schedulers were provided. The default is the completely fair queueing (*cfq*)
algorithm, which has origins in network scheduling. Each process has its own logical queue, and requests at the front of each queue are batched, sorted and served.

The *deadline* scheduler was designed to limit response time variance. Each request sits in two queues, one sorted by CSCAN order, one FIFO, and each has a deadline. The CSCAN order is used, unless a deadline would be violated, and then FIFO is used. As with many algorithms, reads are separated from and given priority over writes because the requesting read process has usually suspended to await I/O completion, and so there are actually four (logical) queues. The *anticipatory* scheduler is no longer supported, and the *noop* scheduler is essentially FIFO, which delivers poor performance on almost all workloads, and thus these two will not be discussed.

A fundamental departure from greedy algorithms, scanning algorithms, and $O(N)$ mixtures thereof was offered by Geist and Ross [15]. They observed that, over the preceding decades, CPU speeds had increased by several orders of magnitude while disk speeds remained essentially unchanged. They suggested that $O(N^2)$ algorithms might be competitive and offered a statically optimal solution that was based on Bellman's *dynamic programming* [16], in which a table of size $O(N^2)$ containing optimal completion sequences was constructed. Although their algorithm was shown to deliver excellent performance in tests on a real system, there were two easily-identifiable problems. It ignored the dynamics of the arrival process, and it ignored the effects of any on-board disk cache.

More recently, Geist, Steele, and Westall [17] partially addressed the issue of arrival dynamics. Although at first glance counter-intuitive, it is often beneficial for disk schedulers with a non-empty queue of pending requests to do nothing at all [18]. The process that issued the most recently serviced request is often likely to issue another request for a nearby sector, and that request could be served with almost no additional effort. They added a so-called *busdriver* delay, to mimic the actions of a bus driver who would wait at a stop for additional riders, to the table-based, dynamic programming algorithm of Geist and Ross, and showed that it delivered excellent performance, superior to any of the four schedulers distributed with Linux 2.6, for a fairly generic, web file-server workload designed by Barford and Crovella [19].

3   A New Scheduler

We now propose an extension of Geist-Steele-Westall algorithm to capture the effects of the on-board cache. On-board caches are common, although their effect on the performance of standard workloads is often minimal. All UNIX-derivative operating systems allocate a significant portion of main memory to I/O caching. In Linux this is called the page buffer cache. The file systems also issue readahead requests when they detect sequential reads of a file's logical blocks. Since the page buffer cache is usually an order of magnitude larger than any on-board disk cache, most of the benefits of caching are captured there. Nevertheless, users can force individual processes to avoid the page buffer cache, and so there are cases where the on-board disk cache could have significant effect.

We model the effects of the on-board cache with just three parameters, the number of segments, the number of sectors per segment, and the pre-fetch size, also in sectors. We assume the cache is fully associative with FIFO replacement, and, should a request exceed the segment size, we assume wrap-around. Although manufacturers are often secretive about the operations of on-board disk caches, the minimal information we require can usually be obtained from the SCSI mode page commands. The sdparm utility [20] is a convenient tool for accessing such.

With this information we maintain a shadow cache within the scheduler. Upon entry to the scheduler, even if a previously computed optimal sequence is still valid and no $O(N^2)$ table-building would be required, we check the entire arrival queue of pending requests against the shadow cache for predicted cache hits. If any request is found to be a predicted hit, it is scheduled immediately.

The shadow cache comprises pairs of integers denoting the start and end sectors of the span assumed to be contained within each segment. When a request is dispatched, the pair (start sector, end sector + pre-fetch) is written into the shadow cache at the current segment index, and the current segment index is advanced. To check for a predicted read hit, the requested span is compared against the spans contained in each segment.

Finally, we use a form of soft deadline to control response time variance. Reads are given priority over writes, and only reads use the table-building algorithm. Writes are served in CSCAN order. If the oldest pending read request exceeds a maximum READDELAY parameter, we forgo table-building and serve reads in CSCAN order as well, until the oldest reader no longer exceeds this age. The deadline is soft because it only guarantees service within the next sweep.

4   A Virtual Performance Throttle

We now describe how the real performance of these algorithms can be predicted using a virtual machine. We start by specifying a service time model for the targeted physical drive. A highly accurate model would require a detailed mapping of logical to physical blocks, which, for modern disks, is often serpentine in nature [21]. Within a single track, consecutive logical sectors usually map directly to consecutive physical sectors, but the mapping of logical tracks to physical tracks is considerably more complex because cylinder seek time is now less than head switching time. As a result, logical tracks are laid out in bands that allow multiple cylinder seeks per head switch when the disk is read in a logically sequential way. Nevertheless, as we will illustrate in the next section, at the macroscopic level, where seek distances are measured in millions of sectors, a linear model will suffice. Thus we assume we have a service time model of the form $X_r = R_r/2 + S_r(d_r/D_r)$, where R_r is rotation, S_r is maximum seek time, and D_r is maximum seek distance.

The idea of the virtual performance throttle (VPT) is to insert an *iprobe* into the SCSI path of the virtual system to force virtual service times that are proportional to real ones, with an identifiable constant of proportionality. If the virtual seek distance is d_v with maximum D_v, then we want the virtual system to deliver a service time of kX_r, where $d_r/D_r = d_v/D_v$, for some system-wide constant, k. Instead, it will deliver a service time of X_v. Clearly, we can achieve our goal by delaying the completed virtual request by $kX_r - X_v$, except that k is unknown to us. Further, due to factors outside our control, such as server load and network congestion on the hardware supporting the virtual machine, an appropriate k may change during the course of our tests. Thus we need a self-scaling system.

We specify an initial value of k and allow the VPT to constrain the flow of disk requests being served based on this value and the linear service time model. The VPT uses two kernel probes in the generic SCSI driver: a *jprobe* in the down path to record when requests leave for the virtual disk and an *iprobe* in the up path to intercept and delay completed requests upon return from the virtual disk. The *jprobe* calculates the target completion time of the request and passes it to the *iprobe*, which then determines how long the request should be delayed after its completion but before returning it to the requesting process. The *iprobe*'s queue of delayed completions is checked periodically with a timer. Once the target completion time has passed, the request is injected back into the SCSI generic path.

As noted, the service time on the virtual disk is subject to change, and so it is possible that a request will arrive to the *iprobe* after its target completion time. This means that the scale factor, k, is too small. The *iprobe* increases it and passes the new value to the *jprobe*. Similarly, if the VPT develops a large queue of delayed requests, we know that the target performance is an underestimate, and the *iprobe* can decrease k to shorten the overall length of the simulation. The *iprobe* reports the current value of k on each change. In practice, we use preliminary tests with a dynamic k to find stable values and then fix k at a stable value for each real test.

5   Platform and Workload

The real test platform used in our study was a Linux (2.6.30) system with two, Intel Xeon 2.80GHz processors, 1 GB main memory, a Western Digital IDE system drive and two external Seagate Cheetah 15K.4 SCSI drives, each with its own Adaptec 39320A Ultra320 SCSI controller. Tests were restricted a single Cheetah drive. The disk is a model ST373454 with 4 recording surfaces and a formatted capacity of 73.4 GBytes. It rotates at 15,000 rpm yielding a rotation time of 4 ms. The disk has 50,864 tracks per recording surface and was formatted at 512 bytes per sector.

To approximate the linear service time at the macroscopic level, we disabled the on-board cache, opened */dev/sda* using the *O_DIRECT* mode, which forced a bypass of the page buffer cache, and read 100,000 randomly selected pages. The time required to read each page and the distance in sectors from the previously read page were captured. We sorted this data in order of increasing distance and plotted distance versus time. The result was a reasonably linear band of noise approximately 4 ms in width. The data was then smoothed using a filter that replaced each point with the average of the (up to) 1001 points centered at the

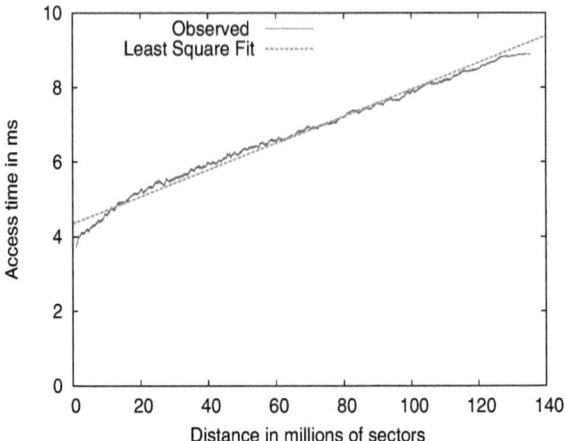

Fig. 1. Stochastic sector to sector costs

point in question. The filtered data and the least squares approximation to it are plotted in Figure 1. The linear model is $X_r = 4.25 + 5.25(d_r/D_r)$.

The Cheetah manual [22] indicates that 7,077KB is available for caching, which yields 221 512-byte sectors per segment. We assumed that a single span of requested sectors, plus the pre-fetch, would be cached in each segment. Single request spans larger than 157 sectors (221-64) were rare in our workloads, but for those cases we assumed a wrap-around with over-write within the segment.

The KVM-based virtual machine was hosted on an IBM 8853AC1 dual-Xeon blade. It was configured with a 73GB virtual SCSI disk, for which the available emulator was an LSI Logic / Symbios Logic 53c895. The virtual disk image was stored on a NetApp FAS960c and accessed via NFS.

We compared the performance of our cache-aware, table-scheduling (CATS) algorithm with *cfq* and *deadline* on both real and virtual platforms. The CATS *busdriver* delay was set to 7*ms* and the maximum READDELAY to 100*ms*.

We used two workloads in this study. Both were similar, at the process level, to that used by Geist, Steele, and Westall [17], which was based on the approach used by Barford and Crovella [19] in building their Scalable URL Reference Generator (SURGE) tool. Each of 50 processes executed an ON/OFF infinite request loop, shown in pseudo-code in Figure 2.

The Pareto(α,k) distribution is a heavy-tailed distribution,

$$F_X(x) = \begin{cases} 1 - (k/x)^\alpha & x \geq k \\ 0 & elsewhere \end{cases} \tag{1}$$

The discrete Zipf distribution is given by

$$p(i) = k/(i+1), \qquad i = 0, 1, ..., N \tag{2}$$

```
forever{
    generate a file count, n, from Pareto(α₁,k₁);
    repeat (n times){
        select filename using Zipf(N);
        while (file not read){
            read page from file;
            generate t from Pareto(α₂,k₂);
            sleep t milliseconds;
        }
    }
}
```

Fig. 2. ON/OFF execution by each of 50 concurrent processes

where k is a normalizing factor, specifically, the reciprocal of the $N+1^{st}$ harmonic number. A continuous approximation,

$$F_X(x) = \frac{log(x+1)}{log(N+2)} \qquad 0 \le x \le N+1 \qquad (3)$$

suffices for our study.

File count parameters were taken directly from the Barford and Crovella study, $(\alpha_1, k_1) = (2.43, 1.00)$. The shape parameter of the sleep interval, $\alpha_2 = 1.50$, was also taken from this study, but we used a different scale parameter, $k_2 = 2.0$, because our sleep interval was milliseconds per block rather than seconds per file.

Both the virtual SCSI drive and the Cheetah drive were loaded with 1 million files in a two-level directory hierarchy where file sizes were randomly selected from a mixture distribution also suggested by Barford and Crovella. This mixture distribution is lognormal(9.357,1.318) below 133 KB and Pareto(1.1,133K) above, where the lognormal(μ,σ) distribution function is given by:

$$F_Y(y) = \int_0^y e^{-\frac{(log_e t - \mu)^2}{2\sigma^2}} /(t\sigma\sqrt{2\pi})dt \quad y > 0 \qquad (4)$$

To induce reasonable fragmentation, we erased a half million files, selected at random, and then added back a half million files with different, randomly selected sizes. Due to the relatively small capacity of these drives, we chose to truncate at 100 MB any files that would have exceeded that size.

For each algorithm, for each test run, we captured the arrival times, service initiation times, and service completion times of 50,000 requests. We captured these time stamps by directly instrumenting the kernel outside of the schedulers, and, during each test run, we stored the time stamps to a static kernel array. The time stamp data was extracted from the kernel array after the test run by using a custom system call.

The two workloads were identical at the process level but decidedly different at the drive level. For the first, each file was opened with mode flag *O_DIRECT*, which, as noted earlier, forced the associated I/O to by-pass the main memory page buffer cache. The second workload differed from the first only in that the *O_DIRECT* flag was not used. The page buffer cache, a standard feature of UNIX-derivative operating systems, is dynamic and can grow to become quite large. For tests described here, 65MB was often observed.

Finally, although the Cheetah drive supports tagged command queueing (TCQ), we disabled it for all tests. We found that, for all schedulers, allowing re-scheduling by the drive hardware <u>decreased</u> performance. We would have thought this to be an anomaly, but we have observed the same result for other SCSI drives on other Linux systems.

6   Results

The results for the first workload, using *O_DIRECT*, are shown in Table 2. We see that the virtual system uniformly predicted higher mean service, higher mean response, and lower throughput than was found from measurements of the real system. Nevertheless, on all three measures, the predicted performance rank of the three algorithms was correct: CATS performs better than *deadline*, which performs better than the Linux default, *cfq*. Thus algorithm selection could be made solely on the basis of the virtual system predictions.

Table 2. Performance on *O_DIRECT* workload

	real			virtual ($k=8$)		
algorithm	cats	deadline	cfq	cats	deadline	cfq
mean service (ms)	1.96	2.71	1.39	2.58	3.24	2.36
variance service	8.51	9.76	5.85	9.03	8.23	7.78
mean response (ms)	37.35	59.87	124.70	53.79	78.27	117.13
variance response	6961.50	561.15	839270.49	16641.07	633.28	28651.71
throughput (sectors/ms)	8.19	6.08	2.19	6.15	5.06	3.38

We also gauged the effectiveness of the shadow cache in predicting real cache hits by placing record markers within the captured time stamp trace of the CATS scheduler on all records that were predicted by the shadow cache to be hits. We examined the service time distribution of a large collection of single-sector requests, independent of any trace, and found a prominent initial spike at 250 microseconds. We then processed the CATS trace and marked any record with service time below 250 microseconds as an actual hit. We found that the shadow cache correctly predicted 97% of the 31,949 actual hits observed.

When I/O is staged through the main memory page buffer cache (the second workload), the results are decidedly different, as shown in Table 3. Again the virtual system uniformly overestimated mean service time and mean response time and underestimated throughput, but again it correctly predicted the performance rank of all three algorithms on all three measures.

Table 3. Performance on non-*O_DIRECT* workload

algorithm	real			virtual (k=8)		
	cats	*deadline*	*cfq*	*cats*	*deadline*	*cfq*
mean service (ms)	6.53	7.41	7.80	7.15	7.60	8.57
variance service	11.13	8.80	17.62	6.22	6.05	10.30
mean response (ms)	114.91	121.87	179.17	189.45	198.33	258.75
variance response	8080.16	3296.87	35349.39	19292.52	6839.66	65796.33
throughput (sectors/ms)	12.00	12.04	9.08	11.44	11.68	8.82

7 Conclusions

We have suggested a method for predicting the performance of disk scheduling algorithms on real machines using only their measured performance on virtual machines. The method uses a dynamically loaded kernel intercept probe (*iprobe*) to adjust low-level virtual device timing to match that of a simple model derived from the real disk device. We used this method to predict the performance of three disk scheduling algorithms, one of which is new. Although the virtual system was seen to uniformly underestimate performance, it correctly predicted the relative performance of the three algorithms as measured on a real system under two workloads.

It it fair to charge that we are simply using a virtual operating system as an elaborate simulator. Nevertheless, this simulator provides almost all the subtle nuances of a real operating system and yet requires almost no programming effort on our part. A low-level model of device service time performance and the drop-in *iprobe* are all that is required.

We are currently working on methods to achieve more accurate absolute predictions. As yet we have not accounted for measurement overhead inherent in our method. We believe that this is one of the causes of the uniformly underestimated performance. Another potential source of error is the model service time ascribed to a cache hit, which we fix at 250 microseconds, even though this only an observed maximum. Unfortunately, reducing this value, X_r, will require an increase in scale factor, k, so that $kX_r - X_V$ remains non-negative, and increasing k increases run-time. We may be faced with a trade-off between accuracy and (simulation) run-time. Nevertheless, we believe that accounting for these factors will result in more accurate predictions.

References

1. Creasy, R.: The origin of the VM/370 time-sharing system. IBM Journal of Research & Development 25, 483–490 (1981)
2. Popek, G.J., Goldberg, R.P.: Formal requirements for virtualizable third generation architectures. Commun. ACM 17, 412–421 (1974)
3. Robin, J.S., Irvine, C.E.: Analysis of the Intel®Pentium's™ability to support a secure virtual machine monitor. In: Proc. of the 9th Conf. on USENIX Security Symposium (SSYM 2000), pp. 10–10. USENIX Assoc., Berkeley (2000)

4. VMWare, Inc.: Understanding full virtualization, paravirtualization, and hardware assist (2007),
   http://www.vmware.com/files/pdf/VMware_paravirtualization.pdf
5. Barham, P., Dragovic, B., Fraser, K., Hand, S., Harris, T., Ho, A., Neugebauery, R., Pratt, I., Warfield, A.: Xen and the art of virtualization. In: Proc. 19th ACM Symp. on Operating System Principles, Bolton Landing, New York, pp. 164–177 (2003)
6. Kivity, A., Kamay, Y., Laor, D., Lublin, U., Liguori, A.: kvm: the Linux virtual machine monitor. In: Proceedings of the Linux Symposium, Ottawa, Ontario, Canada, pp. 225–230 (2007)
7. Geist, R., Jones, Z., Westall, J.: Virtualizing high-performance graphics cards for driver design and development. In: Proc. 19th Annual Int. Conf. of the IBM Centers for Advanced Studies (CASCON 2009), Toronto, Ontario, Canada (2009)
8. Geist, R., Jones, Z., Westall, J.: Virtualization of an advanced course in operating systems. In: Proc. 3rd Int. Conf. on the Virtual Computing Initiative (ICVCI3), Raleigh, North Carolina (2009)
9. Jones, Z.H.: A Framework for Virtual Device Driver Development and Virtual Device-Based Performance Modeling. PhD thesis, Clemson University (2010)
10. Coffman, E., Hofri, M.: On the expected performance of scanning disks. SIAM J. on Computing 11, 60–70 (1982)
11. Coffman Jr., E.G., Gilbert, E.N.: Polling and greedy servers on a line. Queueing Syst. 2, 115–145 (1987)
12. Geist, R., Daniel, S.: A continuum of disk scheduling algorithms. ACM TOCS 5, 77–92 (1987)
13. Worthington, B.L., Ganger, G.R., Patt, Y.N.: Scheduling algorithms for modern disk drives. In: Proceedings of the 1994 ACM SIGMETRICS Conference on Measurement and Modeling of Computer Systems, Nashville, TN, USA, pp. 241–251 (1994)
14. Pratt, S., Heger, D.A.: Workload dependent performance evaluation of the Linux 2.6 i/o schedulers. In: Proc. of the Linux Symposium, Ottawa, Ontario, vol. 2, pp. 425–448 (2004)
15. Geist, R., Ross, R.: Disk scheduling revisited: Can $o(n^2)$ algorithms compete? In: Proc. of the 35th Annual ACM SE Conf., Murfreesboro, Tennessee, pp. 51–56 (1997)
16. Bellman, R.E.: Dynamic Programming. Dover Publications (2003) (incorporated)
17. Geist, R., Steele, J., Westall, J.: Enhancing webserver performance through the use of a drop-in, statically optimal, disk scheduler. In: Proc. of the 31st Annual Int. Conf. of the Computer Measurement Group (CMG 2005), Orlando, Florida, pp. 697–706 (2005)
18. Iyer, S., Druschel, P.: Anticipatory scheduling: a disk scheduling framework to overcome deceptive idleness in synchronous i/o. In: SOSP 2001: Proceedings of the Eighteenth ACM Symposium on Operating Systems Principles, pp. 117–130. ACM, New York (2001)
19. Barford, P., Crovella, M.: Generating representative web workloads for network and server performance evaluation. In: Proc. ACM SIGMETRICS Measurement and Modeling of Computer Systems, pp. 151–160 (1998)
20. Douglas Gilbert: sdparm utility 1.03 (2008),
    http://sg.danny.cz/sg/sdparm.html
21. Qian, J., Meyers, C.R., Wang, A.I.A.: A Linux implementation validation of track-aligned extents and track-aligned raids. In: USENIX Annual Technical Conference, Boston, MA, pp. 261–266 (2008)
22. Seagate Technology LLC: Product Manual Cheetah 15K.4 SCSI. Pub. no. 100220456, rev. d edn., Scotts Valley, CA (2005)

Modeling Wireless Sensor Networks Using Finite-Source Retrial Queues with Unreliable Orbit[*]

Patrick Wüchner[1], János Sztrik[2], and Hermann de Meer[1]

[1] Faculty of Informatics and Mathematics,
University of Passau, Innstraße 43,
94032 Passau, Germany
patrick.wuechner@uni-passau.de
[2] Faculty of Informatics, University of Debrecen,
Egyetem tér 1. Po.Box 12, 4010 Debrecen, Hungary
jsztrik@inf.unideb.hu

Abstract. Motivated by the need for performance models suitable for modeling and evaluation of wireless sensor networks, we introduce a retrial queueing system with a finite number of homogeneous sources, unreliable servers, orbital search, and unreliable orbit. All random variables involved in model construction are assumed to be independent and exponentially distributed. Providing a generalized stochastic Petri net model of the system, steady-state analysis of the underlying continuous-time Markov chain is performed and steady-state performance measures are computed by the help of the MOSEL-2 tool. The main novelty of this investigation is the introduction of an unreliable orbit and its application to wireless sensor networks. Numerical examples are derived to show the influence of sleep/awake time ratio, message dropping, and message blocking on the senor nodes' performance.

Keywords: performance evaluation, unreliable finite-source retrial queue, wireless sensor network, energy efficiency, self-organization.

1 Introduction

Wireless sensor networks (WSNs, [1, 12]) are communication networks with harsh resource constraints. They need lightweight, energy-efficient, and self-organizing communication protocols.

In this paper, we propose a model that allows to discuss the trade-off between the energy efficiency and performance of WSNs by showing the positive and

---

[*] This research is partially supported by the German-Hungarian Intergovernmental Scientific Cooperation (HAS&DFG, 436 UNG 113/197/0-1), by the New Hungary Development Plan (TÁMOP 4.2.1./B-09/1/KONV-2010-0007), by the AutoI project (STREP, FP7 Call 1, ICT-2007-1-216404), by the ResumeNet project (STREP, FP7 Call 2, ICT-2007-2-224619), by the SOCIONICAL project (IP, FP7 Call 3, ICT-2007-3-231288), and by the EuroNF Network of Excellence (FP7, IST 216366).

K.A. Hummel et al. (Eds.): PERFORM 2010 (Haring Festschrift), LNCS 6821, pp. 73–86, 2011.

negative effects of message dropping and blocking for variable sleep/awake time ratios. The model is based on retrial queueing systems [7] in which arriving jobs that find all servers unavailable do not line up in a queue, but join a so-called orbit. An orbit is a buffer from where the jobs retry to get service until they are successfully served. In contrast to ordinary queueing systems, the server(s) might be idle even if the buffer contains jobs.

Due to their broad practical applicability, e.g., in the field of communication networks, and due to their non-triviality, retrial queues have been receiving wide interest in the scientific community. The interested reader is referred to [7] for a recent introduction and summary of main methods, results, and applications.

Here, we focus on finite-source retrial queues. The arrival process is then non-Poisson and depends on the number of customers already staying at the system (see [7, p. 32]). While several variants of finite-source retrial queues have been studied in related work (e.g., in [2–4, 6, 17–19]), the authors are not aware of any discussion of the orbit's unreliability, not even in the infinite-source case.

Our main contribution is presenting and discussing a generalized model of finite-source retrial queues with *unreliable orbit* that also takes unreliable servers (cf. [5, 15, 3, 17]) and orbital search (cf. [8, 11, 13, 18, 19]) into account.

By the help of this model, we discuss the influence of the sleep/awake time ratio, message dropping, and message blocking on the senor nodes' mean response time, serving probability, and blocking probability.

The paper is organized as follows. We introduce the investigated WSN scenario in Sect. 2. In Sect. 3, we present the full model description in form of a generalized stochastic Petri net (GSPN), discuss the underlying continuous-time Markov chain (CTMC), and present the main performance measures. Numerical results, conveniently derived using the MOSEL-2 tool, are presented and their implications on WSN design are discussed in Sect. 4. We conclude by summarizing the paper and giving directions for future work in Sect. 5.

2   Use Case: Wireless Sensor Network

An example WSN scenario is shown in Fig. 1. Immobile sensor nodes (circles) are deployed in a two-dimensional (x, y) area. The nodes are labeled with their distance to the sink measured by the number of communication hops. The sink node (solid black circle) is located at coordinate $(7, 5)$.[1]

Due to harsh resource constraints and the resulting limited transmission range, each node i (located at (x_i, y_i)) is only able to communicate directly with its immediate neighbors, i.e., with all nodes j where $|x_j - x_i| \leq 1$ and $|y_j - y_i| \leq 2$. For example, the node located at $(6, 4)$ is able to exchange messages directly with the nodes located at $(6, 6)$, $(5, 5)$, $(5, 3)$, $(6, 2)$, $(7, 3)$, and also with the sink. We further assume that each node is aware of its own distance to the sink measured in the number of hops.

---

[1] The presented concepts also hold, mutatis mutandis, for mobile sensor nodes. The nodes' labels then need to be updated regularly. In this paper, however, we limit ourselves to immobile nodes for the sake of conciseness.

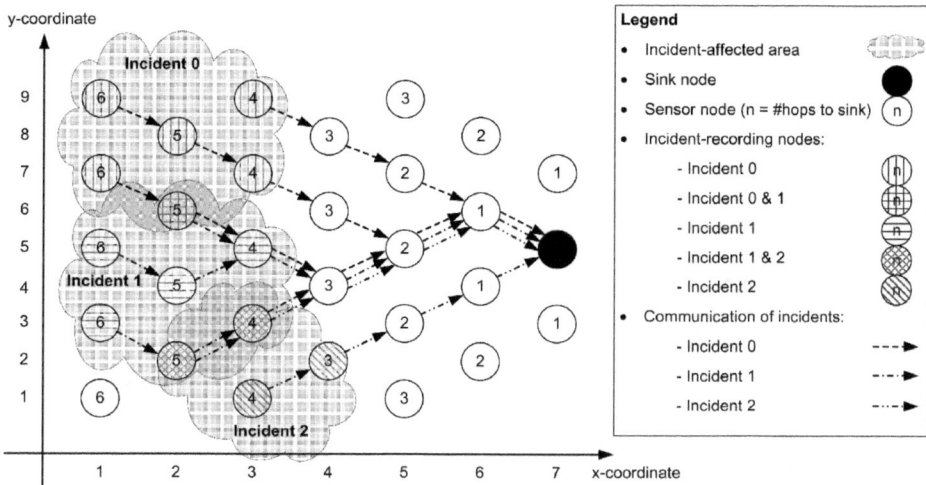

Fig. 1. WSN example scenario

The purpose of the given sensor network is to monitor the covered area, record incidents, and communicate the appearance of incidents to the sink in a multi-hop fashion. In a real system, such incidents could be, e.g., the recognition of an intrusion, temperatures or humidity exceeding or under-running predefined thresholds, or the detection of fire, gas, vibration, movement, noise, etc. We assume that each node may be equipped with several sensors. Hence, each node may detect several but a finite number of distinct incidents.

In the example scenario sketched in Fig. 1, three incidents (Incidents 0, 1, and 2) can be detected by the sensor network. For example, node $(2,6)$ detects Incidents 0 and 1, immediately generates a message for each incident, and tries to send both messages toward the sink. Node $(2,6)$ has six neighbors of which two are closer to the sink: node $(3,7)$ and node $(3,5)$.

In principle, all neighbors can be aware of the transmission due to the broadcast character of the wireless air interface. Hence, each neighbor may serve as next-hop node. However, nodes are self-organizing and, for saving energy, may decline to receive new messages, store messages, or (re-)send messages depending on the node's energy status, the sender's distance to the sink, etc.

By sending an acknowledgment, a receiver agrees to accept the message and to take care of forwarding it further. If none of the neighbors accepts the message, node $(2,6)$ stores it locally and *retries* to forward it later. If in the meanwhile node $(2,6)$ receives further messages reporting the same incident, it merges the messages. Hence, each incident is only stored once at the node.

As soon as the message has successfully been transferred to the next hop, the latter takes care of forwarding the message toward the sink. Messages that reach the sink leave the sensor network.

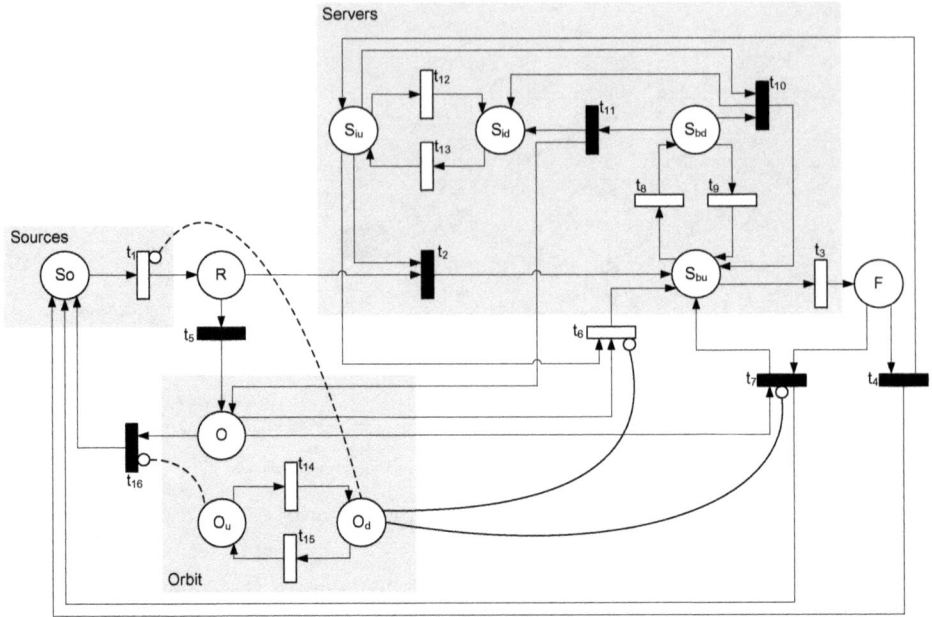

Fig. 2. GSPN of finite-source retrial queue with unreliable servers, unreliable orbit, and orbital search

3   Model and Performance Measures

As a first step toward assessing the influence of system parameters on the performance and energy efficiency of the WSN scenario introduced in Sect. 2, we focus on a single node and its neighbors in this paper.

In this section, we present a model representing such a group of nodes. After constructing the model in the form of a GSPN (cf. [10, p. 64]) and discussing the underlying CTMC (cf. [10, p. 96]), performance measures are derived based on the model's steady-state probabilities.

3.1   Generalized Stochastic Petri Net Model

In Figure 2, the model is graphically represented in form of a GSPN. The model is a generalization of finite-source retrial queues with unreliable servers (cf. [3, 17]) and finite-source retrial queues with orbital search (cf. [18, 19]) by introducing an unreliable orbit, server hopping, and a variable number of server repairmen.

The main model parameters, the places' descriptions, and the transitions' functions are summarized in Tabs. 1, 2, and 3, respectively. Tab. 4 maps the use case's parameters to the model parameters and defines default values used as a basis for numerical evaluations in Sect. 4.

Table 1. Main model parameters

Parameter	Symbol	Range
Number of sources	K	$\mathbb{N}$
Number of servers	S	$\mathbb{N}$
Number of repairmen	S_R	$\mathbb{N}$
Arrival rate	λ	$\mathbb{R}^+$
Service rate	μ	$\mathbb{R}^+$
Retrial rate	ν	$\mathbb{R}^+$
Search probability	p	$(0,1)$
Busy server breakdown rate	δ_{Sb}	$\mathbb{R}^+$
Busy server repair rate	τ_{Sb}	$\mathbb{R}^+$
Idle server breakdown rate	δ_{Si}	$\mathbb{R}^+$
Idle server repair rate	τ_{Si}	$\mathbb{R}^+$
Orbit breakdown rate	δ_O	$\mathbb{R}^+$
Orbit repair rate	τ_O	$\mathbb{R}^+$
Sources blocked (servers down)[2]	β	0, 1
Sources blocked (orbit down)[2]	γ	0, 1
Server hopping[2]	σ	0, 1
Server flushing[2]	ϕ	0, 1
Orbit flushing[2]	ω	0, 1

Table 2. Places used in Fig. 2 with capacity (column "Cap.") and initial marking (column "i.M.")

ID	Description	Cap.	i.M.
So	Active sources	K	K
R	Incoming requests	1	0
S_{bu}	Servers, busy & up	S	0
S_{iu}	Servers, idle & up	S	S
S_{bd}	Servers, busy & down	S	0
S_{id}	Servers, idle & down	S	0
O	Orbit	K	0
O_u	Orbit up	1	1
O_d	Orbit down	1	0
F	Finished requests	1	0

Note that not all model properties can be mapped to the graphical GSPN representation conveniently. For example, the dashed inhibitor arcs are subject to additional guard functions. Please refer to Tab. 3 (IF statements in column *Value*) for a comprehensive list of all guard functions that have to be considered.

The model consists of three main parts (gray boxes in Fig. 2): a finite set of sources, a finite set of servers, and the orbit component.

In our finite-source model, there are K sources represented by K Petri net tokens initially residing in place So. All tokens located in place So represent incidents that are currently not sensed by, reported to, or accepted by the node under investigation. New incidents arrive to the node with arrival rate λ per unreported incident (transition t_1). Remember that incoming reports of incidents that are already currently processed by the node do not imply new tasks which motivates the application of a finite-source model.

Arriving tokens enter place R from where they immediately try to enter the group of S servers. The node immediately tries to forward new messages (represented by tokens in place R) to one of the next hops, represented by the group of S servers. Each server might be idle and up (tokens in place S_{iu}), busy and up (S_{bu}), idle and down (S_{id}), or busy and down (S_{bd}). Busy servers represent next-hop nodes that are currently not able to receive incident messages since they are processing former messages (t_3) with rate μ. A server is considered down when the corresponding next-hop node is sleeping, i.e., in power-saving mode. Only servers that are up and idle are ready to receive tokens. If none of the servers (next hops) is idle and up (awake), arriving tokens (incident messages) move to the orbit O (investigated node's local storage of messages) via t_5.

---

[2] The parameter value 0 refers to *false/off/disabled*, the parameter value 1 refers to *true/on/enabled*.

Table 3. Transitions used in Fig. 2

ID	Type[3]	Description	Value
t_1	E	Request generation	IF $(\beta = 0$ OR $(S_{id} + S_{bd} < S))$ AND $(\gamma = 0$ OR $O_u = 1)$: λSo
t_2	I	Incoming request to server	PRIO 1
t_3	E	Service	μS_{bu}
t_4	I	Served and no orbital search	WEIGHT $1 - p$
t_5	I	Incoming request to orbit	PRIO 0
t_6	E	Retrial	IF $O_u = 1$: νO
t_7	I	Served and orbital search	IF $O_u = 1$: WEIGHT p
t_8	E	Busy server breakdown	$\delta_{Sb} S_{bu}$
t_9	E	Busy server repair	$\tau_{Sb} \min(S_{bd}, S_R)$
t_{10}	I	Server hopping on failure	IF σ: PRIO 1
t_{11}	I	Server flushing on failure	IF ϕ: PRIO 0
t_{12}	E	Idle server breakdown	$\delta_{Si} S_{iu}$
t_{13}	E	Idle server repair	$\tau_{Si} \min(S_{bi}, \max(0, S_R - S_{bd}))$
t_{14}	E	Orbit breakdown	δ_O
t_{15}	E	Orbit repair	τ_O
t_{16}	I	Orbit flushing on failure	IF ω AND $O_u = 0$

Table 4. Default parameters based on use case

Parameter	Symbol	Default[4]
Number of incident types (sources)	K	10
Mean inter-arrival time per unreported incident (1/arrival rate)	λ^{-1}	1 min
Mean retrial time per stored incident message (1/retrial rate)	ν^{-1}	5 ms
Number of potential next hops (servers)	S	5
Mean processing time at next hop (1/service rate)	μ^{-1}	20 ms
Probability that node is aware of next hop's service completion (orbital search)	p	0.1
Ratio of sleep/awake time	α	10
All nodes' mean awake time (1/failure rate)	δ^{-1}	50 ms
All nodes' mean sleeping time (1/repair rate)	τ^{-1}	$\alpha\delta^{-1}$
Message transfer from sleeping to operational next hop (server hopping)	σ	0 (off)
Drop message on falling asleep (server/orbit flushing)	ϕ, ω	0 (off)
Block incoming messages if next hops are down or re-forwarding is disabled (source blocking on server/orbit failure)	β, γ	0 (off)

Idle servers fail (t_{12}), i.e., idle next hops fall asleep, with rate δ_{Si}. Each failed idle server gets repaired (t_{13}), i.e., idle next hop wakes up, with a rate of τ_{Si} (if $S_R \geq S$). Similarly, busy servers fail (t_8) with rate δ_{Sb} and get repaired (t_9) with τ_{Sb} each (if $S_R \geq S$). If $\delta_{Si} = \delta_{Sb} > 0$, we call the breakdowns *independent*, and if $\delta_{Si} = 0 < \delta_{Sb}$, we call them *active* (cf. [17]). In the following, we assume that all breakdowns and repairs are independent with rates $\delta := \delta_{Si} = \delta_{Sb}$ and $\tau := \tau_{Si} = \tau_{Sb}$, respectively. We call δ^{-1} and τ^{-1} the next hops' *mean awake and sleeping time*, respectively.

For keeping the model general and being able to compare the model to related work, we also allow the number of server repairmen S_R to be smaller than the number of servers. If $0 < S_R < S$, failed busy servers are repaired with a higher priority than idle servers. Hence, S_R is the maximum number of repairmen available for failed busy servers and $\max(0, S_R - S_{bu})$ is the remaining maximum

---

[3] Type of transition: I: immediate transition $\Rightarrow$ column *Value* denotes priority (PRIO) or weight (WEIGHT); E: timed transition $\Rightarrow$ column *Value* denotes firing rate.

[4] Unless otherwise stated, these values are used as a basis for numerical results presented in Sect. 4.

number of repairmen available for failed idle servers (see column *Value* in Tab. 3 for t_9 and t_{13}). In the following, we assume that $S_R = S$.

The model allows server *hopping* (t_{10}) and *flushing* (t_{11}) on server failure if the parameters σ and ϕ are set to *true* (i.e., 1), respectively. Server hopping enables tokens to be directly transferred from a failing server to any operational idle server. If server flushing is active, tokens are moved from a failing server to the orbit. If both options are enabled, server hopping has higher priority. Workload at failing servers is resumed after repair if both options are disabled. In the following, we assume that hopping is always disabled ($\sigma = 0$) and server flushing is referred to as *next-hop dropping* since incident messages are dropped from failing next hops if $\phi = 1$.

Tokens located in the orbit (O) represent incident messages stored in the node under investigation for retransmission. Each token located in the orbit (O) is retrying (t_6) to enter the group of servers after an exponentially distributed retrial time with mean $1/\nu$.

Representing the node's potential to refrain from storing incoming messages and from sending stored messages for power-saving reasons, the orbit is subject to failure (t_{14}) with rate δ_O. A failed orbit gets repaired (t_{15}) with rate τ_O. Depending on parameter ω, the failing orbit may discard all stored tokens (*node dropping*, $\omega = 1$) via t_{16} or keep them for later resumption. In the following, we assume $\delta_O = \delta$ and $\tau_O = \tau$.

Via parameters β and γ, *blocked sources* (cf., [17]) can be modeled (see guard function of t_1). In the blocked case, sources do not generate new calls if all servers ($\beta = 1$) and/or the orbit ($\gamma = 1$) is down. In the unblocked case ($\beta = \gamma = 0$), sources are aware neither of server nor orbit failures. Blocked sources represent the node's ability to decline incoming messages since no next hop is available (*next-hop blocking*, $\beta = 1$) or since the node itself is in power-saving mode (*node blocking*, $\gamma = 1$). Here, power-saving of the investigated node refers to the case when it still might forward messages to next hops but refrains from storing and re-sending new incident messages that cannot be sent immediately.

With probability p, a busy server on service completion (token at place F) informs the (operational) orbit of its upcoming idleness and, if available, directly receives (t_7) the next token from the orbit (*orbital search*). With a probability of $1 - p$, orbital search is not performed (t_4). For $p \approx 1$, the performance of a reliable retrial queue resembles the performance of a classical first-come-first-served queue. In the following, we assume that the investigated node gets aware of a next hop's idleness with a probability of 10%, i.e., $p = 0.1$.

3.2   Underlying Markov Chain

The described GSPN can be mapped to the five-dimensional stochastic process $X(t) = (S_{bu}(t), S_{bd}(t), S_{iu}(t), O(t), O_u(t))$, where $0 \le S_{bu}(t) \le \min(S, K)$, $0 \le S_{bd}(t) \le \min(S, K)$, $0 \le S_{iu}(t) \le S$, $0 \le O(t) \le K$, and $O_u(t) \in \{0, 1\}$ are the number of tokens in places $S_{bu}, S_{bd}, S_{iu}, O$, and O_u, respectively, at time $t \ge 0$. Note that $S_{id}(t) = S - (S_{bu}(t) + S_{bd}(t) + S_{iu}(t))$, $S_o(t) = K - (O(t) + S_{bu}(t) + S_{bd}(t))$, and

$O_d(t){=}1{-}O_u(t)$. Places R and F do not have to be considered because all states where $R(t) > 0$ or $F(t) > 0$ are vanishing due to enabled immediate transitions.

Since all involved random variables are exponentially distributed, $X(t)$ constitutes a CTMC. We denote the state space of $X(t)$ with $\mathbf{X}$. Since $\mathbf{X}$ is both finite and irreducible, the CTMC is ergodic for all positive values of the arrival rate λ. From now on, the system is assumed to be in steady state, i.e., $t \to \infty$. Due to space limitations and the high complexity of the underlying CTMC, we refrain from visualizing it here. We also refrain from giving equations for the size of $\mathbf{X}$ based on K and S which is a tedious combinatorial problem. Note that for a less complex model, this was achieved in [19].

3.3   Main Performance Measures

The stationary probabilities of the CTMC discussed in Sect. 3.2 are

$$P(s_{bu}, s_{bd}, s_{iu}, o, o_u)$$
$$= \lim_{t \to \infty} P(S_{bu}(t) = s_{bu}, S_{bd}(t) = s_{bd}, S_{iu}(t) = s_{iu}, O(t) = o, O_u(t) = o_u).$$

The stationary probabilities can conveniently be derived using the software tool MOSEL-2 (see Sect. 4). Knowing the stationary probabilities, the main performance measures of the finite-source retrial queue can be obtained as follows:

– *Mean number of tokens at operational servers $\overline{S_{bu}}$:*

$$\overline{S_{bu}} = \sum_{\forall (s_{bu}, s_{bd}, s_{iu}, o, o_u) \in \mathbf{X}} s_{bu} P(s_{bu}, s_{bd}, s_{iu}, o, o_u).$$

– *Mean number of tokens at failed servers $\overline{S_{bd}}$:*

$$\overline{S_{bd}} = \sum_{\forall (s_{bu}, s_{bd}, s_{iu}, o, o_u) \in \mathbf{X}} s_{bd} P(s_{bu}, s_{bd}, s_{iu}, o, o_u).$$

– *Mean number of tokens at the orbit $\overline{O}$:*

$$\overline{O} = \sum_{\forall (s_{bu}, s_{bd}, s_{iu}, o, o_u) \in \mathbf{X}} o P(s_{bu}, s_{bd}, s_{iu}, o, o_u).$$

– *Probability that all servers are down P_{S_d}:*

$$P_{S_d} = \sum_{\forall (0, s_{bd}, 0, o, o_u) \in \mathbf{X}} P(0, s_{bd}, 0, o, o_u).$$

– *Probability that orbit is down P_{O_d}:*

$$P_{O_d} = \sum_{\forall (s_{bu}, s_{bd}, s_{iu}, o, 0) \in \mathbf{X}} P(s_{bu}, s_{bd}, s_{iu}, o, 0).$$

– *Mean number of blocked active sources* $\overline{So_b}$:

$$\overline{So_b} = \sum_{\substack{\forall(0,s_{bd},0,o,o_u)\in\mathbf{X}\ |\ \beta=1;\\ \forall(s_{bu},s_{bd},s_{iu},o,0)\in\mathbf{X}\ |\ \gamma=1;}} (K - (s_{bu} + s_{bd}))P(s_{bu}, s_{bd}, s_{iu}, o, o_u).$$

– *Utilization of servers:* $\rho = \frac{\overline{S_{bu}}}{S}$.

– *Mean number of busy servers:* $\overline{S_b} = \overline{S_{bu}} + \overline{S_{bd}}$.

– *Mean number of tokens flushed on orbit failure:* $\overline{O_f} = \omega\overline{O}$.

– *Mean number of tokens at service or orbit:* $\overline{M} = \overline{S_b} + \overline{O}$.

– *Mean number of active sources:* $\overline{So} = K - \overline{M}$.

– *Probability that a specific source is blocked:* $P_b = \frac{\overline{So_b}}{\overline{So}}$.

– *Mean number of unblocked active sources:* $\overline{So_a} = \overline{So} - \overline{So_b}$.

– *Overall arrival rate:* $\lambda_{in} = \lambda\overline{So_a}$.

– *Departure rate of served tokens:* $\lambda_s = \mu\overline{S_{bu}}$.

– *Departure rate of unserved tokens:* $\lambda_u = \lambda_{in} - \lambda_s$.

– *Probability of an incoming token getting served:* $P_s = \frac{\lambda_s}{\lambda_{in}}$.

– *Mean waiting time:* $\overline{W} = \frac{\overline{O}}{\lambda_{in}}$.

– *Mean response time:* $\overline{T} = \frac{\overline{M}}{\lambda_{in}}$.

4   Numerical Results

Instead of deriving the underlying CTMC and solving the system of global balance equations manually, we directly formulate the GSPN shown in Fig. 2 using the Modeling, Specification and Evaluation Language (MOSEL-2). MOSEL-2's evaluation environment then derives the performance measures by automatic generation and numerical evaluation of the underlying CTMC. Due to space limitations, we refer the interested reader to [9, 16] for details on MOSEL-2. The corresponding MOSEL-2 model is available on request. The scalability of MOSEL-2 in the context of finite-source retrial queues is discussed in [18].

The validity of the proposed model can be shown by comparing its results in the case of a reliable orbit to numerical results obtained in [3] and [17] (both partly based on the Pascal program provided in [14, p. 272–274]), and [18]. For all comparable parameter settings, our results perfectly match the reference results.

Unless stated otherwise, the model parameters are chosen according to Tab. 4 in the following.

On the x-axis of all result graphs, the parameter α (*sleep/awake ratio*) is given. As already indicated in Tab. 4, $\alpha = \frac{\delta}{\tau} = \frac{\tau^{-1}}{\delta^{-1}}$, where $\tau^{-1} := \tau_O^{-1} = \tau_{Sb}^{-1} = \tau_{Si}^{-1}$ is the mean time the nodes are in power-saving mode, and $\delta^{-1} := \delta_O^{-1} = \delta_{Sb}^{-1} = \delta_{Si}^{-1}$ is the mean time the nodes (investigated node and its idle or busy next hops) are awake. Hence, α is the proportion of time the investigated nodes are in power-saving mode. The higher α, the longer nodes sleep in comparison to the time they are awake and the less energy they use in the course of time.[5]

On the y-axis of the presented result graphs, we focus on the mean response time $\overline{T}$, source blocking probability P_b, and probability of getting served P_s.

4.1   Influence of Sleep/Awake Ratio and Mean Awake Time

Fig. 3 shows the effect of modifying the sleep/awake ratio α (x-axis) and the mean awake time δ^{-1} (curves) on the mean response time $\overline{T}$ (y-axis).

Increasing the energy efficiency (sleep/awake ratio) clearly results in a higher mean response time. Increasing the mean awake time while keeping the mean sleep/awake ratio constant also increases the mean response time, because the sleeping time is a multiple of the awake time. Unfortunately, decreasing the mean awake time is not arbitrarily possible, since switching state often comes with additional overhead not (yet) handled by the model.

An important result is that accepting higher response times, i.e., increased delay tolerance, considerably eases the design of energy-efficient WSNs.

For sleep/awake ratio $\alpha \approx 0$, the probability that a node sleeps is very small. For all values of the mean awake time δ^{-1}, the mean response time $\overline{T}$ is then close to the processing time $\mu^{-1} \approx 20ms$ of the next hop. Since the inter-arrival time $\lambda^{-1} = 1min$ is very large in comparison to the processing time, the probability that at arrival of a new incident all next hops are busy is close to zero. Consequently, the probability for having to wait (and to retry) is also close to zero which results in a negligible waiting time.

In the following, we set the mean awake time to $\delta^{-1} = 50ms$ and discuss whether the mean response time can be decreased by dropping messaged from sleeping nodes.

4.2   Influence of Sleep/Awake Ratio and Message Dropping

Fig. 4 shows of the effect of the sleep/awake ratio α (x-axis) and incident message dropping by sleeping nodes $\phi = \omega$ (curves) on the mean response time $\overline{T}$ (y-axis, top) and on the serving probability P_s (y-axis, bottom).

It can be seen at the top of Fig. 4 that dropping stored incident messages from sleeping nodes significantly decreases the mean response time, even below the reliable case. This is because messages that are dropped may spend less time in the system than messages that are successfully served.

Therefore, we also need to have a look at the probability P_s that an accepted incident message can be successfully processed. This is shown at the bottom of Fig. 4. Dropping stored messages clearly implies smaller serving probabilities.

---

[5] Note that in this paper we do not yet consider energy used by the process of switching between awake and sleep states.

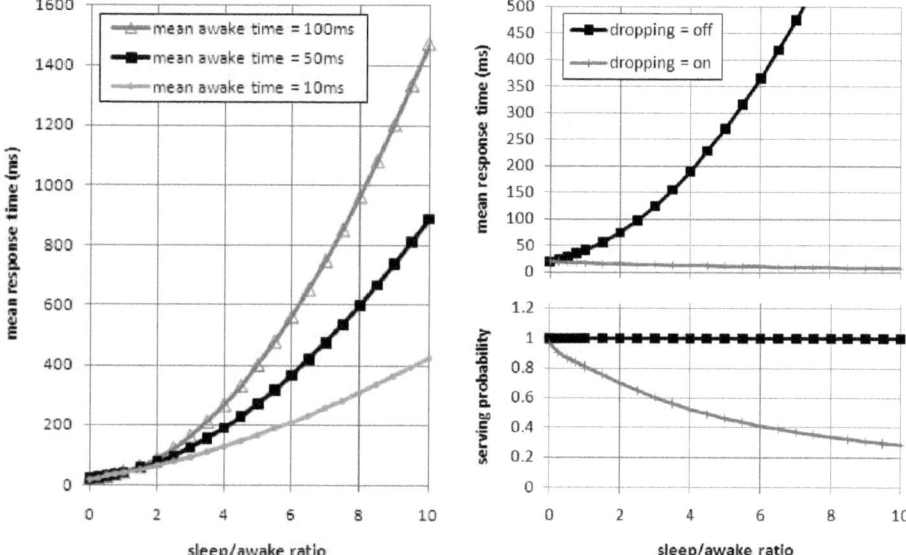

Fig. 3. Effect of sleep/awake ratio α (x-axis) and mean awake time δ^{-1} (curves) on mean response time $\overline{T}$ (y-axis)

Fig. 4. Effect of sleep/awake ratio α (x-axis) and message dropping $\phi = \omega$ (curves) on mean response time $\overline{T}$ (y-axis, top) and on serving probability P_s (y-axis, bottom)

Therefore, dropping should be used for outdated messages only. However, the age of tokens cannot be considered by the presented model.

Note that even for $\alpha \approx 0$, i.e., the probability of the nodes being awake is close to 1, the serving probability is not necessarily close to 1. The serving probability also highly depends on the absolute value of the awake time since after each awake time, stored incident messages are dropped (regardless how short the sleep time is afterwards).

In Sect. 4.3, we aim at increasing the serving probability by blocking incoming messages when nodes are sleeping.

4.3 Influence of Sleep/Awake Ratio and Message Blocking

Fig. 5 shows the effects of the sleep/awake ratio α (x-axis) and of the message blocking $\beta = \gamma$ (curves) on the mean response time $\overline{T}$ (y-axis, top), the serving probability P_s (y-axis, middle), and the blocking probability P_b (y-axis, bottom).

First, we discuss the serving probability shown in the middle of Fig. 5. Blocking incoming messages has a positive effect on the serving probability. For sleep/awake ratio $\alpha > 3$, the effect of next-hop blocking (——) has significantly more effect than node blocking (——).

In comparison to the non-blocking case (——), the following improvements can be observed:

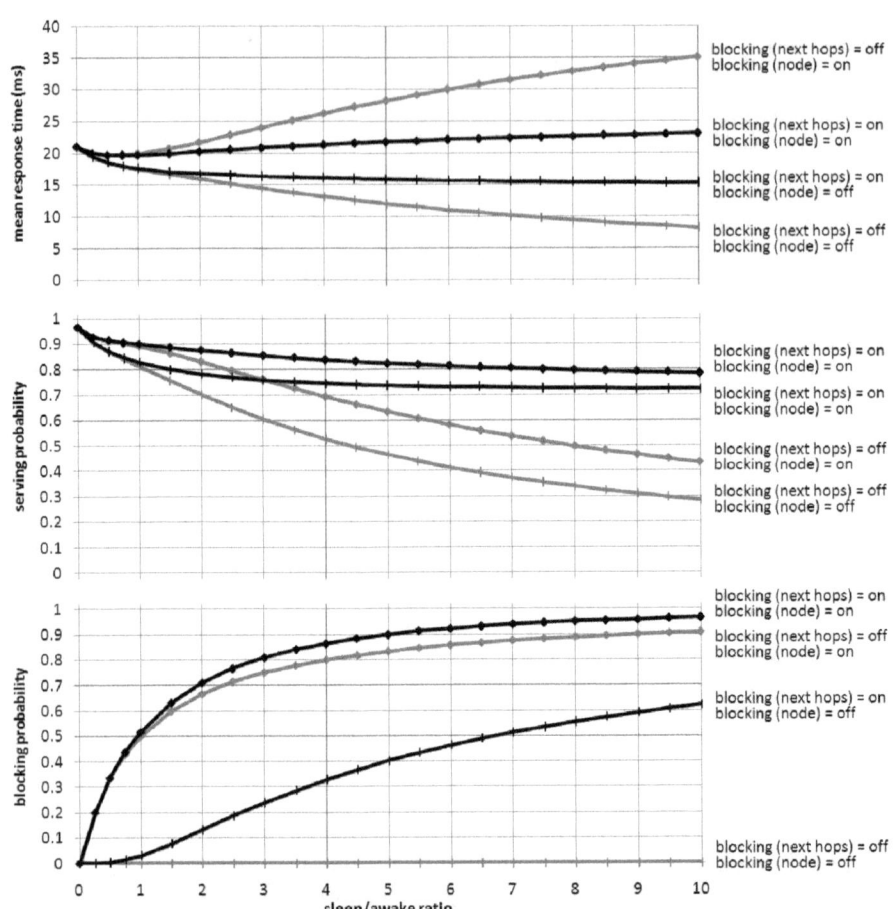

Fig. 5. Effect of sleep/awake ratio α (x-axis) and message blocking β & γ (curves) on mean response time $\overline{T}$ (y-axis, top), serving probability P_s (y-axis, middle), and blocking probability P_b (y-axis, bottom)

- Some improvement (—◆—): Node blocking prevents arriving messages from being dropped immediately when the node is in power-saving mode and immediate forwarding is not possible due to busy or sleeping next hops at the same time.
- Good improvement (—+—): If next-hop blocking is enabled, messages do not join the node when no next hop is accepting messages. This reduction of the number of messages stored at the node increases the serving probability, since all stored messages are dropped later with a high probability.
- Best improvement (—◆—): The combination of both blocking mechanism results in the best improvement of the serving probability.

Focusing on the two best cases with respect to the serving probability (next-hop blocking is enabled, curves —◆— and —+—), we now discuss which is the better alternative regarding the mean response time given at the top of Fig. 5.

Curve —+— refers to the case when node blocking is disabled. As seen before, the serving probability drops significantly when the sleep/awake ratio is increased, because it is more likely that a message needs to wait in the node and hence may be dropped if storage gets disabled before an operational and free next hop is found. Since dropped messages tend to stay less time in the system than served messages, the mean response time decreases.

In the scenario of curve —+—, the probability that an incoming message gets immediately dropped is higher than in the scenario of curve —●—, where this situation is not possible due to node blocking.

Curve —●— follows the same behavior as —+— for small sleep/awake ratios ($\alpha < 0.5$) but is then dominated by another effect that can be explained as follows. Messages are only accepted when they can at least be stored locally. Hence, if they cannot be forwarded immediately on arrival, they will (in contrast to scenario —+—) always experience waiting time. Additionally, since the next-hop nodes follow the same sleep/awake ratio as the investigated node, stored messages have to wait longer before being successfully transferred to an available next hop with increasing sleep/awake ratio.

Hence, while —●— is more attractive than —+— with respect to the serving probability, it is less attractive regarding the mean response time. Moreover, it can be seen at the bottom of Fig. 5 that —+— has a significantly lower probability of blocking incoming incident messages than —●—.

We can therefore conclude that next-hop blocking is more attractive than node blocking. It should be noted, however, that next-hop blocking is relying on local knowledge about the next hops' state. How this knowledge can be obtained in a self-organized manner and which overhead might come with this is not yet covered by this paper.

5 Conclusions

A generalization of finite-source retrial queueing systems is studied. In addition to unreliable servers, the model can be used to evaluate the effects of an unreliable orbit. The model is used to discuss the trade-off between energy efficiency and performance in a wireless sensor network use case. The model evaluation is carried out using the MOSEL-2 tool. The numerical results are discussed in detail and show the positive and negative effects of message dropping and message blocking.

Our future work aims at finding self-organized mechanisms to provide the sensor nodes with the necessary local knowledge. We also need to discuss in more detail how single-node results can be aggregated suitably to evaluate multiple-hop scenarios and larger network topologies. Validation of model results by comparison with test-bed results is considered. Finally, the presented model should assist in developing energy-efficient and self-organizing lightweight communication protocols for wireless sensor networks.

References

1. Akyildiz, I.F., Vuran, M.C.: Wireless Sensor Networks. John Wiley & Sons (July 2010)

2. Almasi, B., Bolch, G., Sztrik, J.: Heterogeneous finite-source retrial queues. Journal of Mathematical Sciences 121(5), 2590–2596 (2004)
3. Almasi, B., Roszik, J., Sztrik, J.: Homogeneous finite-source retrial queues with server subject to breakdowns and repairs. Mathematical and Computer Modelling 42, 673–682 (2005)
4. Amador, J.: On the distribution of the successful and blocked events in retrial queues with finite number of sources. In: Proc. of the 5th Int'l Conf. on Queueing Theory and Network Applications, pp. 15–22 (2010)
5. Artalejo, J.R.: New results in retrial queueing systems with breakdown of the servers. Statistica Neerlandica 48, 23–36 (1994)
6. Artalejo, J.R.: Retrial queues with a finite number of sources. J. Korean Math. Soc. 35, 503–525 (1998)
7. Artalejo, J.R., Gómez-Corral, A.: Retrial Queueing Systems: A Computational Approach. Springer, Heidelberg (2008)
8. Artalejo, J.R., Joshua, V.C., Krishnamoorthy, A.: An M/G/1 retrial queue with orbital search by the server. In: Artalejo, J.R., Krishnamoorthy, A. (eds.) Advances in Stochastic Modelling, pp. 41–54. Notable Publications Inc., NJ (2002)
9. Begain, K., Bolch, G., Herold, H.: Practical Performance Modeling – Application of the MOSEL Language. Kluwer Academic Publishers (2001)
10. Bolch, G., Greiner, S., de Meer, H., Trivedi, K.: Queueing Networks and Markov Chains, 2nd edn. John Wiley & Sons, New York (2006)
11. Chakravarthy, S.R., Krishnamoorthy, A., Joshua, V.C.: Analysis of a multi-server retrial queue with search of customers from the orbit. Performance Evaluation 63(8), 776–798 (2006)
12. Dressler, F.: Self-Organization in Sensor and Actor Networks. John Wiley & Sons (2007)
13. Dudin, A.N., Krishnamoorthy, A., Joshua, V.C., Tsarenkov, G.V.: Analysis of the BMAP/G/1 retrial system with search of customers from the orbit. Eur. J. Operational Research 157(1), 169–179 (2004)
14. Falin, G., Templeton, J.: Retrial Queues. Chapman & Hall (1997)
15. Wang, J., Cao, J., Li, Q.: Reliability analysis of the retrial queue with server breakdowns and repairs. Queueing Systems 38, 363–380 (2001)
16. Wüchner, P., de Meer, H., Barner, J., Bolch, G.: A brief introduction to MOSEL-2. In: German, R., Heindl, A. (eds.) Proc. of 13th GI/ITG Conference on Measurement, Modelling and Evaluation of Computer and Communication Systems (MMB 2006), GI/ITG/MMB. University of Erlangen, VDE Verlag (2006)
17. Wüchner, P., de Meer, H., Bolch, G., Roszik, J., Sztrik, J.: Modeling finite-source retrial queueing systems with unreliable heterogeneous servers and different service policies using MOSEL. In: Al-Begain, K., Heindl, A., Telek, M. (eds.) ASMTA 2007 Conference, Prague, Czech Republic, pp. 75–80 (June 2007)
18. Wüchner, P., Sztrik, J., de Meer, H.: Homogeneous finite-source retrial queues with search of customers from the orbit. In: Proc. of 14th GI/ITG Conference on Measurement, Modelling and Evaluation of Computer and Communication Systems (MMB 2008), Dortmund, Germany (March 2008)
19. Wüchner, P., Sztrik, J., de Meer, H.: Finite-source M/M/S retrial queue with search for balking and impatient customers from the orbit. Computer Networks 53(8), 1264–1273 (2009)

Markov Chains and Spectral Clustering

Ning Liu and William J. Stewart

Department of Computer Science
North Carolina State University, Raleigh, NC 27695-8206, USA
nliu@ncsu.edu,
billy@ncsu.edu

Abstract. The importance of Markov chains in modeling diverse systems, including biological, physical, social and economic systems, has long been known and is well documented. More recently, Markov chains have proven to be effective when applied to internet search engines such as Google's PageRank model [7], and in data mining applications wherein data trends are sought. It is with this type of Markov chain application that we focus our research efforts. Our starting point is the work of Fiedler who in the early 70's developed a spectral partitioning method to obtain the minimum cut on an undirected graph (symmetric system). The vector that results from the spectral decomposition, called the Fiedler vector, allows the nodes of the graph to be partitioned into two subsets. At the same time that Fiedler proposed his spectral approach, Stewart proposed a method based on the dominant eigenvectors of a Markov chain — a method which was more broadly applicable to nonsymmetric systems. Enlightened by these, somewhat orthogonal, results and combining them together, we show that spectral partitioning can be viewed in the framework of state clustering on Markov chains. Our research results to date are two-fold. First, we prove that the second eigenvector of the signless Laplacian provides a heuristic solution to the NP-complete state clustering problem which is the dual of the minimum cut problem. Second, we propose two clustering techniques for Markov chains based on two different clustering measures.

Keywords: spectral clustering, graph partitioning, Markov chains, eigenvector.

1 Introduction

The aim of clustering is to group objects together on the basis of similarity measurement criteria. Roughly speaking, clustering algorithms can be divided into two classes: hierarchical and partitioning [4]. In this paper, we focus on graph partitioning which provides clustering information for graph nodes [1,6].

A graph $G = (V, E)$, where V is a set of vertices and E is a set of edges, can be partitioned into two disjoint subsets V_1 and V_2, $V_1 \cup V_2 = V$, by deleting those edges that connect V_1 and V_2. A common objective is to partition the graph in such a way that the partition has minimum cost, e.g., the minimum number of

K.A. Hummel et al. (Eds.): PERFORM 2010 (Haring Festschrift), LNCS 6821, pp. 87–98, 2011.

edges cut. In a weighted graph this involves ensuring that the sum of the weights on the removed edges is as small as possible. In graph theoretic language, this is called the minimum cut. Given two sets of vertices V_1 and V_2 which partition a graph, the numerical value assigned to the cut, called the "cut value", is given by

$$\text{cut}(V_1, V_2) = \sum_{i \in V_1, j \in V_2} w_{ij}$$

where w_{ij} is the weight on the edge that connects vertex i to vertex j. Spectral graph partitioning is an important and effective heuristic approach for finding good solutions to the minimum cut problem. It was introduced in the early 1970s [3], and popularized in the early 1990s [11].

The starting point for applying spectral partitioning on a graph is to create a matrix representation of the graph, e.g., the Laplacian matrix. Fiedler first derived well known representations of the Laplacian to show its connection to the minimum cut problem [3] i.e., minimizing the total weight of edges cut. Besides minimum cut, a number of different graph partitioning objectives have also been proposed, such as maximum association [12] and normalized cut (balanced minimum cut) [12]. Maximum association and minimum cut can be viewed as dual problems of each other. In this paper we introduce the signless Laplacian matrix [2] to model the maximum association in spectral clustering, and show that the eigenvalues of the signless Laplacian have important physical interpretations.

Markov chains are widely used in modeling diverse systems and in data mining applications, e.g., applied to internet search engines such as Google's PageRank model [7]. Spectral based methods can also be applied to a Markov chain for the purposes of state clustering analysis. The dominant eigenvectors of the transition probability matrix of a Markov chain allow the states of the Markov chain to be arranged into meaningful groups [13]. Most recently, an interesting observation of spectral clustering provided by Meila et al.[8] is that spectral clustering can be depicted in the framework of Markov random walks on graph structure. It would appear that the results from Fiedler and Stewart have been unified, but in fact this is not the case as we shall explain later.

The rest of this paper is organized as follows. Section 2.1 reviews the basic procedure of spectral clustering by the Laplacian matrix. Section 2.2 introduces a new spectral method to cluster graph nodes with the objective of maximum association, the dual problem of minimum cut. Section 2.3 shows how to construct the stochastic matrix of a random walk from a graph so as to perform spectral clustering. In Section 2.4, we introduce two clustering measures on the states of Markov chains, first the normalized cut measure and second the distance measure of states from the steady state, and we propose a novel technique of clustering on graph nodes which is based on incorporating these two clustering techniques. Section 3 describes an application of NCD Markov chains while Section 4 presents our conclusions.

2   Spectral Graph Partitioning and State Clustering

2.1   Minimum Cut via Eigen-Analysis

In this section, we first review the basic procedure of spectral clustering. The starting point is to create the *Laplacian matrix* of a graph. Representations of the Laplacian show connections to the cut problem [3]. The Laplacian matrix L of an undirected graph $G = (V, E)$ with n vertices and m edges is the $n \times n$ matrix whose elements are as follows.

$$L_{ij} = \begin{cases} \sum_k w_{ik}, & \text{if } i = j \\ -w_{ij}, & \text{if } i \neq j, \ i \text{ and } j \text{ are adjacent} \\ 0, & \text{otherwise,} \end{cases}$$

The matrix L can be obtained as $L = D - A$, where A is the adjacency matrix of G and D is a diagonal matrix whose i^{th} diagonal element is $D_{ii} = \sum_k w_{ik}$. It is well known that the Laplacian matrix of a graph has a number of interesting properties [3].

1. L is symmetric positive semi-definite. As such its eigenvalues are all real and non-negative. Furthermore the eigenvectors of L constitute a full set of n real and orthogonal vectors.
2. $Le = 0$, where e is a column vector whose elements are all equal to 1. Thus 0 is the smallest eigenvalue of L and e is its corresponding eigenvector.
3. For any vector x, we have

$$x^T L x = \sum_{\{i,j\} \in E} w_{ij} (x_i - x_j)^2. \tag{1}$$

Given a partition of V into V_1 and V_2 ($V_1 \cup V_2 = V$), a partition vector p is defined as

$$p_i = \begin{cases} +1, & \text{vertex } i \in V_1, \\ -1, & \text{vertex } i \in V_2. \end{cases} \tag{2}$$

Clearly we see that $p^T p = n$. Given a Laplacian matrix L and a partition vector p, we have, from Equation (1),

$$p^T L p = \sum_{\{i,j\} \in E} w_{ij} (p_i - p_j)^2.$$

Observe from Equation (2), that the weight of edges within each set V_1 or V_2 is not counted in this sum, while the weight of each edge connecting a vertex of V_1 to a vertex in V_2 is multiplied by a factor of 4. Given that we have defined the cut value as $\text{cut}(V_1, V_2) = \sum_{i \in V_1, j \in V_2} w_{ij}$, it follows that $p^T L p = 4 \text{cut}(V_1, V_2)$ and the Rayleigh quotient is

$$\frac{p^T L p}{p^T p} = \frac{1}{n} \cdot 4 \text{cut}(V_1, V_2). \tag{3}$$

A well known result from matrix computation [5,10] is that the maximum and minimum of the Rayleigh quotient can be obtained as the largest and smallest eigenvalues of the Laplacian matrix L:

$$\lambda_{max} = \max_{x \neq 0} \frac{x^T L x}{x^T x} \quad \text{and} \quad \lambda_{min} = \min_{x \neq 0} \frac{x^T L x}{x^T x}, \tag{4}$$

where λ_{max} and λ_{min} are the largest and smallest eigenvalues of L respectively, and the x that achieves the maximum or minimum is the corresponding eigenvector of λ_{max} or λ_{min}. Therefore, the minimum value of the Rayleigh quotient is zero, which is the smallest eigenvalue of L corresponding to the eigenvector e. This partition vector indicates that all the vertices of the graph are in the same set, which means nothing is cut. This is the trivial partition. It is the second smallest eigenvalue (referred to as the Fiedler value) of L that provides the optimal value, and its corresponding eigenvector (referred to as the Fiedler vector) gives the real valued partition vector for our minimum cut problem.

Once the eigenvector corresponding to the second smallest eigenvalue has been computed, we can partition the vertices into two subsets. In the ideal case, the eigenvector takes on only two discrete values and the signs of the values tell us exactly how to partition the graph. However, the eigenvectors assume continuous values and we need to choose a splitting point to partition it into two parts. Here we take value zero. Actually, there are many other different ways of choosing such a splitting point. One can take the median value as the splitting point or one can search for the splitting point such that the resulting partition has the best cut value. We can also group close values together to partition the eigenvector into several parts.

2.2   Node Clustering in Graphs

Sometimes we wish to determine clustering information for a graph G, i.e., to figure out which vertices in G have strong mutual cohesion. The problem of maximizing the total weight within two clusters is the dual of the minimum cut problem. In our work, we propose a new method for modeling this dual problem by introducing an eigen-analysis of the signless Laplacian matrix M. We shall see that the clustering solution from an eigen-analysis of the signless Laplacian coincides with the solution of the minimum cut problem. We now proceed through the details of this dual problem and describe our approach to solving it.

Equation (1) suggests how we might find the maximum total weight within clusters. If we change the minus sign in Equation (1) and construct a matrix M that satisfies

$$p^T M p = \sum_{\{i,j\} \in E} w_{ij}(p_i + p_j)^2, \tag{5}$$

then we see that the edges connecting the two subsets V_1 and V_2 do not contribute to the value of this equation, but instead the edges within each cluster contribute

4 times their weight. The dual problem can then be solved, and the optimal p which maximizes $p^T M p$ provides our sought-after clustering information.

To obtain the matrix M, we need the unoriented incidence matrix I_G of G. I_G is an $n \times m$ matrix, wherein each row is associated with a vertex and each column is associated with an edge, and the column of I_G corresponding to an edge from vertex i to j has zeros everywhere except in positions i and j: the two nonzero elements are given by $\sqrt{w_{ij}}$.

Given any vector x, let k be the element of $I_G^T x$ that corresponds to the edge $\{i, j\}$. Then,

$$(I_G^T x)_k = \sqrt{w_{ij}}(x_i + x_j).$$

If we now construct the matrix $M = I_G I_G^T$, then

$$x^T M x = x^T I_G I_G^T x = (x^T I_G)_{1 \times m}(I_G^T x)_{m \times 1}, \quad \text{for all} \quad x.$$

This implies that

$$x^T M x = \sum_{\{i,j\} \in E} w_{ij}(x_i + x_j)^2.$$

Replacing x with the partition vector p yields Equation (5). The matrix M is called the signless Laplacian matrix [2]. It can also be obtained as $M = D + A$, where A is the adjacency matrix of the graph G and D is a diagonal matrix with diagonal entry $D_{ii} = \sum_k w_{ik}$. The Rayleigh quotient $p^T M p / p^T p$ provides a quantitative evaluation of the cohesion of vertices within clusters. We can maximize the Rayleigh quotient $p^T M p / p^T p$ to obtain the node clusters in graphs. Actually, because of the relation $M = 2D - L$, maximizing $p^T M p / p^T p$ is equivalent to minimizing $p^T L p / p^T p$ in the minimum cut problem.

By the Perron-Frobenius theorem[10], since M is real-symmetric, nonnegative and irreducible, all its eigenvalues are real and the eigenvector corresponding to the largest eigenvalue is the only eigenvector of M whose elements are all nonzero and positive. We already know that the eigenvalues of the signless Laplacian matrix M can be depicted as the maximum, minimum and intermediate values of the Rayleigh quotient. Since the eigenvector corresponding to the largest eigenvalue of M is positive, this means that all the vertices are in one cluster; it does not satisfy our objective of getting two clusters of vertices. Therefore the second largest eigenvalue of M is the maximum value that we want and it approximately estimates the value of the cohesion of vertices within the clusters. The associated eigenvector generates the clustering information among these vertices of the graph G. The clustering result indicated here should coincide with the minimum cut result.

This leads us to ask why we still need to model the dual problem if the results from the dual and the original are the same. One significant reason is that the eigenvalues of the signless Laplacian have important physical interpretations. They provide a quantitative evaluation of total weight of edges within clusters. Since the eigenvalues of the Laplacian matrix provide a measure of the total weight of edges connecting clusters, the ratio of these eigenvalues can be used to analyze and evaluate the effects of partitioning and clustering.

Spectral clustering not only provides a bi-partitioning result, i.e. two clusters of graph nodes, but also multiple clusters. Roughly, there are two classes of methods in spectral clustering to obtain multiple clusters: (a) recursive bi-partitioning, and (b) using multiple eigenvectors. In Section 3, we will show an example of using multiple eigenvectors to generate more than two clusters.

2.3   State Clustering on Markov Chains and Graphs

Spectral based methods may also be applied to a directed graph viewed as a Markov chain for clustering purposes. In a Markov chain, the state transition probability diagram can be viewed as a graph with directed edges. It is well known that when a Markov chain is irreducible and aperiodic with a finite state space, its stochastic matrix has a single eigenvalue on the unit circle, the eigenvalue 1, and the left-hand eigenvector corresponding to this unit eigenvalue is the stationary distribution vector. In [13] it is shown that the right-hand eigenvector of a stochastic matrix corresponding to the subdominant eigenvalue, i.e., the eigenvalue with second largest modulus closest to, but strictly less than, 1, provides the information necessary to cluster states into coherent groups. This approach is based on the concept of distance (the number of transition steps) from each state to the stationary distribution.

Therefore, the spectral approach can be applied not only in the symmetric case, e.g., Laplacian and signless Laplacian matrices, but also in the asymmetric case such as Markov chains. The clusters of graph nodes in spectral clustering are based on the minimum cut sense, while the clusters of Markov chains states are in the sense of distance from each state to the steady state. Actually these two clustering measures can be connected in the framework of a random walk on a graph topology. In other words, spectral clustering can be viewed in the framework of state clustering on Markov chains [8].

It has been shown that the adjacency matrix A of a graph can be converted to a transition probability matrix P to generate a random walk on a graph as:

$$P = D^{-1}A.$$

D is a diagonal matrix whose diagonal elements are the vertex degree. Using the Laplacian matrix $L = D - A$ and the signless Laplacian matrix $M = D + A$, we obtain the following formulas showing their relationship with P:

$$I - P = D^{-1}L \quad \text{and} \quad I + P = D^{-1}M$$

Proposition 1. If an eigenpair (λ, v) is a solution of the eigenvalues/vectors problem $Pv = \lambda v$, then the pair (λ, v) is also a solution of the generalized eigenvalues/vectors problems $(1 - \lambda)Dv = Lv$ and $(1 + \lambda)Dv = Mv$.

Proof: Given a pair (λ, v) which satisfies the eigenvalues/vectors problem $Pv = \lambda v$, since $P = D^{-1}A$ and $L = D - A$, we have [8]

$$Pv = \lambda v \quad \Rightarrow \quad D^{-1}Av = \lambda v \quad \Rightarrow \quad D^{-1}(D - L)v = \lambda v$$

$$\Rightarrow \quad Iv - D^{-1}Lv = \lambda v \quad \Rightarrow \quad (1 - \lambda)Dv = Lv \tag{6}$$

It is the same for the signless Laplacian matrix $M = D + A$:

$$Pv = \lambda v \quad \Rightarrow \quad D^{-1}Av = \lambda v \quad \Rightarrow \quad D^{-1}(M - D)v = \lambda v$$

$$\Rightarrow \quad D^{-1}Mv - Iv = \lambda v \quad \Rightarrow \quad (1 + \lambda)Dv = Mv \tag{7}$$

It is known that the eigenvectors of the generalized eigenvalue problem $(1 - \lambda)Dv = Lv$ provide a heuristic solution for the minimum balanced cut on a graph [12]. Therefore, from the proposition above, we can assert that the right eigenvectors of P also provide the balanced cut solution, just like the eigenvectors of the generalized problem on M.

Before we discuss the details of clustering information given by the right-hand eigenvectors of P, let us first look at an interesting property of P. This property is only true for the P of random walk generated on a graph, and not for all transition probability matrices.

Proposition 2. All the eigenvalues of the probability matrix P derived from a random walk on a graph are real.

Proof: $P = D^{-1}A$, though not symmetric does has a symmetric structure. Since

$$D^{1/2}PD^{-1/2} = D^{-1/2}AD^{-1/2},$$

the matrix P and the symmetric matrix $D^{-1/2}AD^{-1/2}$ are similar. Similar matrices share the same eigenvalues. Therefore all eigenvalues of P are real.

Proposition 2 provides and alternative way to calculate the eigenvectors and eigenvalues of P: it is easier to compute the eigenvectors/eigenvalues of a symmetric matrix than an unsymmetric one.

Let λ be an eigenvalue of P generated from a graph, and let x_R and x_L be the corresponding right and left eigenvectors of P respectively. Then

$$Px_R = \lambda x_R \quad \text{and} \quad P^T x_L = \lambda x_L.$$

First, let us focus on x_R. For the eigenvalue problem of P:

$$Px_R = \lambda x_R \quad \Rightarrow \quad D^{-1}Ax_R = \lambda x_R.$$

If we premultiply $D^{1/2}$ on both sides, we obtain

$$D^{-1/2}AD^{-1/2}(D^{1/2}x_R) = \lambda(D^{1/2}x_R).$$

Therefore, the value λ and the vector $D^{1/2}x_R$ are respectively an eigenvalue and eigenvector of the matrix $D^{-1/2}AD^{-1/2}$. Second, let us now consider x_L. Since A is symmetric:

$$P^T x_L = \lambda x_L \quad \Rightarrow \quad AD^{-1}x_L = \lambda x_L.$$

If we now premultiply $D^{-1/2}$ on both sides we obtain

$$D^{-1/2}AD^{-1/2}(D^{-1/2}x_L) = \lambda(D^{-1/2}x_L).$$

Hence the value λ and the vector $D^{-1/2}x_L$ are respectively an eigenvalue and eigenvector of the matrix $D^{-1/2}AD^{-1/2}$. Now we see that x_R and x_L can be obtained from calculating the eigenvectors of the matrix $D^{-1/2}AD^{-1/2}$. Let v be the eigenvector of $D^{-1/2}AD^{-1/2}$ corresponding to the eigenvalue λ. Then

$$x_R = D^{-1/2}v \quad \text{and} \quad x_L = D^{1/2}v.$$

Let us now return to the clustering information provided by the probability matrix P of a random walk on a graph. Spectral clustering is in the measure of minimum balanced cut, while Markov chains clustering is in the measure of distance of states from the steady state. Now a question rises: do these two kinds of clustering results coincide in the right eigenvectors of P? The answer is not simply "yes" or "no", but instead depends on the eigenvalues of P.

Clustering on Markov chains uses the right eigenvector of P corresponding to the eigenvalue with modulus closest to 1. If the eigenvalue closest to the unit circle is positive, then it will be shifted to the eigenvalue closest to 0 in the generalized eigenvalue problem of L in Equation (6) which is used in spectral clustering, and the corresponding eigenvector gives the minimum balanced cut on the graph. In this case the two clustering results coincide. However if the eigenvalue of P closest to the unit circle is negative, then it will be shifted to the eigenvalue closest to 2 in Equation (6) and so the corresponding eigenvector will not give the clustering information based on the minimum balanced cut of a graph. Now the two clustering results are distinct.

2.4   Two Clustering Techniques for Markov Chains

After generating a random walk on a graph via $P = D^{-1}A$, we can perform a clustering analysis on the Markov chain of this graph. If the Markov chain is ergodic, then spectral decomposition of the transition probability matrix P provides two clustering measures of graph nodes.

The first measure is based on the minimum balanced cut. If we sort the eigenvalues of the probability matrix P in descending order, the eigenvector associated with the second largest eigenvalue (should be positive)— which is shifted to the second smallest one in the generalized eigenvalue problem (6) and the second largest one in (7) — gives clusters of states based on the minimum balanced cut from the spectral clustering point of view. There is much material available concerning the use of the eigenvectors of P having positive eigenvalues to carry out spectral clustering [9].

The second measure is based on the "distance" of each state to the steady state. If we sort the modulus of eigenvalues in descending order, the eigenvector associated with the second largest one (not necessarily positive) gives clusters of states based on this distance measure from a Markov chain point of view. If the second largest one is positive, then the two clustering measures coincide; otherwise they do not. Previous research has paid much more attention to the first measure than to the second, due primarily to the fact that clustering using the first measure is closely associated with the structure of real world networks. However the second measure is also related to the graph structure and we should not simply ignore the information it provides.

Let us focus on the case when the eigenvalue having the second largest modulus is negative. How should we identify the clustering information from the corresponding eigenvector? An element with small amplitude in the eigenvector indicates the corresponding state is close to the steady state and perhaps we can say that it belongs to a group of states which are more closely linked to each other. On the other hand, an element with large amplitude indicates the corresponding state is far from the steady state and possibly belongs to a group in which states are not closely linked to each other. That means the clustering information should be identified through close amplitude values of elements in the eigenvector. The eigenvector associated with a negative eigenvalue whose modulus is very close to 1 indicates the role or position of states (graph nodes) in each cluster and also the cluster structure.

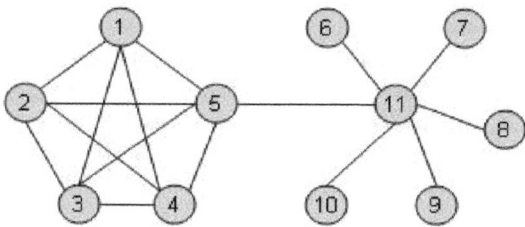

Fig. 1. Joint complete and star graph

A concrete example may help in understanding the clustering techniques for Markov chains. Figure 1 is an unweighted graph with 11 vertices. After forming the transition probability matrix P of the associated random walk on this graph, we find that second largest eigenvalue of P is $\lambda_1 = 0.8852$ while the eigenvalue with second largest modulus is $\lambda_2 = -0.9336$. The corresponding eigenvectors are v_1 and v_2.

$$(\mathbf{v_1}, \mathbf{v_2}) = \begin{pmatrix} -0.1423 & 0.0105 \\ -0.1423 & 0.0105 \\ -0.1423 & 0.0105 \\ -0.1423 & 0.0105 \\ -0.0770 & -0.0707 \\ 0.2581 & -0.3085 \\ 0.2581 & -0.3085 \\ 0.2581 & -0.3085 \\ 0.2581 & -0.3085 \\ 0.2581 & -0.3085 \\ 0.2285 & 0.2880 \end{pmatrix}.$$

Based on the first clustering measure from v_1, we obtain cluster 1 {vertices 1,...,5} and cluster 2 {vertices 6,...,11}. Moving to the second clustering measure, we make two observations:

(a) Eigenvector v_2 shows that the first five elements associated with cluster 1 have relatively small modulus while others associated with cluster 2 have relatively large modulus. This means that cluster 1 has a structure in which states are closely linked to each other, while cluster 2 has the opposite structure. This observation can be confirmed by the figure. The left part of figure is a clique and right part is a star.

(b) If we sort the values of elements in v_2, then the value of vertex 5 is closer to cluster 2 than other vertices in cluster 1 while the value of vertex 11 is closer to cluster 1 than other vertices in cluster 2. This is reasonable, because vertex 5 and 11 have the possibility of transitioning to the other cluster in a single step; they are the connecting vertices between clusters. This can also be confirmed by the observation that vertex 5 in v_2 has a modulus which is relatively large compared with other vertices in cluster 1; vertex 11 has relatively smaller modulus in cluster 2.

Therefore, the eigenvector associated with a negative eigenvalue whose modulus is very close to 1 indicates firstly, the cluster structure and secondly, the role or position of states (graph nodes) in each cluster. When we perform a clustering analysis on P, it is appropriate to combine both clustering techniques, using not only the eigenvectors with positive eigenvalues but also those with negative eigenvalues to obtain more comprehensive information concerning each cluster of states.

There are two applications of Markov chains associated with the two types of clustering techniques. One is associated with nearly completely decomposable (NCD) Markov chains, and the other with nearly periodic Markov chains. In NCD Markov chains, there are several positive eigenvalue very close to 1. In this case, the eigenvectors corresponding to this kind of eigenvalue will provide the first clustering measure information. In nearly periodic Markov chains, there are multiple eigenvalues close to the unit circle but not close to 1. In this case, the corresponding eigenvectors provide the second clustering measure information.

To summarize, spectral decomposition of P not only provides the clusters of nodes satisfying the minimum balanced cut, the same as that obtained from the Laplacian and the signless Laplacian matrix, but also it provides additional clustering information concerning each cluster's structure and the role or position of graph nodes in each cluster.

3   NCD Markov Chains

The right-hand eigenvectors of the transition probability matrix P provide clustering information concerning the states of a Markov chain — especially the eigenvector corresponding to the second largest eigenvalue of P. This idea can be applied to a Markov chain with NCD structure to find nearly uncoupled blocks. It is frequently the case that the state space of a Markov chain can be partitioned into disjoint subsets, with strong interaction between states within subsets but with weak interaction between subsets themselves. Such problems are called nearly completely decomposable (NCD) Markov chains.

For example, given the following transition probability matrix P (Courtois [13]) with NCD structure as indicated,

$$
P = \left(\begin{array}{ccc|ccc|ccc}
0.85 & 0.0 & 0.149 & 0.0009 & 0.0 & 0.00005 & 0.0 & 0.00005 \\
0.1 & 0.65 & 0.249 & 0.0 & 0.0009 & 0.00005 & 0.0 & 0.00005 \\
0.1 & 0.8 & 0.0996 & 0.0003 & 0.0 & 0.0 & 0.0001 & 0.0 \\
\hline
0.0 & 0.0004 & 0.0 & 0.7 & 0.2995 & 0.0 & 0.0001 & 0.0 \\
0.0005 & 0.0 & 0.0004 & 0.399 & 0.6 & 0.0001 & 0.0 & 0.0 \\
\hline
0.0 & 0.00005 & 0.0 & 0.0 & 0.00005 & 0.6 & 0.2499 & 0.15 \\
0.00003 & 0.0 & 0.00003 & 0.00004 & 0.0 & 0.1 & 0.8 & 0.0999 \\
0.0 & 0.00005 & 0.0 & 0.0 & 0.00005 & 0.1999 & 0.25 & 0.55
\end{array}\right),
$$

we compute the eigenvalues to be

Eigenvalues $= \left(\, 1.0\ 0.9998\ 0.9985\ 0.7500\ 0.5501\ 0.4000\ 0.3007\ -0.1495 \right).$

The eigenvalues $\lambda_2 = 0.9998$ and $\lambda_3 = 0.9985$ are very close to the unit eigenvalue 1, and very close to each other. From the discussion in Section 2.4, the second and third right-hand eigenvectors are approximately of equal importance, and hence the third cannot be ignored. Both of them provide clustering information. The corresponding right-hand eigenvectors of λ_2 and λ_3 are

$$
\mathbf{v_2, v_3} = \left(\begin{array}{cc}
-0.3536 & -0.4876 \\
-0.3536 & -0.4878 \\
-0.3536 & -0.4883 \\
-0.3536 & 0.3783 \\
-0.3536 & 0.3777 \\
0.3536 & 0.0073 \\
0.3536 & 0.0073 \\
0.3536 & 0.0073
\end{array}\right),
$$

These two eigenvectors can be used to obtain the NCD blocks for the Markov chain. Based on v_2, the first five states form one cluster, the remaining states form the other cluster. After checking v_3, it is obvious that the first five states can yet again be partitioned into two clusters corresponding to values that are close to one another This leads to three blocks of states for this NCD Markov chain, namely $\{1,2,3\}$, $\{4,5\}$ and $\{6,7,8\}$.

4   Conclusion

In this paper, we introduced the concept of modeling the dual problem of graph partitioning, i.e., graph clustering using a spectral method applied to the signless

Laplacian matrix. This spectral algorithm generates clusters values (eigenvalues) that can be compared with cut values. Furthermore, we proposed two clustering techniques for Markov chains. These two clustering measures obtained from the probability matrix allow us to obtain more comprehensive information of clusters on graph nodes.

References

1. Brand, M., Huang, K.: A unifying theorem for spectral embedding and clustering. In: 9th International Conference on Artificial Intelligence and Statistics (2003)
2. Cvetkovic, D., Rowlinson, P., Simic, S.K.: Signless Laplacians of finite graphs. Linear Algebra and its Applications 423, 155–171 (2007)
3. Fiedler, M.: Algebraic connectivity of graphs. Czecheslovak Mathematical Journal 23, 298–305 (1973)
4. Filippone, M., Camastra, F., Masulli, F., Rovetta, S.: A survey of kernel and spectral methods for clustering. Pattern Recognition 41(1), 176–190 (2008)
5. Golub, G.H., Van Loan, C.F.: Matrix computations, 3rd edn. Johns Hopkins University Press (1996)
6. Higham, D.J., Kibble, M.: A unified view of spectral clustering. University of Strathclyde Mathematics Research Report (2004)
7. Hilgers, P.V., Langville, A.N.: The five greatest applications of Markov Chains (2006)
8. Meila, M., Shi, J.: Learning segmentation by random walks. In: NIPS, pp. 873–879 (2000)
9. Meila, M., Shi, J.: A random walks view of spectral segmentation (2001)
10. Meyer, C.D.: Matrix analysis and applied linear algebra. SIAM (2000)
11. Pothen, A., Simon, H., Liou, K.-P.: Partitioning sparse matrices with eigenvectors of graphs. SIAM Journal on Matrix Analysis and Applications 11(3), 430–452 (1990)
12. Shi, J., Malik, J.: Normalized Cuts and Image Segmentation. IEEE Transactions on Pattern Analysis and Machine Intelligence 22, 888–905 (2000)
13. Stewart, W.J.: Introduction to the Numerical Solution of Markov Chains. Princeton (1994)

On the Analysis of Queues with Heavy Tails: A Non-Extensive Maximum Entropy Formalism and a Generalisation of the Zipf-Mandelbrot Distribution

Demetres D. Kouvatsos and Salam A. Assi

Networks and Performance Engineering Research Group (NetPEn),
Informatics Research Institute (IRI), University of Bradford, Bradford, BD7 1DP, UK
D.Kouvatsos@Bradford.ac.uk, S.A.Assi@Leeds.ac.uk

Abstract. A critique of a non-extensive maximum entropy (NME) formalism is undertaken in conjunction with its application into the analysis of queues with heavy tails that are often observed in performance evaluation studies of heterogeneous networks exhibiting traffic burstiness, self-similarity and/or long range dependence (LRD). The credibility of the NME formalism, as a method of inductive inference, for the study of non-extensive systems with long-range interactions is explored in terms of four consistency axioms of extensive systems with short-range interactions. Focusing on a a general physical system and, as a special case, a single server queue with finite capacity, it is shown that the NME state probability is characterised by a generalisation of the Zipf-Mandelbrot (Z-M) type distribution depicting heavy tails and asymptotic power law behaviour. Typical numerical experiments are employed to illustrate the adverse combined impact of traffic burstiness and self-similarity on the behaviour of the queue. A reference to open issues relating to the NME formalism and open queueing networks is included.

Keywords: Performance evaluation, extensive maximum entropy (EME) formalism, non-extensive maximum entropy (NME) formalism, generalised exponential (GE), traffic burstiness, self-similarity, short-range dependence (SRD), long-range dependence (LRD), queueing systems, Zipf-Mandelbrot (Z-M) distribution.

1 Introduction

Empirical traffic characterisation studies in networks of diverse technology and the Internet have shown that traffic flows often exhibit burstiness, self-similarity and/or long-range dependence (LRD) causing performance degradation and the formation of queues with heavy length tails and asymptotic power law behaviour (e.g., [20]).

Traffic distributions with such properties are often employed to generate workloads in simulation modelling for performance evaluation studies of high speed networks. These, however, tend to be rather inflexible, computationally expensive and may display unusual characteristics (e.g., [6,19]). Some analytic mechanisms for estimating the tail index of Internet traffic with heavy tails were reported by Rezaul and Grout [19], based on the Pareto distribution. Further analytic studies can be found in [1,10,14,15]. These

K.A. Hummel et al. (Eds.): PERFORM 2010 (Haring Festschrift), LNCS 6821, pp. 99–111, 2011.

works are based on the optimisation of generalisations of the classical Boltzmann-Gibbs-Shannon entropy function (c.f.,[4,21]), such as those proposed in the diverse fields of Quantification Theory [8], Statistical Physics [23] and Information Theory [18].

This paper explores a non-extensive maximum entropy (NME) formalism for the study of queueing systems with long range interactions and heavy tails. The methodology is based on the optimization of Havrda-Charvat-Tsallis generalised entropy function (c.f., [8,23]), subject to suitable mean value constraints. The credibility of the NME formalism, as a method of inductive inference for non-extensive systems with long range interactions, is assessed in terms of four potential consistency axioms for extensive systems proposed in [22]. Moreover, the NME formalism is applied into the analysis of a general system Q and, as a special case, a single server queue with finite capacity, where formal connections are made between the NME state probabilities and a generalisation of the Zipf-Mandelbrot (Z-M) type distribution [16] depicting heavy queue tails and asymptotic power law behaviour.

The concepts of the Boltzmann-Gibbs-Shannon entropy and Havrda-Charvat-Tsallis entropy are reviewed in Section 2. The NME formalism and a generalisation of the Z-M distribution are highlighted in Section 3. The suitability of the NME formalism, as a method of inductive inference, is addressed in Section 4. The NME state probability of a single server queue with finite capacity and heavy tails is reviewed in Section 5. Typical numerical experiments are included in Section 6. Concluding remarks and a reference to open issues relating to the NME analysis of open queueing network models follow in Section 7.

2   On the Interpretation of the Classical and Generalised Entropies

2.1   The Classical Entropy Function

For a general system Q with an integer number of possible (microscopic) configurations or states $N(N > 0)$ and 'short-range interactions', the classical entropy function, $H^*(p_N)$ [4,21] is defined by

$$H^*(p_N) = -c \sum_{n=1}^{N} p_N(n) \log p_N(n) \tag{1}$$

where $c(c > 0)$ is a positive constant and $\{p_N(n), n = 1, 2, ..., N\}$ are the associated event or state probabilities. The entropy function $H^*(p_N)$ can be interpreted as a measure of uncertainty or information content that is implied by p_N about a physical system Q with short-range interactions such as "the formation of chemical bonds" and "holding matter together" [7] in Statistical Physics. In this context, quantities such as 'energy and entropy' are considered as 'extensive' variables in the sense that the overall energy of the system Q with short-range interactions is "proportional to the system size" [7].

By analogy, traffic flows in queues with short-range dependence (SRD), such as those represented by Poisson (regular), compound Poisson (bursty) [12] and batch renewal (BR) (correlated) [13] processes, influence the creation of single queues and networks, where the state and entropy variables are extensive. In this context, the extensive maximum entropy (EME) formalism (c.f., [9]), as a method of inductive inference (c.f., [22]) for determining the form of the EME state probability distribution, is based on the maximisation of the extensive entropy function, $H^*(p_N)$, subject to normalisation and suitable mean value constraints. The implementation of the EME formalism is achieved by applying the method of Lagrange's undetermined multipliers leading to the characterisation, for example, of modified geometric (Geo) and generalised Geo (GGeo) (e.g., [12,13]) types of state probabilities for single server queues.

2.2 A Generalised Entropy Function

For a general system Q with an integer number of possible (microscopic) configurations or states $N(N > 0)$ and 'long-range interactions', such as gravity in Statistical Physics, "energy and entropy are no longer extensive quantities" [7]. This increases the complexity of the physical system for which the state probability distribution associated say, with energy, can no longer be determined by maximizing the Boltzmann-Gibbs-Shannon entropy function, $H^*(p_N)$.

To address this problem, Tsallis [23] proposed a generalisation of the extensive entropy function, $H^*(p_N)$ to a 'non-extensive' entropy measure, $H^*(p_{q,N})$, namely

$$H^*(p_{q,N}) = c(1 - \sum_{n=1}^{N} p_{q,N}(n)^q)/(q-1) \qquad (2)$$

where $c(c > 0)$ is a positive constant, q is a real number known as the 'non-extensivity' parameter, which measures the degree of long-range interactions and $\{p_{q,N}(n), n = 1, 2, ..., N\}$ are the associated state probabilities of system Q. As $q \to 1$, $H^*(p_{q,N})$ reduces to the Boltzmann-Gibbs-Shannon entropy, $H^*(p_N)$.

The non-extensive $H^*(p_{q,N})$ entropy function proposed in [23] is identical to the one devised earlier by Havrda-Charvat [8]in the field of Quantification Theory for classification processes. Maximizing $H^*(p_{q,N})$ in the context of the canonical(statistical) ensemble in Statistical Physics, subject to the normalization and mean (generalised internal) energy constraint, it was shown (c.f., [23]) that the NME state probability is of the Z-M type distribution [16], having heavy tails and asymptotic power law behaviour.

By analogy, self-similar input traffic processes, such as fractional Brownian (fBm), influence the formation of queues with LRD traffic flows and 'long-range interactions'. In this case, the state and entropy variables are non-extensive leading, respectively, to NME state probabilities of the Z-M and generalised Z-M (G-Z-M) types, subject to appropriate constraints (c.f., [1,10,14,15,23]).

NME solutions for infinite and finite capacity queues were first established in [1,14] by maximising the extensive Havrda-Charvat entropy [8] and other generalised entropy measures (e.g., [18]), subject to normalisation and generalised exponential (GE) type queueing theoretic mean value constraints. The original NME solution of the Z-M type derived by Tsallis [23] in Statistical Physics was adopted by Karmeshu and Sharma [10] for the NME study of a single server queue with infinite capacity, subject to the normalisation and mean queue length (MQL) constraint. The latter was based on a formula devised by Norros [17] for the estimation of the buffer capacity of a simple storage system with a fBm input process. This NME solution, however, is invalid for single server queues at equilibrium as it violates the Little's law at the service centre of the queue. More recently, NME solutions of a G-Z-M type were determined in [15] for single server queues with infinite and finite capacities, based on the reinterpretation of a proposed heuristic generalisation of the Norros formula [17].

3   A Non-Extensive Maximum Entropy Formalism and a Generalisation of the Z-M Distribution

Consider a general non-extensive system Q with long-range interactions that has a set S of possible discrete states $\{S_0, S_1, S_2, ...\}$, which may be finite or countable infinite and state $\{S_n, n = 0, 1, ...\}$ may be specified arbitrarily. Suppose the available information about Q places a number of constraints on the state probabilities $\{p_q(S_n), S_n \in S\}$, belonging in closed convex set Ω, where q is the non-extensivity parameter. Without loss of generality, it is assumed that the constraints take the form of mean values of several suitable functions $\{f_{1,q}(S_n), f_{2,q}(S_n), ..., f_{m,q}(S_n)\}$, where m is less than the number of feasible states.

A NME framework can be established to determine the form of $p_q(S_n)$ that maximises the Havrda-Charvat-Tsallis entropy, $H^*(p_q)$, namely

$$H^*(p_q) = c(1 - \sum_{S_n \in S} p_q(S_n)^q)/(q - 1) \tag{3}$$

subject to normalisation and the mean value constraints, namely

$$\sum_{S_n \in S} p_q(S_n) = 1 \tag{4}$$

$$\sum_{S_n \in S} f_{k,q}(S_n) p_q(S_n) = F_{k,q} \tag{5}$$

where $c(c > 0)$ is a positive constant and $\{F_{k,q}, k = 1, 2, ..., m\}$ are the prescribed mean values defined on the set of functions $\{f_{k,q}(S_n), k = 1, 2, ..., m\}$. Note that $H^*(p_q)$ can be described as a low-order truncation of Renyi's information theoretic entropy [18].

By employing the Lagrange's method of undetermined multipliers, the maximisation of $H(p_q)$, subject to mean value constraints (4)-(5), determines a least biased NME solution for the state probabilities, namely

$$p_q(S_n) = \frac{1}{Z_q} \left[1 + (1 - q) \sum_{k=1}^{m} \beta_k f_{k,q}(S_n) \right]^{\frac{1}{q-1}} \tag{6}$$

where $\{\beta_k\}, k = 1, 2, ..., m\}$ are the Lagrangian multipliers corresponding to the constraints (5), $Z_q = \exp\{\beta_0\}$ is the normalizing constant and β_0 is the Lagrangian multiplier determined by the normalisation constraint.

The NME solution (6) can be interpreted as a G-Z-M type distribution depicting heavy queue tails and asymptotic power law behaviour.

4   The NME Formalism as a Method of Inductive Inference for Non-Extensive Systems

The principles of EME [9] and minimum relative entropy (MRE), a generalisation, were shown in [22] to be uniquely correct methods of inductive inference for extensive systems, subject to a prior probability estimate and new information given in the form of suitable mean values.

The approach adopted in [22] was based on the fundamental assumption that the use of the EME and MRE principles, as methods of inductive inference for extensive systems, should lead to consistent results when there are different ways to solve a problem by taking into account the same information. This fundamental requirement was formalised in terms of four consistency axioms, namely uniqueness, invariance, system independence and subset independence. It was shown that optimizing any function other than entropy or relative entropy will lead to inconsistencies unless the function in question and the entropy or relative entropy share, respectively, identical maxima or minima. In other words, given new constraint information, there is a unique distribution satisfying these constraints that can be chosen by a procedure based on EME and MRE formalisms complying with the consistency axioms.

The relevance of the four consistency axioms for extensive systems (c.f., [22]) on the credibility of the NME formalism, as a method of inductive inference, is investigated below.

4.1   Uniqueness

Adopting the notation of Section 3, let $f_{q,N}, h_{q,N} \in \Omega$ be two distinct probability distributions defined on the set of states S of the non-extensive system Q having the same non-extensive entropy functions, namely $H^*(f_{q,N}) = H^*(h_{q,N})$.

Using the NME solution (6), it can be shown that

$$H^*(f_{q,N}) = H^*(h_{q,N}) = \\ \alpha H^*(f_{q,N}) + (1-\alpha)H^*(h_{q,N}) \le H^*(\alpha f_{q,N} + (1-\alpha)h_{q,N}) \tag{7}$$

Since the set Ω is convex, there is a distribution (i.e., weighted average) given by $\alpha f_{q,N} + (1-\alpha)h_{q,N}$, which belongs to Ω and has an extended maximum entropy greater than $H^*(f_{q,N}) = H^*(h_{q,N})$. Therefore, there cannot be two distinct probability distributions $f_{q,N}, h_{q,N} \in \Omega$ having the same maximum entropy in Ω. Thus, the Havrda-Charvat-Tsallis NME formalism satisfies the axiom of uniqueness (c.f., [22]).

4.2   Invariance

Following the analytic methodology in [22], let Γ be a coordinate transformation from state $S_n \in S, n = 1, 2, ..., N$ to state $R_n \in R, n = 1, 2, ..., N)$, where R be a transformed set of N possible discrete states $\{R_n, n = 1, 2, ..., N\}$ with $(\Gamma p_{q,N})(R_n) = J^{-1}p_{q,N}(S_n)$, where J is the Jacobian $J = \partial(R_n)/\partial(S_n)$. Moreover, let $\Gamma\Omega$ be the closed convex set of all probability distributions $\Gamma p_{q,N}$ defined on R such that $\Gamma p_{q,N}(R_n) > 0$ for all $R_n \in R, n = 1, 2, ..., N$ and $\sum_{n=1}^{N} \Gamma p_{q,N}(R_n) = 1$.

It can be clearly seen that, transforming of variables from $S_n \in S$ into $R_n \in R$, the Havrda-Charvat-Tsallis extended entropy function (c.f., (6)) is transformation invariant, namely

$$H^*(p_{q,N}) = H^*(\Gamma p_{q,N}) \tag{8}$$

Thus, since the minimum in $\Gamma \Omega$ corresponds to the minimum in Ω, the NME formalism satisfies the axiom of invariance.

4.3 System Independence

Consider two general non-extensive systems Q and R each of which having a finite set of $N, N > 0$ possible discrete states $\{x_n, n = 1, 2, ..., N\}$ and $\{y_n, n = 1, 2, ..., N\}$, respectively. Moreover, let X and Y be the random variables describing the state of the systems Q and R, respectively, with corresponding state probabilities $f_{q,N}(x_n) = Pr\{X = x_n\}$ and $g_{q,N}(y_n) = Pr\{Y = y_n\}$, respectively.

Assuming that Q and R are independent systems, then using the joint probability, $h_{q,N}(x_k, y_n) = Pr(x_k, y_n), k, n = 1, 2, ..., N$ and the definition of (6), it clearly follows that

$$H^*[h_{q,N}] \neq H^*(f_{q,N}) + H^*(g_{q,N}) \tag{9}$$

The inequality (9) implies that, in information theoretic terms, the joint NME state probability distribution of two independent non-extensive systems Q and R defies, due to the presence of long-range interactions, the axiom of system independence (c.f., [22]). Thus, this attribute of the NME formalism, as a method of inductive inference, is clearly most suitable for the quantitative studies of non-extensive dynamic systems with heavy queue tails and asymptotic power law behaviour.

Note that in the case of $q \to 1$ limit, equation (9) becomes

$$H^*[h_{q,N}] = H^*(f_{q,N}) + H^*(g(Y_{q,N})) \tag{10}$$

The equality (verifies that the joint EME state probability distribution, as expected, satisfies the axiom of system independence (c.f., [22]). This is an appropriate property of the EME formalism as a method of inductive inference for the study of extensive systems with short-range interactions.

4.4 Subset Independence

Consider a general non-extensive system Q that has wlog a finite number, $L(L > 0)$, of disjoint sets of discrete states $\{S_i^*, i = 1, 2, ..., L\}$, whose union is S. Let $\{x_{ij}, i = 1, 2, ..., L; j = 1, 2, ..., L_i\}$ be a conditional state belonging to the set $\{S_i^*, i = 1, 2, ..., L_i\}$, where L_i is the finite number of possible conditional states in S_i^*. Moreover, let ξ_i be the probability that a state of the system Q is in the set $\{S_i^*, i = 1, 2, ..., L_i\}$ and let probability $f_{q,i}(x_{ij}) \in \Omega_i$, where Ω_i, is the closed convex set of all probability distributions on S_i^*. Moreover, let x be an aggregate state of system Q and probability $f_q(x) \in \Omega$, where Ω is the closed convex set of all probability distributions on S.

Using the definition of the extended entropy function (c.f.,(6)), it can be shown that

$$H^*(f_q) = \sum_i \xi_i H_i^*(f_{q,i}) \tag{11}$$

where $H_i^*(f_{q,i})$ is the extended conditional entropy function defined on the set of states $S_i, i = 1, 2, ..., L$. Therefore, maximising the generalised aggregate entropy function, $H^*(f_q)$, subject to an aggregate set of available constraints, it is equivalent to maximising each generalised conditional entropy function, $H_i^*(f_{q,i})$, individually, subject to a conditional set of available constraints. Thus, the Havrda-Charvat-Tsallis NME formalism satisfies the axiom of subset independence [22].

5   A Finite Capacity Queue with Heavy Length Tails

In this section, the NME solution for the state probability distribution of a single server finite capacity queue is devised, as a special case of the NME solution (6), in terms of the normalisation, server utilisation, mean queue length and full buffer state probability constraints. As it became evident from the earlier analysis of queues with short-range interactions, based on the classical ME formalism (c.f., [12]), the selection of these constraints is motivated by the fact that they may, generally, capture the main system characteristics and they can be expressed in terms of known input system parameters, such as the mean arrival rate λ, the squared coefficient of variation (SCV) of the interarrival times, (Ca^2), the mean service rate μ and the SCV of the service times, (Cs^2). By analogy, this behaviour has also been observed in the analysis of queues with long range interactions (c.f., [1,10,14,15]), where the constraints of the associated NME solutions include, in addition, the non-extensivity parameter, q (c.f., [8,23]), or, equivalently, the Hurst self-similarity parameter, H (c.f., [3]) such that $q = 1.5 - H, 1/2 < q, H < 1$ (c.f., [10,15,19]).

Note that a heuristic generalization of Norros formula [17], for the estimation of the buffer capacity of a storage system (under a fBm input process), was conjectured in [15]. This formula was reinterpreted as a MQL constraint, L_H, in the NME analysis of a stable single server gS-S/GE/1 queue with infinite capacity, an abstact generalised self-similar (gS-S) arrival process with parameters H, λ and Ca^2 and a GE-type service time distribution completely defined by μ and Cs^2. This MQL heuristic expression is given by

$$L_H = \frac{\rho^{\frac{1}{2(1-H)}}}{2^{\frac{1}{2(1-H)}}} \left(\frac{\left[1 - \rho + Ca^2 + \rho Cs^2\right]^{\frac{1}{2(1-H)}}}{(1-\rho)^{\frac{H}{1-H}}} \right) \tag{12}$$

where H takes values in the interval $\frac{1}{2} < H < 1$ [15]. Note that the heuristic formula (12) takes explicitly into account the adverse combined impact of traffic burstiness (via the SCVs Ca^2 and Cs^2) and self-similarity (via pararameter H) on queueing system performance. The expresion (12) reduces correctly to the Norros formula [17] when $Ca^2 = Cs^2 = 1$. Moreover, for $H = \frac{1}{2}$ (i.e., $q \to 1$) equation (12) yields the mean queue length formula of a stable GE/GE/1 queue [12].

5.1   An NME State Probability as a G-Z-M Type Distribution

Consider a single server gS-S/GE/1/N queue with i) finite capacity, N ii) a gS-S arrival process with mean arrival rate λ, interarrival time SCV, Ca^2 and Hurst self-similarity parameter, H and iii) a GE-type service time distribution with mean service rate, μ and service time SCV, Cs^2. Moreover, at any given time, let $p_{q,N}(n), n = 0, 1, .., N$, be the state probability of having n, say, messages in the queue.

Suppose that the prior information about the state probability can be expressed in terms of the following mean value constraints (c.f., [14,15]): Normalisation, server utilisation $Uq, N = 1 - p_{q,N}(0)$, MQL, $L_{q,N} = \sum_{n=0}^{N} np_{q,N}(n)$ and full buffer state probability, $\phi_{q,N} = p_{q,N}(N)$, where $\phi_{q,N}$ satisfies the flow balance condition, namely

$$\lambda(1 - \pi_{q,N}) = \mu U_{q,N} \tag{13}$$

where $\pi_{q,N}$ is the blocking probability that an arrival message find a full capacity queue.

The form of the NME queue length distribution, $p_{q,N}(n), n = 0, 1, ..., N$ of the gS-S/GE/1/N queue can be characterised by maximising the Havrda-Charvat-Tsallis nonextensive entropy,

$$H^*(p_{q,N}) = c(1 - \sum_{n=0}^{N} p_{q,N}(n)^q/(q-1) \tag{14}$$

subject to the normalisation and the aforementioned constraints: server utilisation, MQL and full buffer state probability.

The NME state probability of the gS-S/GE/1/N can be clearly reduced from the generic G-Z-M solution (6 and is given by

$$p_{q,N}(n) = \frac{1}{Z_{q,N}}$$
$$[1 + \alpha_N(1 - q)n + \beta_N(1 - q)h_{q,N}(n) + \gamma_N(1 - q)s_{q,N}(n)]^{\frac{1}{q-1}} \tag{15}$$

where β_N, α_N and γ_N are the Lagrangian multipliers corresponding to the server utilisation, MQL and full buffer state probability constraints. Moreover, $h_{q,N}(n)$ and $s_{q,N}(n)$ are auxiliary functions clearly defined by

$$h_{q,N}(n) = \begin{bmatrix} 0, n = 0 \\ 1, n \neq 0 \end{bmatrix}. \tag{16}$$

$$s_{q,N}(n) = \begin{cases} 0, n < N \\ 1, n = N \end{cases} \tag{17}$$

and $Z_{q,N}$ is the normalizing constant expressed by

$$Z_{q,N} = \sum_{n=0}^{N} [1 + \alpha_N(1 - q)n + \beta_N(1 - q)h_{q,N}(n) + \gamma_N(1 - q)s_{q,N}(n)]^{\frac{1}{q-1}}$$
$$= \zeta \left[\frac{1}{1-q}, \frac{1 + \beta_N(1-q)h_{q,N}(n) + \gamma_N(1-q)s_{q,N}(n)}{\alpha_N(1-q)} \right] \tag{18}$$

where $\frac{1}{1-q} > 1, q > 0$ and ζ denotes the Hurwitz-Zeta function (c.f., [2]).

Note that a direct derivation of the NME solution (15), can be seen in [15]. As $N \rightarrow \infty$, the NME solution, $p_{q,N}(n)$ (15) becomes identical to that of the corresponding infinite capacity queue with heavy length tails [14,15]. Moreover, at the limit $q \rightarrow 1$, $p_{q,N}(n)$ reduces to the ME solution for the state probability $p_{q,N}(n)$ of the GE/GE/1/N queue [12].

For $q < 1, \rho < 1$ and for large number of messages n, $p_{q,N}(n)$ follows asymptotically, as expected, a power law, which turns out, as expected, to be identical to the one obtained for a stable gS-S/GE/1 (c.f., [15]) and also to the one associated with Tsallis original solution (c.f., [10]), namely

$$p_N(n) \sim n^{\frac{-1}{1-q}}, \quad \frac{1}{2} < q < 1 \qquad (19)$$

The Newton-Raphson numerical method can be applied, as in [10] and [15], to implement the NME solutions of single server queues with infinite and finite capacities, respectively. For illustration purposes, the computational implementation of the NME solution is simplified by assuming that the Lagrange multipliers α_N and β_N are asymptotically invariant to the buffer size, N i.e., $\alpha_N = \alpha$ and $\beta_N = \beta$, where α and β are the Lagrange multipliers of the corresponding infinite capacity queue. Moreover, the blocking probability, π, can be computed by using the flow balance condition (13).

6 Numerical Results

This section presents four typical numerical experiments (c.f., Figs. 1 - 3(b)) illustrating the credibility of the NME solutions and also assess the combined impact of bursty and self-similar traffic flows on the performance of the queue.

A plot of the queue length distribution $p_{q,N}(n)$ of a finite capacity gS-S/GE/1/N queue versus state n for different values of non-extensivity parameter $q = 1.5 - H$ is shown in Fig. 1. It can be seen that decreasing values of q impose gradually, as expected, heavier long tails on the state probabilities, $p_{q,N}(n)$.

The queue length distribution $p_{q,N}(n)$ versus n for $q = 0.6$ and different values of Ca^2 is shown in Fig. 2. It can be observed that, for small states n, higher input traffic burstiness (i.e., variability) has no much influence, as expected, on the tails of the state probabilities. However, as the values of state n increase beyond a small threshold value, traffic burstiness imposes progressively, in the presence of high self-similarity, heavier tail behaviour on the state probabilities.

The relationship between the utilisation, $U_{q,N}$ and ρ for $Ca^2 = 3, 20$ and different values of q is plotted in Figs. 3(a) and 3(b). It can be observed in Fig. 3(a) that for smaller values of ρ under moderate traffic burstiness (variability) at $Ca^2 = 3$, the utilisation $U_{q,N}$ for smaller values of non-extensivity parameter $\frac{1}{2} < q < 1$ is progressively decreasing. This indicates that increasing self-similarity in traffic flows does not have an adverse effect on queue performance when the server is underutilised. Note that, as it was observed earlier in [5], the acute transition from low to high utilisation for lower values of q, as the parameter ρ attains increasing values, is a typical attribute of self-similarity and/or LRD network traffic with moderate variability (c.f., [10,15]).

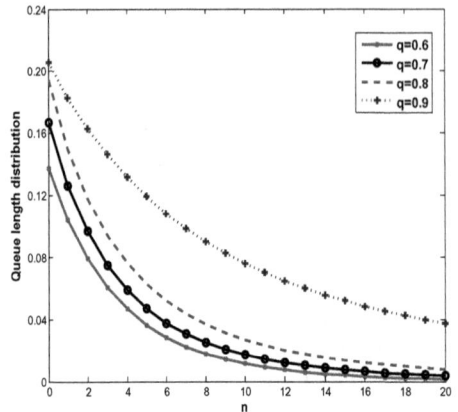

Fig. 1. The relation between $p_{q,N}(n)$ and n for a gS-S/GE/1/N queue with $Ca^2 = 4$, $Cs^2 = 9$, $N = 20$, $\lambda = 0.8$, $\mu = 1.0$ and $\{q = 0.6, 0.7, 0.8, 0.9\}$

Fig. 2. The relation between $p_{q,N}(n)$ and n for a gS-S/GE/1/N queue with $Cs^2 = 9$, $N = 5$, $q = 0.6$, $\lambda = 0.8$, $\mu = 1.0$ and $\{Ca^2 = 1, 3, 8, 16\}$

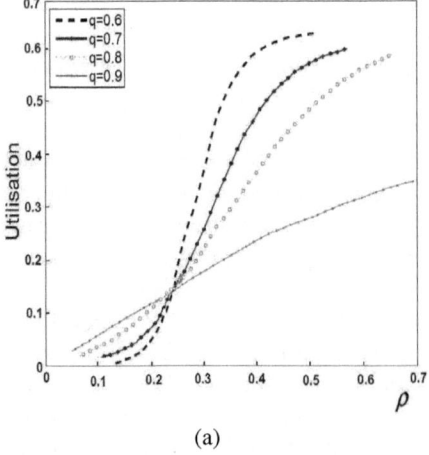

(a)

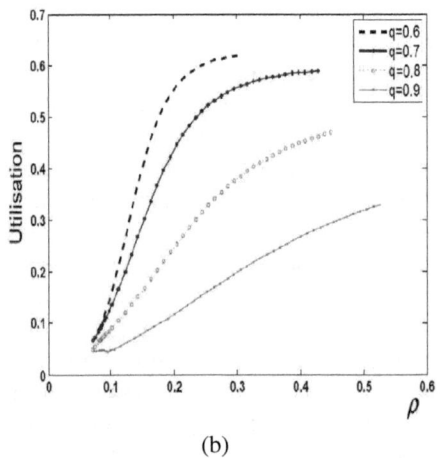

(b)

Fig. 3. The relation between $U = 1 - p_{q,N}(0)$ and ρ for a gS-S/GE/1/N queue with $Cs^2 = 4$, $N = 20$, $\{q = 0.6, 0.7, 0.8, 0.9\}$ and (a) $Ca^2 = 3$ or (b) $Ca^2 = 20$

However, this utilisation 'anomaly' of the plot $U_{q,N}$ versus ρ in Fig. 3(a) is no longer valid as the traffic displays 'burstier' characteristics with Ca^2 being equal to the increased value from 3 to 20. As it can be seen in Fig. 3(b), the relationship between $U_{q,N}$ and ρ reaches a more distinguished pattern as the adverse impact of higher traffic burstiness on utilisation $U_{q,N}$, even for small values of ρ, is quite evident. In particular, the curves of Fig. 3(b) have a much lower intersection point towards $\rho \to 0$ whilst those in Fig. 3(a) are quite independent from each other up to a higher threshold value of ρ. The extremal type of behaviour displayed by the curves of Fig. 3(b) illustrates the adverse impact of traffic flows with combined high burstiness and self-similarity on the performance of the queue.

The numerical experiments of Figures (Figures 1 - 3(b)) demonstrate the credibility and robustness of the power-type NME queue length distributions and assess effectively the adverse combined impact of traffic variability and self-similarity on queue performance.

7   Conclusions

An exploration of the NME formalism for the study of non-extensive systems with long range interactions was undertaken, based on the maximization of the Havrda-Charvat-Tsallis entropy function, subject to new information in the form of suitable mean value constraints. The credibility of NME formalism, as a method of inductive inference, was formally assessed in terms of four consistency axioms for the EME and MRE formalisms for extensive systems (c.f., [22]), namely uniqueness, invariance, system independence and subset independence. It was established that the NME solution satisfies three of these consistency axioms, namely uniqueness, invariance and subset independence, but it did not comply with the axiom of system independence, due to the long-range interactions in non-extensive systems. Thus, the NME formalism was shown to be a most suitable method of inductive inference for the study of dynamic systems with heavy queue tails and asymptotic power law behaviour. Furthermore, it was determined that the NME state probability distribution of a general physical system, Q is characterised by a G-Z-M type distribution depicting bursty and heavy tails with asymptotic power law behaviour. Typical numerical experiments focusing on the special case of a gS-S/GE/1/N queue with finite capacity were included to highlight the credibility and robustness of the NME solution and related performance metrics and also verify the adverse combined impact of traffic burstiness and self-similarity on the performance of the queue.

The NME formalism for single server gS-S/GE/1/N queues with or without finite capacity provides simple but efficient analytic building blocks towards further advances into the theoretical characterisation of underlying bursty and S-S arrival processes and service time distributions associated with exact/approximate NME state probabilities, as appropriate. Although exact NME product form solutions for arbitrary open queueing network models (QNMs) are not possible (c.f., [11]), nevertheless analytic NME product-form approximations and queue-by-queue decomposition algorithms may be established. The latter can be linked, to a tolerable accuracy, with approximate marginal state probabilities and related performance metrics for QNMs of converging multi-service heterogeneous networks taking into account merging, splitting and departing flow steams of bursty and S-S traffic processes.

References

1. Assi, S.A.: An Investigation into Generalised Entropy Optimisation with Queueing Systems Applications. MSc Dissertation, Dept. of Computing, School of Informatics, University of Bradford (2000)
2. Bateman, H.: Higher Transcendental Functions, vol. 1. McGraw-Hill, New York (1953)
3. Beran, J.: Statistics for Long-Memory Processes. Chapman & Hall (1994) ISBN 0-412-04901-5
4. Chakrabarti, C.G., Kajal, D.: Boltzmann-Gibbs Entropy: Axiomatic Characterisation and Application. Internat. J. Math. & Math. Sci. 23(4), 243–251 (2000)
5. Choudhury, G.L., Whitt, W.: Long-tail Buffer-content Distributions in Broadband Networks. Performance Evaluation 30, 177–190 (1997)
6. Crovella, M.E., Lipsky, L.: Long-lasting Transient Conditions in Simulations with Heavy-Tailed Workloads. In: Proc. of Winter Simulation Conference, pp. 1005–1012 (1997)
7. Tsallis Statistics, Statistical Mechanics for Non-extensive Systems and Long-Range Interactions. Notebooks (23:22 January 29, 2007),
   http://www.cscs.umich.edu/~crshalizi/notabene/tsallis,html
8. Havrda, J.H., Charvat, F.: Quantification Methods of Classificatory Processes: Concept of Structural Entropy. Kybernatica 3, 30–35 (1967)
9. Jaynes, E.T.: Information Theory and Statistical Mechanics. Physical Review 106, 620–630 (1957)
10. Karmeshu, Sharma, S.: Long Tail Behaviour of Queue Lengths in Broadband Networks: Tsallis Entropy Framework. Technical Report, School of Computing and System Sciences, J. Nehru University, New Delhi, India (August 2005)
11. Karmeshu, Sharma, S.: q-ExponentiaL Product-Form Solution of Packet Distribution in Queueing Networks: maximisation of Tsallis Entropy. IEEE Communication Letters 10(8), 585–587 (2006)
12. Kouvatsos, D.D.: Entropy Maximization and Queueing Network Models. Annals of Operation Research 48, 63–126 (1994)
13. Kouvatsos, D.D., Awan, I., Fretwell, R., Dimakopoulos, G.: A Cost-Effective Approximation for SRD Traffic in Arbitrary Multi-Buffered Networks. Computer Networks 34, 97–113 (2000)
14. Kouvatsos, D.D., Assi, S.A.: An Investigation into Generalised Entropy Optimisation with Queueing System Applications. In: Merabti, M. (ed.) The Proceedings of the 3rd Annual Postgraduate Symposium on the Convergence of Telecommunications, Networking and Broadcasting (PGNet 2002), pp. 409–414. Liverpool John Moores University Publisher (2002)
15. Kouvatsos, D.D., Assi, S.A.: On the Analysis of Queues with Long Range Dependent Traffic: An Extended Maximum Entropy Approach. In: Proceedings of the 3rd Euro-NGI Conference on Next Generation Internet Networks - Design and Engineering for Heterogeneity, Trodheim, Norway, pp. 226–233 (May 2007)
16. Mandelbrot, B.B.: The Fractal Geometry of Nature. W.H. Freeman, New York (1982)
17. Norros, I.: A Storage Model with Self-similar Input. Queueing Systems and their Applications 16, 387–396 (1994)

18. Renyi, A.: On Measures of Entropy and Information. In: Proceedings of the 4th Berkely Symposium Math Stat And Probability, vol. 1, pp. 547–561 (1961)
19. Rezaul, K.M., Grout, V.: A Comparison of Methods for Estimating the Tail Index of Heavy-tailed Internet Traffic. In: Innovative Algorithms and Techniques in Automation, Industrial Electronics and Telecommunications, pp. 219–222. Springer, Dordrecht (2007)
20. Sahinoglu, Z., Tekinay, S.: On Multimedia Networks: Self-similar Traffic and Network Performance. IEEE Communication Magazine 37, 48–52 (1999)
21. Shannon, C.E.: A Mathematical Theory of Communication. Bell Syst. Tech. J. 27, 379–423, 623–656 (1948)
22. Shore, J.E., Johnson, R.W.: Axiomatic Derivation of the Principle of ME and the Principle of Minimum Cross-Entropy. IEEE Transaction on Information Theory IT-26, 26–37 (1980)
23. Tsallis, C.: Possible Generalisation of Boltzmann-Gibbs Statistics. Journal of Statistical Physics 52(1-2), 479–487 (1988)

Performance Evaluation with Hidden Markov Models*

E. de Souza e Silva[1], R.M.M. Leão[1], and Richard R. Muntz[2]

[1] Federal Univ. of Rio de Janeiro - COPPE/PESC
P.O. Box 68511, RJ 21941-972, Brazil
{edmundo,rosam}@land.ufrj.br
[2] UCLA - CS Departament
4732 Boelter Hall, LA, CA 90024, USA
muntz@cs.ucla.edu

Abstract. The use of hidden Markov models (HMMs) has found widespread use in many different areas. This chapter focuses on HMMs applied to the performance evaluation of computer systems and networks. After presenting a brief review of background material on HMMs, applications such as channel delay and loss characteristics, traffic modeling and workload generation are surveyed. The power of HMMs as predictors of performance metrics is also highlighted. We conclude by presenting a few features of the module of the Tangram-II performance evaluation tool that is targeted to HMMs.

Keywords: hidden Markov models, performance evaluation, network applications.

1 Introduction

Hidden Markov models (HMM) have been used in a myriad of applications that include speech recognition, signal processing, artificial intelligence, computational biology, finance, image processing, and medical diagnosis, to name a few. Reference [2] provides a comprehensive bibliography of over 300 articles on these applications. The first major successful application of HMMs was in speech recognition [13] and was followed by application in a widespread set of application areas. The HMM framework provides a rich theoretical basis that can be adapted to many different applications. Despite the large body of literature on HMM applications, its use in performance evaluation and computer communication is relatively new. The introduction of the application of HMMs in performance modeling motivates the current chapter.

In performance modeling, the analyst usually has a clear understanding of how the system to be analyzed works. If a Markovian model is the paradigm of choice, the main task is to select the appropriate system state variables and the range of possible values for each. For instance, if the system can be modeled by a

---

* This work is supported in part by grants from CNPq, NSF and FAPERJ.

K.A. Hummel et al. (Eds.): PERFORM 2010 (Haring Festschrift), LNCS 6821, pp. 112–128, 2011.

single server queue and the goal is to predict the waiting time for a given arrival rate, the state variable of choice is the number of customers queued for service. Other state variables may be introduced, for instance, to represent a phase-type service time distribution (instead of the exponential distribution) or to represent changes in the mean arrival rate with time.

Another kind of model that falls in this category is availability modeling. In this case, state variables represent components that can be operational or in different modes of failure. Among the questions one may answer is the fraction of time the system is available to the user and the probability that the system fails in a given time interval.

Usually, models like a simple queueing system or dependability model, are parameterized from some prior knowledge of the system behavior. A key point to have in mind is that, in these models, one takes the point of view that the system state is directly observable. On the contrary, with HMM the system state of the underlying process is assumed to *not* be directly observable. Rather one can observe some values that are a probabilistic function of the state of the underlying Markov process. (Thus the origin of the name *Hidden* Markov Model.)

To clarify these ideas, consider the following simple example. Suppose that one is interested in predicting the behavior of a customer who is browsing through the web pages of an online book store. Assume that we would like to determine the probability that a specific customer is ready to order an item based on their past behavior within their session. Suppose that, after studying the behavior of many customers, the analyst decides to build a simple 4-state Markovian model. Assume that the customer's states are chosen as follows: *just browsing* (JB), searching for items in the store, *interested in a product* (IP) and *ready to order an item* (RO). For convenience we also introduce a *leaving* (LV) state to the model to indicate that a customer can leave after being in any of the previous states. (A transition to state LV to corresponds to a session ending and a new session starts in this specific example with a transition to state JB.)

To estimate the transition probabilities, we might for example, follow the trajectory of a set of customers, where the customers indicate along the way their current state of intent. Then, with this training data, one might estimate the state transition probabilities from this example data using the relative frequency of transitions from the chosen states. Figure 1(a) illustrates one possible Markov chain and its transition probabilities.

Since we assume that a state change occurs with each page click, one might ask the following question: what is the probability that a customer is in the ready to order state after visiting a particular sequence of pages? This might be used, for example, to provide the user with some specific promotion information. Clearly, the efficacy of the model depends on its accuracy to answer the questions such as this.

In the above model it is assumed that we have access to data that includes the *intention* of a customer as each page is visited. If available, this can be used to estimate transition probabilities between states and also the conditional probabilities for the page type given the state of user intent. However this state

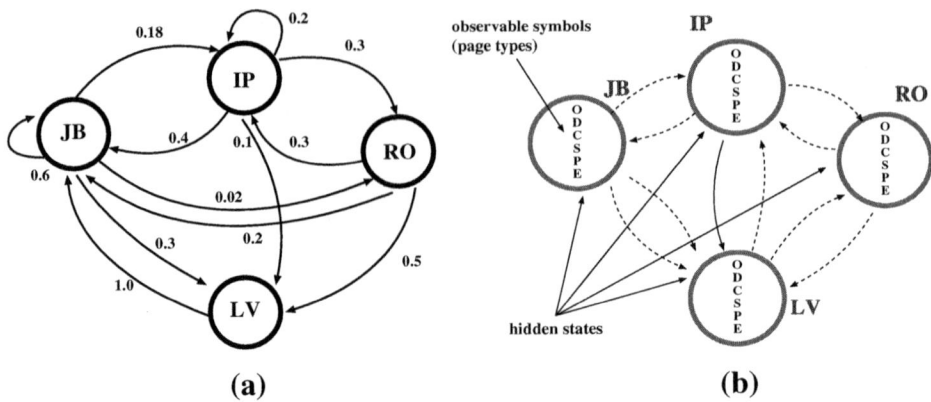

Fig. 1. An example of (a) a simple Markovian model of customer behavior; (b) an application example of HMMs

information may only be available under special circumstances such as when collecting training data. After model construction, when applying the model to online customers, the user's state of intent is not directly observable. An even more challenging circumstance in building a HMM would be when the training data only includes the type of pages a customer visits and not their state of intent. For example, assume that there are 7 different types of pages in the web site: product overview (O), product details (D), set of products within a category (C), shopping cart (S), purchase (P), exit (E). It is easy to record the type of pages the user clicks on, but we cannot know the customer's intent directly. As a consequence, we cannot directly observe the state of the customer's intent (JB, IP, RO). But it is intuitive that the states are correlated with the sequence of observations. In other words, the Markov states JB, IP, RO are *hidden* in the training data but the sequence of page types being visited is known and provides some evidence of of what the user's state of intent is.

Assume that, somehow, we are able to determine the transition probabilities of the hidden chain and, for each hidden state, the conditional probability of emitting an observable *symbol*. (In this example, the probability of clicking on a particular page type (O, D, etc.). Figure 1(b) shows the hidden states and observable symbols of the model. A typical question to be asked is what is the most probable hidden state (that is, the customer intent) given the observed sequence of pages visited up to the current time.

However, as mentioned above, in general we do not know a priori the transition probabilities of the hidden chain. In fact, in most problems, we do not even know how many states the hidden chain contains. (Imagine that you not only cannot perform an experiment, as described earlier, where a set of customers voluntarily provide the information of their intent but you may not even a priori know the number of or interpretation of the "states of intent".) In this case, you know little about the hidden MC and can only record the sequence of page clicks for a number of sessions. But to answer the question of the previous paragraph we must first build a completely specified hidden MC (including all transition probabilities) and also the probability of clicking a page of each type conditioned on the current state of the hidden MC. The only information available may be the sequence of page clicks. In this case, the problem is to find the unknown model parameters that maximize the probability that the observed sequence is generated by the model. (In general this will depend on (a) an a priori probability distribution over the space of possible models and (b) the conditional probability of the observed sequence(s) given a specific model.)

Now suppose that many of the customers of the book store are registered and they provide their age upon registration. Suppose you divide the customers into 2 groups: young and mature folks. Then you build, as above, two different models optimized for each type of customer. (Here one assumes that data to build the models consists of observation sequences where one knows the type of the customer (young or mature) for each sequence.) After building the two models from such example sequences, a third typical type of question is, given a specific user session and the sequence of page clicks observed (but not the customer type), what is the probability that the customer is young versus mature? This is equivalent to asking which of the two models better describes the sequence of page clicks observed.

It should be clear from the above examples that there are different questions that the analyst can ask from a HMM as compared to the traditional Markov models. In this chapter we intend to provide a few examples where HMMs have been used in performance evaluation (including computer communication). After presenting, in section 2, the notation used and a brief review of introductory material to provide the mathematical foundation for the subsequent sections, in section 3 we survey a few models that have been proposed in the literature. In section 4 we outline the main features of the Tangram-II module specifically tailored to deal with HMMs and in section 5 we conclude the chapter.

2   Preliminaries

In this section we first summarize the notation to be used in the rest of this chapter. In the previous section we described a typical type of application of HMMs and the types of question that the model may be used to answer in the context of that application. Here we summarize, independent of a particular application, a more formal definition of a HMM and the most common categories of problems across applications. Later sections will discuss particular applications in the context of performance evaluation.

2.1   Notation the Main HMM Problems

Below we introduce the notation for the parameters of a HMM.

1. The number of states, N, in the underlying (hidden) Markov chain. The i-th state is denoted by s_i. (In the example of section 1 $N = 4$.)

2. The (hidden) state at time t is denoted by q_t. So, the sequence of states up to time T is $q_1 q_2 \ldots q_T$.
3. The number M, of distinct observations. The observed values are denoted by v_k for $1 \leq j \leq M$. ($M = 6$ in the example of section 1.)
4. The observed value at time j is denoted O_j. So, the sequence of observations up to time T is $O_1 O_2, \ldots O_T$. A shorthand notation for the sequence of observations up to time T is $\mathcal{O}_T$.
5. The state transition probabilities: $\mathbf{P} = [p_{i,j}]$, where $p_{i,j} = P[q_{t+1} = s_j | q_t = s_i]$ for all t.
6. The conditional distribution for observed value conditioned on the state of the underlying MC: $b_j(k) = P[O_t = v_k | q_t = s_j]$. $\mathcal{B} = \{b_j(k)\}$ denotes the set of all such conditional probability distributions.
7. $\boldsymbol{\pi}^{(1)}$ will denote the initial state probability vector, i.e., $\boldsymbol{\pi}^{(1)} = [\pi_1^{(1)}, \pi_2^{(1)}, \ldots, \pi_N^{(1)}]$ where $\pi_i^{(1)} = P[q_1 = s_i]$.
8. The model $\mathcal{M}$ is defined by the combination of $\boldsymbol{\pi}$, $\mathbf{P}$ and $\mathcal{B}$, and we use the notation $\mathcal{M} = (\mathbf{P}, \mathcal{B}, \mathbf{p})$ to indicate the complete set of model parameters.

Suppose that know the parameters of a HMM, for instance, the parameters for the HMM of Figure 1(b) that models the page types accessed by an user. Assume that we want to generate a sequence of page types visited by an user. From the model, a sequence of observable symbols can be easily generated as follows.

1. Choose an initial state according to $\pi^{(0)}$.
2. Set $t = 1$.
3. Choose $O_t = v_k$ according to the conditional probability distribution for state q_t, i.e., if $q_t = s_i$, choose v_k with probability $b_i(k)$.
4. Choose the next state according to the state transition probabilities.
5. Set $t = t + 1$ if $t < T$, else terminate.

In section 1 we describe an example used to introduce typical problems the analyst is faced with. For instance, if the model $\mathcal{M}$ is known, from a given (observable) sequence of page accesses $\mathcal{O}_T$, what is the probability that the model is in each of the hidden states at T? In addition, how to estimate the model parameters $(\mathbf{P}, \mathcal{B}, \boldsymbol{\pi})$ if the only information available are sequences of page accesses? Formally, we can identify four basic problems:

Problem 1:
Given the observation sequence $\mathcal{O}_T = O_1 O_2 \ldots O_T$ and a model $\mathcal{M}$, compute the probability of observing the output sequence $\mathcal{O}_T$ given the underlying model $\mathcal{M}$, i.e., $P[\mathcal{O}_T | \mathcal{M}]$.

Problem 2:
Given the observation sequence $\mathcal{O}_T = O_1 O_2 \ldots O_T$ and a model $\mathcal{M}$, how to determine a *best* state sequence $\mathcal{Q}_T = q_1, q_2, \ldots, q_T$ which *best explains* the output sequence $\mathcal{O}_T$. In order words, in this problem we want to unveil the hidden states in the model. To have a well specified problem here we need to define more formally what is meant by *best*, i.e., define the specific objective

function. Two possibilities are: (i) for each time t what is the most probable state of the underlying MC, and (ii) what is the most probable sequence of states. Note that these are different in general. For example, the answer to (i) could have $s_t = s_i$ and $s_{t+1} = s_j$ but if $p_{i,j}$ is 0 the sequence of states $\langle s_i, s_j \rangle$ could never occur.

Returning to the example of section 1, if the transition JB to RO is zero (instead of 0.02 as in Figure 1) the sequence *just browsing* to *ready to order* could never occur, but one can calculate the probability that the customer is *ready to order* after observing a sequence of page visits.

Problem 3:
Given the observation sequence $\mathcal{O}_T = O_1 O_2 \ldots O_T$, construct an underlying model $\mathcal{M}$, such that $P[\mathcal{O}_T|\mathcal{M}]$ is maximized.

Problem 4:
Given several possible models $\mathcal{M}_i$ $1 \leq i \leq K$ and a sequence of observed values, what is the probability that $\mathcal{M}_j$ is the actual model, i.e. $P[\mathcal{M}_j|\mathcal{O}]$. In the example of customers browsing an online bookstore, the two models could be one for a young customer and one for a mature customer.

2.2   A Brief Discussion of Algorithms for Solving These Basic Types of Problems

Problem 1. $P[\mathcal{O}_T|\mathcal{M}]$:
An iterative algorithm exists for computing $P[\mathcal{O}_T|\mathcal{M}]$ which has complexity $N^2 T$ [13].

Define $\alpha_t(\mathcal{O}_t, q_t = s_i|\mathcal{M}) = P[\mathcal{O}_t, q_t = s_i|\mathcal{M}]$ where $\mathcal{O}_t = O_1 O_2 \ldots O_t$, i.e., the first t observed values. Thus $\alpha_t(\mathcal{O}_t, q_t = s_i)$ is the probability of the first t observations **and** ending up in state s_i at time t. For convenience in notation we may sometimes drop the explicit reference to the model $\mathcal{M}$ but it is assumed unless otherwise stated. So we use the shorthand, $\alpha_t(\mathcal{O}_t, q_t = s_i)$ and sometimes even $\alpha_t(\mathcal{O}_t, s_i)$. Then: $\alpha_{t+1}(\mathcal{O}_{t+1}, q_{t+1} = s_j|\mathcal{M}) = \sum_{i=1}^{N} \alpha_t(\mathcal{O}_t, q_t = s_i|\mathcal{M}) p_{i,j} b_j(O_{t+1})$

Starting with $\alpha_1(O_1, s_i) = \pi^{(0)}[s_i] b_i(O_1)$, the above expression is used to calculate the values of α at time $t+1$ from the values for α at time t. It should be clear that:

$$P[\mathcal{O}_T|\mathcal{M}] = \sum_{i=1}^{N} \alpha(\mathcal{O}_T, q_T = s_i|\mathcal{M})$$

The computational complexity for each time increment is clearly N^2 since there are N values to calculate at each increment and calculating each value takes on order N calculations.

Similarly an iterative backward algorithm exists using an auxiliary function $\beta_t(\mathcal{O}_{t+1,T}, q_t = s_i|\mathcal{M}) = P[\mathcal{O}_{t+1,T}|q_t = s_i, \mathcal{M}]$ where $\mathcal{O}_{t+1,T} = O_{t+1} O_{t+2} \ldots O_T$, i.e., the observed values from $t+1$ to T. (Again, see [13] for more details.)

Problem 2.

There is an algorithm for a specific version of Problem 2: computing $P[q_T = s_i | \mathcal{M}, \mathcal{O}]$. (Note that, for convenience, the notation here has dropped the subscript T from $\mathcal{O}_T$.)

From the solution to Problem 1 we know how to compute $P[\mathcal{O}, q_T = s_i | \mathcal{M}]$ and since $P[\mathcal{O}|\mathcal{M}] = \sum_{\text{states}, i} P[\mathcal{O}, q_T = s_i | \mathcal{M}]$ we have an expression for $P[\mathcal{O}|\mathcal{M}]$ as well. Since:

$$P[q_T = s_i | \mathcal{O}, \mathcal{M}] = \frac{P[q_T = s_i, \mathcal{O}|\mathcal{M}]}{P[\mathcal{O}|\mathcal{M}]} = \frac{P[q_T = s_i, \mathcal{O}|\mathcal{M}]}{\sum_{\text{states}, s_i} P[\mathcal{O}, q_T = s_i | \mathcal{M}]}$$

In terms of the $\alpha_T(\mathcal{O}_T, s_i)$ functions that were introduced earlier, we have:

$$P[q_T = s_i | \mathcal{O}, \mathcal{M}] = \frac{\alpha_T(\mathcal{O}, s_i)}{\sum_{\text{states}, s_j} \alpha_T(\mathcal{O}, s_j)}$$

Problem 3.

One important problem is to determine the model parameters to fit the observed data. In other words, how to estimate the model parameters in order to maximize the probability of the observation sequence. Here we assume that the number of states in the underlying MC (N) is known. Roughly, the procedure for adjusting the model is based on the *maximum likelihood estimation* (MLE) method and a technique derived from the *Expectation-Maximization* (EM) algorithm known as Baum-Welch iterative procedure to obtain a local maximum for the likelihood function [18]. For HMMs, the Expectation step evaluates an auxiliary function (based on the likelihood function) $L(\overline{\mathcal{M}}, \mathcal{M}) = \sum_{\mathcal{O}_T} \log P[\mathcal{O}_T, \mathcal{Q}_T | \mathcal{M}] P[\mathcal{Q}_T | \mathcal{O}_T, \overline{\mathcal{M}}]$, where $\overline{\mathcal{M}}$ is the current estimated model parameters, $\mathcal{Q}_T$ is the sequence of hidden states during $(1, T)$ and $P[\mathcal{O}_T, \mathcal{Q}_T | \mathcal{M}]$ is the joint likelihood of observable variables and hidden states. The Maximization step obtains the new model parameter values which maximizes $L(\overline{\mathcal{M}}, \mathcal{M})$.

Maximizing the above function leads to simple formulas for HMMs, as follows. Let $\mathcal{M}^{(k)}$ denote the model parameters after the k-th iteration. $\mathcal{M}^{(0)}$ is the initial estimate.

Define $\phi_t^{(k)}(i, j) = P[q_t = s_i, q_{t+1} = s_j | \mathcal{O}, \mathcal{M}^{(k)}]$, i.e., conditioned on the observation sequence and the k-th estimate of the model parameters. These quantities can be computed from the forward and backward computations as introduced earlier for Problem 1.

The next iteration of model parameters is based on the EM algorithm. For example for the $(k + 1)$-st model iteration, the estimate for the transition probabilities are given by:

$$p^{(k+1)}(i, j) = \frac{\sum_{t=1}^{T-1} \phi_t^{(k)}(i, j)}{\sum_{t=1}^{T-1} \sum_{j=1}^{N} \phi_t^{(k)}(i, j)}$$

Similar expressions are used to find the next iteration of values for initial state probabilities and the conditional probabilities for observation values conditioned on the current MC state. It can be shown that this iteration converges to at least a local maximum likelihood solution for the model parameters [13]. Note that one interesting feature of the procedure is that model parameters can be re-estimated based on new observations.

Problem 4.
Basically this is a classification problem and one simply applies Bayes Theorem.

$$P[\mathcal{M}_j|\mathcal{O}] = \frac{P[\mathcal{M}_j]P[\mathcal{O}|\mathcal{M}_j]}{\sum_{i=1}^{K} P[\mathcal{M}_i]P[\mathcal{O}|\mathcal{M}_i]}$$

The a priori probabilities $P[\mathcal{M}_i]$ also have to be known or estimated. The remaining components in the above expression can be computed using the procedure described for Problem 1.

2.3   Remarks

In the sections above we briefly summarized introductory material on HMMs. In what follows we make a few additional remarks which are useful to understand the remaining sections.

The choice of N. When building a model, the analyst may not be able to associated some significance to each hidden state nor have a clear idea on the number of hidden states that should be employed. In the example of section 1, the description of the problem leads naturally to hidden states associated with the user intent, and the number of states N naturally follows from this. However, in many application models, there may be no meaning that can be attributed to the hidden states and, therefore, N must be guessed. In addition, even when one can think of some meaning to the hidden states, the value of N may not be evident. Referring to the online book store example, one can easily imagine other user intents. For instance, one may choose to break the *interested in a product* state into two, such as *showing some interest* and *very interested* in a product. What would be the "best" model to use?

Deciding on the value of N depends largely on the analyst experience. This issue is somewhat similar to that of determining which state variables to use when one is building a Markov chain model. We should be aware that the choice of N impacts the accuracy of the model to obtain the desired answer as well as the cost to solve the model. There is some formalism that can be used to help the analyst. This problem is treated in [14]. There, the estimator proposed is based on the entropy for a stationary stochastic process. A function of the model entropy is selected and evaluated for different values of N. The smallest value of N is chosen such that, by increasing N, no significant improvement in the function value is observed.

The hidden Markov chain structure. Another issue is the choice of a proper model structure. In order to apply the EM-algorithm, the analyst has to guess initial values for the entries of the state transition matrix $\mathbf{P}$. In general, one may choose $p_{ij} \neq 0$ for all i and j. However, depending on the particular problem, the analyst may have a good grasp on the structure of the underline Markov chain. For instance, in the online bookstore example, one may know a priori that the probability of jumping from the *just browsing* state to the *ready to order* state is so small that it can be neglected. It is easy to see that, if $p_{ij}^{(0)} = 0$, then $p_{ij}^{(k)} = 0$ for all the subsequent steps of the EM-algorithm. Thus, the hidden Markov chain initial structure is preserved throughout the iterations. A good initial structure may lead to a more accurate and efficient parameter estimation result for the problem under investigation.

Similar comments apply to the initial choice of the conditional probabilities of symbol emission, $b_j^{(0)}(k)$'s. From Figure 1(b) the probability of clicking page O (*set of products within a category*) while in state *ready to order* may be very low and perhaps can be ignored. If $b_j^{(0)}(k) = 0$, then $b_j^{(n)}(k) = 0$ for the remaining n steps of the EM-algorithm. These kind of structures may be known from some prior knowledge of the model and can be *forced* during the analysis.

Hierarchical models and Hidden Semi-Markov models. The most common type of HMM is presented in Figure 1(b), where a probability distribution for symbol emission is associate to each of the hidden Markov states. The model operates by outputting a unique symbol per visit of a hidden state.

In some problems, however, it may be helpful to use special structures for symbol emission within a state and to relax the assumption that a unique symbol is emitted per visit to any hidden state. This leads to a hierarchy of HMMs (HHMM) [11]. (In [11], examples drawn from the areas of analysis of natural text and cursive handwriting are given.)

One simple hierarchy is to associate a discrete time Markov chain to each hidden state. In this case, a symbol is output at every visit of the *lower level* Markov chain. Section 3 presents examples of structures proposed for two different applications. Figure 2 shows one such structure. One issue is the choice of the number of symbols that can be output (or, equivalently, the number of lower level Markov chain states that can be visited) per visit of a hidden state. One can choose, for instance, to fix the number of symbols that can be emitted per hidden state visit. Another choice is to exit the hidden state once the second level Markov chain reaches an "absorbing state". In this second case, the number of symbols emitted may be unbounded, depending on the structure of the second level Markov chain.

It can be shown that that the parameter estimation procedure for HMMs can be extended to handle hierarchical structures [11]. For special structures such as that proposed in [15], the application of the EM-algorithm leads to simple equations for estimating the model parameters and, in this case, it is computationally advantageously to use this hierarchical structure in contrast with general HMMs structures.

There is another generalization of HMMs called hidden semi-Markov models (HSMM). A HSMM is similar to a HMM, except that the duration of a hidden state is explicitly modeled. More specifically, at each hidden state, a *segment* of symbols is emitted according to some distribution. An issue is the way to compute the probability of emitting a particular segment. One possibility is to assume "conditional independence" where each symbol in a sequence is drawn from the same symbol distribution associated with a hidden state, independently of the previous symbols at that state. However, an arbitrary joint distribution for the segment symbols can also be assumed. It is not difficult to see that, depending on the distribution used for a segment, a HSMM can be reduced to a HMM. In addition, there are similarities between a HSMM and a HHMM.

As in the case of HHMM, one must extend the estimation algorithms to handle HSMM. See [23] for one such extension and the references therein.

3   Hidden Markov Models Used in Performance Evaluation

3.1   Modeling Network Channel Losses and Delay Characteristics

HMMs have been used to develop traffic generation models useful, for instance to include in simulation studies where loss and delay characteristics metrics impact the application under investigation. On the other hand, HMMs have also been used as predictors of these metrics for online applications. The goal in this case is to adapt the application behavior to better cope with the predicted channel conditions.

Salamatian and Vaton [14] were among the first to propose the use of a discrete time HMM to model the loss process of an Internet channel. In their model, one out of two symbols can be emitted: the first indicates a packet loss in the path from source to destination and the second, the complimentary event. The choice of N is based on the entropy of the loss process, as outlined in section 2. After parameterizing the model from real loss traces, they concluded that the loss process was accurately modeled using only 4 hidden states. This is a significant improvement from previous works that employed a k-th order Markov model [22].

Since losses and delay have jointly an adverse impact on the performance of some applications and they are usually correlated, models have been proposed that try to capture these two metrics in a single HMM. In [17] a HMM is built from the "observations" collected by probing packets. Each of the first $M - 1$ symbols represent a delay range after the collected delay samples are discretized. The M-th symbol indicates a loss. The number of hidden states was obtained from studies that compared the autocorrelation of the real traces w.r.t. those generated by the model. Since the HMM is basically a discrete process, obtained from the samples collected at discrete points in time, the authors also built a continuous time HMM from the discrete version, to emulate delay and losses at arbitrary points of time. The transformation from $\mathbf{P}$ to a generator matrix uses the relation: $\mathbf{P} = e^{\Delta \mathbf{Q}}$, where Δ is an interval between samples.

The dynamics of the packet loss process in the Internet can be extremely variable over different time scales and it is difficult to capture the process characteristics with a regular HMM model. Layered models, such as HHMM have been used in these cases.

In [16] a 2-layer HHMM model with two time scales is proposed. At the top of the hierarchy, a Markov chain models different channel loss conditions at a coarse time scale. Each MC state h corresponds to loss rates in a specified range such as *no loss*, *minor loss rates*, *serious loss rate*. The second layer contains a number of HMMs, each similar in structure to that proposed by [14]. A HMM $\mathcal{M}_h$ in the lower hierarchy corresponds to state h of the top layer MC. In this paper, each level in the hierarchy is parameterized *independently* of the other and, as such, the parameter estimation method differs from the usual HHMM parametrization where the corresponding optimization problem would include the entire 2-layer model. Therefore, the transition probabilities of the top MC model are immediately estimated from the relative frequencies of transitions between two loss rates in different ranges, independently of the observed loss bursts.

HMMs can be used as predictors of events in the future. For instance, video or voice applications could benefit from estimating future channel loss statistics to better adapt their encoding algorithms to varying network conditions. In addition, a good predictor for channel characteristics could be used by applications that have the choice of selecting two or more distinct paths to communicate with the application peer.

The work of [16] uses the proposed HHMM to infer the length of loss bursts that an application would experience in the immediate future in a source-destination path. Let T be the length of the prediction interval. From the current observable state of the top-layer MC h and its transition probabilities, the most probably state in the next time interval T is obtained and the loss process is assumed to be that modeled by $\mathcal{M}_h$ during that interval. Therefore the statistics on burst length are immediately inferred.

Reference [15] provides an example of a 2-level HHMM used as a packet loss model to predict losses for an online VoIP application. In their model the hidden states can be thought of as modeling different channel loss conditions at a course time scale, similar to the top layer MC of [16]. Each observation consists of a batch of S packet outcomes, and thus models the loss process at a finner time scale. Clearly, the emission of S symbols per hidden state can be modeled as a HMM in which a hidden state emits one of the 2^S possible sample paths for S packets and so, even for moderate values of S, the model becomes intractable. This is an example where the use of HHMM can be used to obtain significant savings. We can restrict the distribution of the observations in a batch by assuming that losses are generated following a 2-state Gilbert model and, as such, only two parameters, for each upper level state, need to be estimated instead of 2^S probabilities. It should be clear that the Gilbert model parameter values are correlated with the hidden state and model the short-term loss process. Therefore, this HHMM can capture loss process changes at different time scales.

The predictive model is re-estimated from continuous real time measures. From the measured loss, the current channel conditions are evaluated. For instance, the current hidden state is inferred by solving Problem 2. For a given current state, and applying transient analysis over the HHMM, loss process statistics are inferred in the future. The work in [15] applies this technique to select the proper FEC scheme at the packet level for VoIP applications, and shows that a measure of voice quality can be significant improved with little packet transmission overhead.

3.2   Traffic Modeling

Traffic modeling has been an important area of research since traffic impacts the dimensioning of the Internet resources. Traffic models have been used in many areas such as admission control, resource allocation, capacity planning, and traffic classification, to name a few. HMMs have received attention as the traffic model of choice for these problems because of their ability to capture important traffic statistical characteristics with only a relatively small number of states. These studies focus on modeling the packet flow either generated by an individual application or that of the aggregate traffic on a single channel.

Often both the inter-arrival packet times and packet sizes from the traffic generated by an application exhibit significant auto-correlation. These two metrics then have been used to construct HMMs for different objectives.

In [1] a HMM is used for identifying distinct applications by observing the traffic on their TCP connection. The authors show that the sizes of just the first few TCP packets provide good evidence for classifying the traffic.

Both the packet size and packet inter-arrival time are employed in [6, 19] to build their HMM. The objective of [6] is to build accurate traffic models for the packet flow generated by each of several applications such as SMTP, HTTP, network games and instant messaging. It was shown that the models capture well the marginal distribution and auto-covariance of the packet inter-arrival time and packet size for the studied applications. In a related work [5] these models are employed for traffic classification, similarly to what is described in section 2, Problem 4. (See also [19] for another similar example of network traffic classification.)

One of the objectives of a packet level traffic model is to help with the task of capacity planning, and so it is important to devise accurate models for the aggregate traffic in a channel. Reference [9], is an example of such application. There, a HMM is used for modeling the aggregate traffic at each link of a given network structure (nodes connected to links). An optimization problem is defined in which the link capacities are obtained such that the total network cost is minimized under the constraint that different QoS metrics (for different services using the network) are satisfied. The QoS metrics employed are the loss probability and the round trip time (RTT) in a source-destination path. The observed traffic rates at each link are discretized in 30 levels and the observed symbols of the HMM correspond to these levels. The HMM proposed has a branch-Erlangian structure with 4 hidden states. From the HMM, for each link, a function was obtained which maps the delay in a link from the link capacity. This function is then used for the optimization problem.

3.3   Workload Generation

In this set of applications, HMMs are used to generate synthetic workload at the application level, instead of the packet level of section 3.2. Similarly to other HMM model applications, the workload at the application level may involve different time scales. For example, when a user clicks on a hyperlink to access a web page, the browser sends requests for that page and its in-line objects. This *ON* period is followed by an *OFF* interval where the user reads the page contents and no requests are sent to the server. Two time scales are then apparent: one that represents the duration of the OFF periods and another that represents the intervals between requests generated during the *ON* period. As mentioned in previous sections, HSMMs and HHMMs are useful tools to represent different time scales.

In [21] a HSMM is proposed to model a web user's behavior. The hidden states are associated with web pages visited by a web user and the duration of a hidden state is related to the number of HTTP requests received by the web server when the users clicks the corresponding web page. Transition between hidden states characterizes the user's browsing behavior from one page to another. The sequence of observations is a vector where the first element is a request for a specific object and the second is the interval between the current and the previous request. In [21] the proposed HSMM is used for anomaly detection. The model is trained using a sequence of requests made by normal web users and an abnormal user is detected by comparing the expected entropy calculated from the model with that from the monitored user.

HSMMs are also used to represent mobile users and their requirements (e.g. bandwidth to transmit a video) in [12]. The observations are attributes of a mobile user such as measured geo-location (which is different from the real location due to measurement errors) and user requirements. The hidden states represent the real location, directions of movement and speed ranges which are not directly observable. The transitions between hidden states define the user path from a source to a destination state. The information obtained from the HSMM model can be used, for instance to characterize traffic streams arriving from mobile users at a wireless device and consequently to determine the user resource allocation and admission control.

In what follows we describe a student behavioral model when accessing a video server of a real distance learning application. The model, proposed in [4], tries to capture the sequence of commands students issue when viewing pre-recorded lectures. The lectures are from the CEDERJ Computer Science graduate course [8] and the model is parameterized from thousands of logs collected automatically at the video server. Each lecture consists of a pre-recorded video, slides synchronized with the video stream and a series of topics which are associated to particular time points in the video. The students have full DVD-like control such as play, pause, jump (forward or backward) at any time.

The proposed model is a hierarchical HMM inspired by the work of [10]. Transitions between hidden states represent transitions between slides during a user session, but there is no one-to-one correspondence between hidden states and the number of slides in a lecture. Figure 2 illustrates the hierarchical HMM. The observed symbols are play, jump forward, rewind, pause, end of session, next slide.

Note that the structure of the discrete time Markov chains for each hidden state is constrained and, as such, invalid sequence of actions cannot be generated as, for instance, two consecutive pauses.

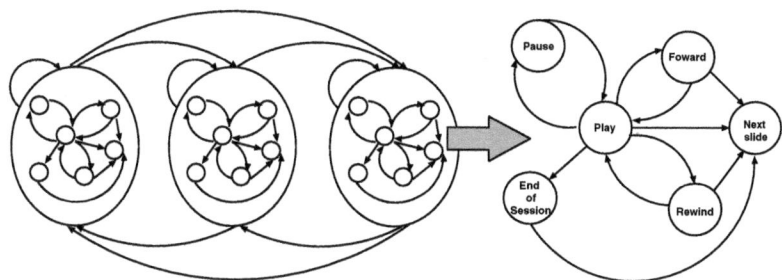

Fig. 2. The hierarchical HMM

The main goal of the proposed HHMM is to generate a synthetic workload used, for instance, to size the video server. The HHMM is capable of generating a sequence of actions but does not include the duration of each action (pause, play, jump forward, etc). The distributions for each action are estimated independently, directly from the trace logs and used jointly with the HHMM to generate the workload.

4   The Tangram-II Tool

Modeling tools are essential to support the development of performance models. Despite the fact that there are a few tools available for handling HMMs, most of them are targeted to specific application areas such as speech recognition and, to our knowledge, none is embedded in a performance evaluation environment. The Tangram-II tool [7,3] has been developed over the last 20 years to provide a useful environment for performance analysts to build and experiment with their models. The tool has a unique modeling paradigm for specifying a model, a rich set of analytical tools, modules to support the calculation of many measures of interest, simulators that include a hybrid fluid and event driven simulator and a traffic generator, all integrated in a single environment.

Concerning HMMs, Tangram-II has a specific module that allows users to work with regular or hierarchical HMMs from the tool's interface. The tool supports three different types of HHMM, each corresponding to specific assumptions for the second layer in the hierarchy. In the first type, a Gilbert Markov model is used for symbol emission, and it is assumed that a fixed number of symbols is emitted at each visit to a hidden state (the state in the top hierarchy). The second type also assumes that the number of symbols emitted per hidden state is fixed, but a general Markov chain can be specified by the user to model the distribution of emissions. The third type of HHMM is similar to the second but the lower layer Markov chain is assumed to contain an absorbing state. When the absorbing state is reached, the top level Markov chain is assumed to make a transition according to the hidden state transition probability matrix **P**. This last type of HHMM allows for the modeling of variable symbol emissions batch sizes for each hidden state.

When creating a HMM one difficulty is to input the initial values for the hidden state transition matrix **P**, especially if the hidden chain contains a non-trivial number of states. Tangram-II facilitates this task by allowing the user to specify the hidden chain with *high level* objects, using the same *Model Specification* module used for regular Markovian models. Any Markov chain model associated with the hidden states can also be specified using the same modeling description paradigm.

Once the model is constructed, the user must specify which state variables are associated with the hidden model and the lower layer model. Figure 3(a) shows a snapshot of one of the tool's interface that allows the user to choose from one of the four types of HMMs. Figure 3(b) shows the interface for an example where a 2-level HHMM is modeled using a Gilbert model for the lower level hierarchy. Finally, Figure 3(c) depicts the parameters that are automatically extracted from the model's specification module. Empty slots are left for the user to include additional initial parameter values, such as the initial probabilities for the hidden states.

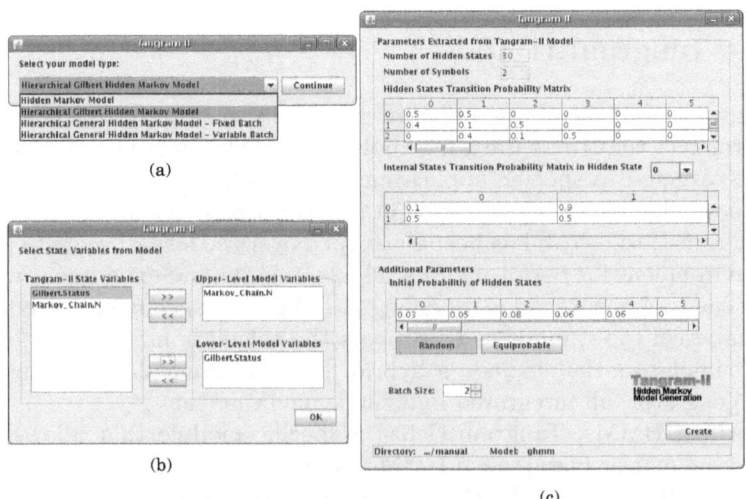

(a)

(b)

(c)

Fig. 3. (a) Model selection interface; (b) State variable selection interface; (c) Additional parameter specification interface

Once the model is specified, the user can select from several solution algorithms. Three of them are associated with the first three HMM problems outlined in section 2. Note that each of the algorithms are adapted to take into account the characteristics of the type of model (regular HMM or HHMM) used.

Before selecting a solution, the user must load the observations collected and stored, for instance, in a trace file. Loading the trace file is done through a special plugin, the MTK HMM plugin. From the data, the user can, for instance, train the model, or calculate the log likelihood of the observation sample. The user can also simulate a previously constructed HMM and generate a sample of symbols. In addition, the tool implements a forecast algorithm, that can predict the probability distribution for the symbols that will follow after a given observed sample. Details can be found in the tool's manual available at [20].

5 Conclusion

Hidden Markov models have proven to be a useful tool for many different areas. In performance evaluation, networking and workload generation have been the predominant application subareas so far. After outlining the mathematical background, we surveyed three applications that served to exemplify the usefulness of HMMs. We also present a modeling tool, Tangram-II that implements the main algorithms for solving the problems surveyed and provides a useful graphical interface to help in the analysis process.

Many modeling issues still remain open, such us the proper choice of the number of hidden states in the model, the choice of the initial parameter values for problem 3 and the initial structure for the underlying Markov chains. Nevertheless, the modeling paradigm is useful when the analyst has little knowledge of how the underlying system works and instead has just access to observations to construct a model that represents the system under study. HMMs have also proven useful for predicting future behavior of measures of interest. We expect many new applications of HMMs to performance problems in the future.

References

1. Bernaille, L., Teixeira, R., Salamatian, K.: Early application identification. In: CoNEXT 2006, pp. 1–12. ACM (2006)
2. Cappé, O.: Ten years of HMMs (March 2001)
3. Carmo, R.M.L.R., de Carvalho, L.R., de Souza e Silva, E., Diniz, M.C., Muntz, R.R.: Performance/Availability Modeling with the TANGRAM-II Modeling Environment. Performance Evaluation 33(1), 45–65 (1998)
4. de Vielmond, C.C.L.B., Leão, R.M.M., de Souza e Silva, E.: A hierarchical HMM for iterative users acessing a multimedia server. In: Brazilian Symposium on Computer Networks, pp. 469–482 (2007) (in Portuguese)
5. Dainotti, A., de Donato, W., Pescape, A., Salvo Rossi, P.: Classification of network traffic via packet-level Hidden Markov Models. In: IEEE GLOBECOM 2008, pp. 1–5 (2008)

6. Dainotti, A., Pescapé, A., Rossi, P.S., Palmieri, F., Ventre, G.: Internet traffic modeling by means of hidden markov models. Computer Networks 52(14), 2645–2662 (2008)
7. de Souza e Silva, E., Figueiredo, D.R., Leão, R.M.M.: The TANGRAM-II integrated modeling environment for computer systems and networks. Performance Evaluation Review 36(4), 64–69 (2009)
8. de Souza e Silva, E., Leão, R.M.M., Santos, A.D., Azevedo, J.A., Machado Netto, B.C.: Multimedia Supporting Tools for the CEDERJ Distance Learning Initiative applied to the Computer Systems Course. In: 22nd ICDE World Conference on Distance Education, pp. 1–11 (2006)
9. de Souza e Silva, E., Leão, R.M.M., Trindade, M.B., da Silva, A.P.C., Ribeiro, B.F., Duarte, F.P., Azevedo, J.A.: A methodology for dimensioning IP networks with QoS using Hidden Markov Models. Technical report, UFRJ-COPPE-PESC (2005)
10. Filho, F.S., Watanabe, E.H., de Souza e Silva, E.: Adaptive forward error correction for interactive streaming over the Internet. In: IEEE Globecom 2006, pp. 1–6 (2006)
11. Fine, S., Singer, Y., Tishby, N.: The hierarchical Hidden Markov Model: Analysis and applications. Machine Learning 32, 41–62 (1998)
12. Kobayashi, H., Yu, S.-Z., Mark, B.L.: An integrated mobility and traffic model for resource allocation in wireless networks. In: 3rd ACM International Workshop on Wireless Mobile Multimedia (WOWMOM 2000), pp. 39–47. ACM (2000)
13. Rabiner, L.R.: A tutorial on hidden Markov models and selected applications in speech recognition. Proceedings of the IEEE 77(2), 257–285 (1989)
14. Salamatian, K., Vaton, S.: Hidden markov modeling for network communication channels. Performance Evaluation Review 29(1), 92–101 (2001)
15. Silveira, F., de Souza e Silva, E.: Predicting packet loss statistics with hidden Markov models. Performance Evaluation Review 35(3), 19–21 (2007), http://www.land.ufrj.br/~edmundo
16. Tao, S., Guerin, R.: On-line estimation of internet path performance: An application perspective. In: IEEE Infocom (2004)
17. Wei, W., Wang, B., Towsley, D.: Continuous-time hidden markov models for network performance evaluation. Performance Evaluation 49(1-4), 129–146 (2002)
18. Welch, L.R.: Hidden markov models and the baum-welch algorithm. IEEE Information Theory Society Newsletter 53(4), 10–14 (2003)
19. Wright, C., Monrose, F., Masson, G.M.: Hmm profiles for network traffic classification. In: The 2004 ACM Workshop on Visualization and Data Mining for Computer Security (VizSEC/DMSEC 2004), pp. 9–15. ACM (2004)
20. The TANGRAM-II manual, http://www.land.ufrj.br
21. Xie, Y., Yu, S.-Z.: A large-scale hidden semi-markov model for anomaly detection on user browsing behaviors. IEEE/ACM Transactions on Networking 17(1), 54–65 (2009)
22. Yajnik, M., Moon, S., Kurose, J., Towsley, D.: Measurement and modelling of the temporal dependence in packet loss. In: IEEE Infocom 2004 (1999)
23. Yu, S.-Z., Kobayashi, H.: An efficient forward-backward algorithm for an explicit-duration hidden markov model. IEEE Signal Processing Letters 10(1), 11–14 (2003)

Network Protocol Performance Bounding Exploiting Properties of Infinite Dimensional Linear Equations*

Ioannis Stavrakakis

Department of Informatics and Telecommunications,
National and Kapodistrian University of Athens,
Ilissia, 157 84 Athens, Greece
ioannis@di.uoa.gr

Abstract. This paper presents a quite versatile and widely applicable performance analysis methodology that has been applied for the study of network resource allocation protocols in the past. It is based on the identification of renewal cycles of the operation of the system and the setting up of recursive equations with respect to quantities-indices defined over the renewal cycles and sessions that appear within. Application of the expectation operator on these equations leads to infinite dimensional systems of linear equations which are shown to posses certain properties leading to rigorous and almost arbitrarily tight bounds on various performance metrics of interest. The special case of a random access protocol is used as an example in order to illustrate the derivation of the recursive equations capturing the protocol dynamics and system inputs. Finally, some other examples of application of the methodology are briefly discussed, illustrating the versatility and powerfulness of the approach. This analysis methodology can be quite useful for understanding the behavior of current complex and large scale networking environments, as well as assessing their scalability, stability and performance.

1   Introduction

The field of computer / communication networking was basically born in the sixties following the inception of packet switching and the development of the first networking infrastructure to enable the transport of information between a few research sites in USA. Since then, the field has boomed and has changed drastically the way society functions in almost all aspects.

As it is the case with any scientific and engineering development, the efficiency and effectiveness of the designed networks has been central to their success. Networking is by definition about the *efficient* interconnection of (geographically) distributed entities, as simply providing (dedicated) links between any two entities would not be feasible or acceptable. The greatest challenge in

---

* This work has been supported in part by the IST-FET project RECOGNITION (FP7-IST-257756).

K.A. Hummel et al. (Eds.): PERFORM 2010 (Haring Festschrift), LNCS 6821, pp. 129–140. 2011

designing an efficient networking system is to devise efficient ways for utilising the available resources. These resources could vary from a single common resource (channel) shared by a number of distributed entities, to the distributed storing/processing/forwarding resources made available by independent mobile nodes.

The efficient (computer) resource utilization problem is one that received the attention of computer system designers before the beginning of data networking. For this reason, approaches and methodologies applied to the design and analysis of computer systems were widely imported into the networking arena. Nevertheless, the distributed nature of the available resources in a networking environment called for the development of new methodologies for designing and analyzing networking structures. Considering the fact that a network is a distributed system whose efficient design requires the incorporation of numerous disciplines and tools, networking was widely viewed as inter-disciplinary and hard to be considered as a separate discipline on its own. It is only very recently that the question as to whether a networking discipline or science can be defined, was posed, [1]. Although still borrowing from an ever increasing number of diverse disciplines (such as anthropology and biology, more recently), the twist of the problems networking is defining and approaches to addressing them are becoming quite distinct from the associated underlying disciplines. In addition, the traditional networking paradigm of connecting sites and transporting messages is rapidly changing into one of content accommodation and delivery in a networking environment collectively contributed through individually-owned disperse networking resources. The key resource is also changing from the transmission resource into a pool of enabling resources such as storage, power, mobility, processing, social, etc, [2].

Returning to the performance evaluation methodologies developed to help design and analyze the early packet networks of the sixties, one can clearly observe a focus on stochastic modelling, queueing theory and operations research, supplemented with mathematical theory and tools. This trend pretty match continued in the seventies and through the nineties when designing multi-user (random) resource access schemes (ALOHA, Ethernet type of networks), integrated networks (ATM) and the Internet. These studies were increasingly supplemented by powerful computer simulations, capturing system details that the analysis often neglects; nevertheless, the insight and guidance through an often immense design space that the analytical modelling provides is still invaluable. The paradigm shift that the Internet has brought towards more autonomic networking clouds, which has been further intensified with the proliferation of autonomic networks (P2P) and the shift of focus towards content generation, accommodation and provisioning, has brought into the performance evaluation arena new methodologies coming from disciplines such as biology, anthropology, social sciences, distributed algorithms, complexity, etc.

The focus of this paper is to present a modelling and performance evaluation approach that yields a tractable analysis methodology to a broad class of networking problems. It is based on a fairly versatile stochastic modelling methodology that leads to the formulation of a typically infinite-dimensional system of linear equations whose proper manipulation yields (in a computationally tractable way) tight lower and upper bounds on performance indices of interest. This approach is presented in the context of the analysis of a specific random access multi-user network that was proposed in the late eighties, since the particular example illustrates the applicability of the methodology to determining - among other performance indices - the stability region of the system, as well.

As already stated, the fundamental problem in networking is that of allocating resources to distributed entities. Nowadays, this competition for resources manifests itself in more complex and generalized manner: bandwidth is only one of a plethora of resources supporting networking; the scale of the resources and competition is global; resources and services are proliferating and are largely autonomic; the networking paradigm, priorities and commodities are changing, etc. For such environment - characterized largely by the generation of unprecedented amounts of content and services - the questions of stability, scalability and performance are central to their efficient design. Although new designs and approaches may be necessary to establish stability and scalability (e.g., by focusing on reducing the irrelevant and useless load (demand for resources) through information processing and effective decision making, [16]), the analysis methodology presented here can be quite useful for understanding the "laws and physics" underlying complex and large scale networking environments today, as well as assessing their scalability, stability and performance.

The paper is organized as follows. Section 2 presents the Limited Sensing Random Access Protocol (LS-RAP) that is used here to illustrate the methodology. In section 3 the stochastic modelling and associated recursive equations capturing the protocol and system dynamics are presented. Section 4 presents the infinite dimensional linear equations and describes how their structure and properties can be exploited by invoking relevant theory to derive bounds on the maximum stable throughput and mean delay. Section 5 presents a discussion on some ways to improve on the bounds, as well as on the applicability of the methodology for studying other protocols and queueing systems in the past. Finally, the contribution of this paper is summarized in section 6.

2   The Limited Sensing Random Access Protocol (LS-RAP)

We consider the classical problem of sharing a common communication channel by a large number of distributed users, as it is the case, for instance, in the classical Ethernet. The environment is assumed to be discrete-time and synchronized in the sense that the users are aware of the beginning of a fixed-size time slot whose length equals the packet transmission time. It is assumed that the unit of information is the message that consists of M packets and that the cumulative message arrival process is Poisson with intensity λ messages per message length. As it is widely understood, a random channel access protocol is appropriate for such a networking environment. As a result, collisions arise when more than one packets are transmitted in overlapping slots. A collision resolution algorithm that belongs to the class of the stack algorithms [5,6], and originally proposed in [3,4] is outlined here and will be denoted by LS-RAP (limited sensing random access protocol).

An active user (i.e., one that has a message to transmit) keeps sensing the channel starting from the slot that follows its message generation instant until this message is successfully transmitted; that is, earlier channel history is not needed (limited sensing). At the end of each slot ternary feedback information (F) is available to all active users revealing the channel status, i.e., whether the channel was idle (I), involved in a successful transmission (S), or in a collision (C). When a packet collision occurs, the sender aborts its transmission before the end of the current slot; thus, only one slot is wasted in a collision. A message transmission is successful once its first packet has been successfully transmitted. The users are geographically separated and their status (i.e., existence or nonexistence of a message to be transmitted) cannot be communicated to the rest of the users.

The concept of the stack is used to illustrate the operation of the algorithm and is not needed for its implementation. The content of a counter (assigned to each user) determines the cell of the stack which the particular user belongs in; it also determines the class of users in which it belongs. Let B_k denote the class of users whose counter content equals k, $k \geq 0$ and let B denote the class of inactive users. B_0 denotes the class of new active users, i.e., those users who have a message to transmit but no message transmission has been attempted so far. B_1 is the class of active users who attempt a packet transmission at the beginning of the slot that follows. Each user's counter content is updated at the end of the slots based on the channel feedback, the steps of the algorithm (see below) and the counter content itself; the new value of the counter determines the class which the user enters. Each user is also assigned a downcounter ω whose initial value is M and which decreases by one unit per slot, starting from the slot in which the first packet of the message was successfully transmitted; this counter determines the time when a successfully transmitting user completes the transmission of all M packets of the message and becomes inactive. LS-RAP can be described by the following updates of the user classes; see [3] for a detailed discussion on the implications of these updates.

1. If $F = S$ then: a) $B_0 \rightarrow B_0$ b) $B_1 \rightarrow B_1$, if $\omega > 0$; $B_1 \rightarrow B$, if $\omega = 0$ c) $B_k \rightarrow B_k$, $k \geq 2$.
2. If $F = C$ then: a) $B_0 \rightarrow B_1$ b) $B_k \rightarrow B_{k+1}$, $k \geq 2$ c) For each $b \in B_1$, $b \rightarrow B_2$ with probability $1 - p$; $b \rightarrow B_1$ with probability p.
3. If $F = I$ then: a) $B_0 \rightarrow B_1$ b) If last nonidle slot was involved in a collision, then for each $b \in B_2$, $b \rightarrow B_1$ with probability $1 - p$; $b \rightarrow B_2$ with probability p; $B_k \rightarrow B_k$, $k \geq 3$ c) If last nonidle slot was involved in a successful transmission, then c_i) if the current slot is the first idle slot after the successful one, then $B_k \rightarrow B_k$, $k \geq 2$ c_{ii}) if the current slot is not as in c_i), then $B_k \rightarrow B_{k-1}$, $k \geq 2$.

3 Renewal Cycles, Sessions and Associated Recursive Equations Capturing Protocol and System Dynamics

Key to the analysis of LS-RAP is the identification of renewal cycles and the computation of their (mean) lengths. This is done by deriving recursive equations with respect to the cycle length and auxiliary sessions (that are stochastically identifiable and contained within a cycle), capturing the protocol dynamics and the system characteristics and inputs. Although these recursive equations are typically infinite dimensional, their structure presents sparse couplings which make their description and (approximate) computation less difficult. Furthermore, one can also define other quantities associated with the renewal cycle and sessions and compute them through a largely similar set of recursive equations. Such equations are presented for the case of the cumulative delay experienced by all packets that were generated and were transmitted during a renewal cycle and specific sessions.

The aforementioned general approach is applied here to the case of LS-RAP. A technical definition of the renewal cycle can be given via the use of an imaginary marker. The marker is originally placed in cell 0 of a conceptual stack. Upon collision, the marker is placed in cell 2. The position of the marker changes in the same way in which the counter content of the users of class B_3 changes, depending on the channel feedback. The slot in which the marker returns to cell 0 is the first one of the renewal cycle that follows and it is always *idle*. When the marker is in cell 0 and a successful transmission occurs the marker is placed in cell 1 and moves up or down as described before. Idle slots do not move the marker from cell 0 and they result in renewal cycles of length one.

From the technical definition of the renewal cycle above it is clear that indeed the system stochastically regenerates itself after the first slot of such a cycle. The number of users k, $k \geq 0$, who attempt a packet transmission in the slot which follows the first idle one of a renewal cycle, determines the multiplicity of the session that follows. The length of a session of multiplicity k is defined as the time required until the conflict of multiplicity k is resolved. All users who attempt packet transmission in that slot plus all those entering the system before the end of this session, transmit successfully during that session. Notice that the sessions of multiplicity k are contained within renewal cycles and their stochastic behaviour is completely determined by the multiplicity k and independent from the position within the renewal cycle they appear.

Let l_k denote the length of a session of multiplicity k (in time slots), $k \geq 0$. The following recursive equations can be written with respect to l_k, $k \geq 0$:

$$l_0 = 1, \quad l_1 = 1 + M + l_{F_M} \tag{1a}$$

$$l_k = 1 + \{1 + l_{k,0}\}I_{\{I_1+F_1=0\}} + \{l_{I_1+F_1} + l_{k-I_1+F_3}\}I_{\{I_1+F_1\neq 0\}}, k \geq 2 \tag{1b}$$

$$l_{k,0} = \{1 + l_{k,0}\}I_{\{I_2+F_2=0\}} + \{l_{I_1+F_1} + l_{k-I_1+F_4}\}I_{\{I_2+F_2\neq 0\}}, k \geq 2 \tag{1c}$$

where $I_{\{.\}}$ is the indicator function and $F_M, F_1, F_2, F_3, F_4, I_1, I_2$ are independent random variables; $F_1, F_2, F_3,$ and F_4 are Poisson distributed over one slot, F_M is Poisson over $M + 1$ and I_1, I_2 are binomial with parameter k and p.

The equations in (1) can be explained as follows. a) The session of multiplicity 0 consists only of the idle slot which marks the beginning of this session. b) The session of multiplicity 1 consists of the following parts. i) The idle slot which is always the first of the session. ii) The M slots involved in the successful transmission of the single packet attempted. iii) The length l_{F_M} which is the same as the length of a session of multiplicity F_M. c) For $k \geq 2$, the session consists of the following. i) The idle slot which is always the first of the session. ii) Since collision occurs we have to distinguish between two cases. Let I_1 be the number of users which remain in class B_1 after the splitting and let F_1 be the number of new messages which arrive in the slot before the collided one. ii_a) If $I_1 + F_1 > 0$ we add $l_{I_1+F_1} + l_{k-I_1+F_4}$ to the length of the session since the original session is split into two with the corresponding multiplicities. ii_b) If $I_1 + F_1 = 0$, we add another slot to the session since no transmission takes place, plus $l_{k,0}$. The latter quantity is equal to the length of a session of multiplicity k without including the slot of the original collision, i.e., $l_{k,0} = l_k - 1$.

Equations (1) are a set of versatile recursive equations with respect to the length of sessions of various multiplicities that appear within a renewal cycle, which are needed to determine the length of a renewal cycle. Equations similar to (1) can be derived with respect to other quantities of interest, whose solution yield performance metrics of interest such as the cumulative delay experienced by all messages generated and transmitted during sessions of various multiplicities and ultimately (through averaging with respect to the session multiplicities) over a renewal cycle. In general, one can define any proper counting metric associated with the sessions (for instance, the number of messages experiencing a delay less than n) whose mean value would ultimately yield a performance metric of importance (message delay distribution). The recursive equations for the cumulative delay experienced by all messages generated and transmitted during sessions of various multiplicities and ultimately over a renewal cycle under LS-RAP may be found in [3,4].

4   Performance Evaluation by Exploiting the Structure and General Properties of Infinite Dimensional Linear Equations

The main contribution of the first part of this paper is the presentation of a methodology for deriving bounds on key performance indices (maximum throughput and delays) by exploiting results from infinite dimensional linear equations. These equations are derived by applying the expectation operator on the recursive equations on key random variables, presented earlier in section 3.

By applying the expectation operator on (1), we obtain the following infinite dimensional linear equations with respect to the expected length of a session of multiplicity k, L_k, where the coefficients h_k and α_{kj} depend on λ, p and M and can be found in [3,4].

$$L_k = h_k + \sum_{j=0}^{\infty} \alpha_{kj} L_j \ , k \geq 0 \ . \tag{2}$$

Similar equations can be obtained with respect to C_k (mean cumulative delay of messages over a session of multiplicity k), with *identical* coefficients α_{kj}.

4.1   Bounds on the Maximum Stable Throughput

Bounds on the mean session length and the mean cumulative message delay over a session of multiplicity k, $k \geq 0$, are needed in order to derive bounds on the maximum achievable throughput and mean message delay. For this reason we proceed with the derivation of such bounds by presenting the foundations and mechanisms for deriving them from the infinite dimensional linear equations as shown in (2). The proofs may be found in [3]. *The maximum stable throughput S_{max} is defined as the maximum over all input traffic rates λ that induce $L_k < \infty$* for $k < \infty$. The following proposition provides for a lower bound on S_{max} (see [3] and [5] for its proof).

**Proposition 1. *Lower bound on S_{max}:* ** *If $\{x_k^u\}_{k=0}^{\infty}$ is an infinite sequence of real numbers which satisfy the following conditions: (1) $0 \leq x_k^u < \infty$, $0 \leq k < \infty$, (2) $h_k + \sum_{j=0}^{\infty} \alpha_{kj} x_j^u \leq x_k^u$, $0 \leq k < \infty$, and (3) $h_k \geq 0$, $\alpha_{kj} \geq 0$, for $k \geq 0$, $j \geq 0$, then:*
(a) the infinite dimensionality linear system of equations $h_k + \sum_{j=0}^{\infty} \alpha_{kj} x_j = x_k$, $0 \leq k < \infty$ has a unique nonnegative solution $\{x_k\}_{k=0}^{\infty}$ that satisfies $0 \leq x_k \leq x_k^u$, $0 \leq k < \infty$ and
(b) If $x_k \equiv L_k$ satisfies the above conditions for some input traffic λ^l, then λ^l is a lower bound on S_{max}.

For $\lambda < \lambda^l$ and for some value of the splitting probability p, $0 \leq p \leq 1$, a quantity $L_k^u = \beta(\lambda, p)k - \gamma(\lambda, p)$ is derived analytically in [3] which satisfies the conditions of Proposition 1 (i.e., $L_k \leq L_k^u$ for $\beta(\lambda, p)$, $\gamma(\lambda, p)$ and k finite).

**Proposition 2. *Upper bound on S_{max}:* ** *An upper bound on S_{max}, λ_N^u, can be obtained as the maximum over all Poisson rates for which the truncated up to N system in (2) has a unique nonnegative solution that satisfies the condition $\lim_{M \to \infty} \max_{N > M} \{\sum_{j=N}^{\infty} \alpha_{kj} L_j\} = 0$. Then, λ_N^u decreases monotonically as N increases and $\lim_{N \to \infty} \lambda_N^u = S_{max}$, [6].*

4.2   Bounds on the Message Delay

As discussed in [3], the total message delay consists of two components: the mean access delay and the mean in system delay. The former reflects the mean time spent by a message from the time it is generated until it makes the first transmission attempt (entering class B_1) and can be calculated easily based on conditional probabilities on the channel state and the input traffic rates (see [3]). The in system delay reflects the remaining time spent in the system till successfully transmitted and is more challenging to compute. We derive bounds on the mean in system message delay, D_s, for input traffic rates $\lambda < \lambda^l$ for which it is guaranteed that the delays will be bounded.

Since the operation of the system follows a renewal process and the multiplicities of the sessions are independent and identically distributed random variables, the following theorem is a direct application of the strong law of large numbers (p. 126, [9]) and is easily proved, [3].

Lemma 1. For $\lambda < \lambda^l$, the mean in system delay D_s is given by $D_s = \frac{C}{\lambda L}$ with probability 1 (wp 1), where $L = E\{L_k\}$, $C = E\{C_k\}$ where $E\{\}$ is over all multiplicities k which are Poisson distributed over one slot.

Proposition 3. _Bounds on the mean in system delay_: Upper and lower bounds on D_s can be obtained by deriving upper and lower bounds on L and C and invoking Lemma 1.

4.2.1   Upper Bounds on L_k and C_k:

A linear with respect to k upper bound on L_k, L_k^u, was presented in section 4.1. The values of $L^u \equiv E\{L_k^u\}$ were found to be very close to those of L^l for $\lambda < 0.8\lambda^l$. As λ approaches λ^l, the upper bound increases rapidly and becomes loose; some approaches to calculating a tighter upper bound are discussed in section 5.

To derive an upper bound on C_k, C_k^u, we follow a procedure similar to the one employed in the derivation of L_k^u. By deriving similar recursive equations with respect to c_k, the cumulative in system delay of a session of multiplicity k, and applying the expectation operator, a similar to (2) set of infinite dimensional linear equations with identical coefficients α_{kj} can be derived. An upper bound on C_k $k \geq 0$, of the form

$$C_0^u = 0, \; C_k^u = v_1 k^2 + v_2 k + v_3, \; k \geq 1$$

was obtained for all input traffic rates $\lambda < \lambda^l$ where v_1, v_2, v_3 are some finite constants depending on λ and p and are analytically derived in [3].

4.2.2   Lower Bounds on L_k and C_k:

The theorem that follows the definition below proves that lower bounds on L_k can be obtained by solving truncated (finite) versions of the infinite dimensional equations in (2).

Definition: Majorant/Minorant systems If $x_k = A_k + \sum_{j=0}^{\infty} B_{kj} x_j, 0 \leq k \leq \infty$ (MAJ) and $y_k = \alpha_k + \sum_{j=0}^{\infty} b_{kj} y_j, 0 \leq k \leq \infty$ (MIN) are infinite dimensionality linear systems of equations with $A_k \geq |\alpha_k|$ and $B_{kj} \geq |b_{kj}|, 0 \leq k \leq \infty$, $0 \leq j \leq \infty$, then we say that the system in (MAJ) is a majorant for the system in (MIN); similarly, the system in (MIN) is a minorant for the system in (MAJ).

Theorem 1. If a majorant for a given system of linear equations has nonnegative solutions x_k, $k \geq 0$, then the given system has the solution y_k which satisfy $|y_k| \leq x_k, 0 \leq k \leq \infty$.

The proof of the theorem can be found in [10]. Note that since the infinite dimensionality linear system of equations in (2) has a nonnegative solution for every $\lambda < \lambda^l$ and it is a majorant for its truncated version, Theorem 1 implies that, for every $\lambda < \lambda^l$, the solutions L_k^l of the truncated system (2) are lower bounds on L_k. Lower bounds on C_k are obtained as for L_l by solving truncated version of the corresponding linear equations.

5   Discussion on the Performance Bounding Approach and Related Work

The methodology presented in sections 3 and 4 is fairly powerful and has been applied for the study of processes, protocols and queueing systems appearing in networking. As long as the operation of the system presents renewal cycles (which is typically the case under stability), one can write recursive equations for the length of statistically identifiable sessions appearing within the renewal cycle. Applying the expectation operator on these recursive equations leads to the infinite dimensional system of the form of (2), with non-negative coefficients and constants. Then, the properties of their solutions and the bounds presented in section 4 can be exploited to obtain performance metrics such as the system stability region and mean delay bounds.

First it should be noted that a truncated version of (2) for large N is computationally tractable, especially if one adopts the solution approach based on successive substitution. That is, if one substitutes an arbitrary non-negative sequence L_k^0, $k \geq 0$, to the right-hand side of (2) and derive the sequence L_k^1 in the left-hand side and then repeat the procedure by using the derived sequence L_k^1 in the right-hand side, and so on. It turns out that the solution is reached fairly fast and is of arbitrary accuracy depending on the value of N. By trying an even larger value of N, one can observe a negligible change in the solution for L_k, $k \geq 0$, especially for small k which are heavier contributing (through the Poisson probability of the multiplicities) to the mean renewal cycle length, L, we are interested in. Thus, L is very well approximated by solving a relatively small number N of linear equations.

Deriving upper bounds on L_k, $k \geq 0$, is more cumbersome. In section 4 we presented a linear with respect to k bound on L_k by calculating its coefficients analytically in [3]; a quadratic bound on the cumulative delay is also analytically derived in [3]. As the solutions of the truncated systems of equations for modest N yield a lower bound that is very close to the actual value, the cumbersome derivation of upper bounds on L_k and C_k is practically not necessary. The upper bound on L_k is necessary though in order to derive a lower bound on the maximum throughput, to determine the minimum support capabilities of the system. Although the lower bound on the throughput is very tight away from the instability region, it becomes very loose when approaching the instability region.

The following approaches can help derive tighter upper bounds on L_k (and similarly on C_k). The basic idea behind the first approach is to set L_k for values of k higher than the truncation value N, equal to the (looser) upper bound L_k^u and then solve the resulting finite system of equations to obtain a tighter bound on L_k, for $0 \leq k \leq N$, which are the L_k's contributing heavier to the mean L. The following theorem proves that the solution of these truncated equations constitute indeed an upper bound.

Theorem 2. *Let* $\{x_k^u\}_{k=0}^{\infty}$ *be a sequence of real numbers which satisfies* (α) $0 \leq x_k^u < \infty$, $0 \leq k < \infty$ *and* (β) $h_k + \sum_{j=0}^{\infty} \alpha_{kj} x_j^u \leq x_k^u$ *with* (γ) $h_k \geq 0$, $\alpha_{kj} \geq 0$, *for* $0 \leq k, j \leq \infty$. *Then the following hold.*
  a) The finite dimensionality system of linear equations

$$x_k^{ut} = h_k + \sum_{j=N+1}^{\infty} \alpha_{kj} x_j^u + \sum_{j=0}^{N} \alpha_{kj} x_j^{ut} \qquad (3)$$

has a nonnegative solution x_k^{ut} *which satisfies* $x_k^{ut} \leq x_t^u$, $0 \leq k \leq N$.
  b) If x_k *is a nonnegative solution of the system* $x_k = h_k + \sum_{j=0}^{\infty} \alpha_{kj} x_j$, $k \geq 0$, *then* $x_k \leq x_k^{ut}$, $0 \leq k \leq N$.

By employing the sequences $\{L_k^u\}_{k=0}^{\infty}$ and $\{C_k^u\}_{k=0}^{\infty}$ in the place of the sequence $\{x_k^u\}_{k=0}^{\infty}$ in the above theorem and solving the resulting finite dimensionality linear systems of equations given by (3), tight upper bounds on L_k ((L_k^{ut})) and C_k (C_k^{ut}) were obtained, for $k \leq N = 24$. By considering these tight upper bounds, tight upper bounds on L (L^{ut}) and C (C^{ut}) are obtained.

The lower bounds on S_{max}, λ^l, derived under Proposition 1 turned out not to be close to the upper bound, λ^u, derived under Proposition 2. The fact that the value of λ^u obtained by solving $N = 5$ linear equations was the same (up to the third decimal point) with that obtained by solving for $N = 24$ linear equations, suggests that λ^u is very close to S_{max} and that the lower bound is very loose. This belief can be substantiated and justified by re-deriving a lower bound as under Proposition 1 but using the following expressions for upper bounding L_k (suggested in [7] for the first time):

$$x_0^u = 1, \ x_k^u = (1 + \epsilon)L_k^l, \ 1 \leq k \leq 7 \qquad (4a)$$
$$x_k^u = \beta(\lambda, p)k - \gamma(\lambda, p), \ 8 \leq k \leq \infty \qquad (4b)$$

where ϵ is an arbitrary small positive number. A sequence $\{x_k^u\}_{k=0}^{\infty}$ as in (4) which satisfies the conditions of Proposition 1, was possible to obtain for $\lambda < \lambda^{lt}$ (where $\lambda^{lt} \approx \lambda^u$) up to the third decimal digit), and thus $S_{max} \approx \lambda^u$.

The analysis methodology presented here has been applied to the study of other protocols and queueing systems in networking. In [11] tight bounds on the maximum stable throughput and mean packet delay are derived under a random access protocol that can accommodate packets from 2 distinct priority classes. In [12,13], the process of the successfully transmitted packets over a random access channel employing LS-RAP is analytically matched to a first and second order markov model, by constructing recursive equations describing the renewal cycles induced by the protocol, as well as equations counting the occurrences of specific blocks of 2 or 3 consecutive slots over a cycle, to be used to calculate the transition probabilities of the approximating Markov model.

Finally, the methodology has been applied for the delay analysis of complex priority queueing disciplines in [14,15], which were developed in order to model and evaluate the performance of the Distributed Queued Dual Bus (DQDB) metropolitan networks. Complex (priority) queueing disciplines can be relatively easily studied by formulating recursive equations for the mean cumulative delay of all packets from a given priority class that were transmitted over a renewal cycle. The mean delay is calculated by the ration $\frac{C^i}{\lambda^i X}$, where X is the mean renewal cycle length and C^i and λ^i are the cumulative delay over a cycle and arrival rate of the packets of priority i. It should be noted that X is known under *any* queueing discipline and is given by $\frac{1}{1-\rho}$, where ρ is the utilization of the queueing system. Thus, the renewal cycle length and stability conditions (given by $\rho \leq 1$) are already available. Lower bounds on C^i can be computed by solving truncated version of the resulting infinite dimensional linear equations. An upper bound on the induced mean packet delay for class i, D^i_{up}, can be obtained (assuming that the priority discipline is non-preemptive and work-conserving) in terms of their lower bounds D^i_{lo}, and the mean delay, D^{FIFO}, induced by the equivalent FIFO queueing system (which is typically known or easy to derive), from the following expression assuming K priority classes [14]:

$$D^i_{up} = \frac{1}{\lambda^i}[\lambda D^{FIFO} - \sum_{k=1, k \neq i}^{K} \lambda^k D^k_{lo}] \tag{5}$$

6   Conclusions

The contribution of this paper is about a comprehensive presentation of a quite versatile and widely applicable methodology for evaluating the performance of resource allocation protocols and queueing systems appearing in networking.

The methodology consists of two steps. First, the renewal cycles of the system under study are identified and recursive equations that invoke properly identified sessions contained within the renewal cycles are derived. In addition to ultimately yielding the length of the renewal cycle, these recursive equations can help construct (similarly) additional ones with respect to random variables that count key events of interest over such sessions, such as the total delay incurred by the packets, the number of packets experiencing a delay exceeding a threshold, etc; performance metrics of interest can then easily be derived by considering the expected values of such random variables.

The second step of the methodology considers a system of infinite dimensional linear equations that is formulated by taking the expectation on the recursive equations developed in the first step. A rich theory on the properties of the solutions of such equations is presented along with techniques to obtain tight bounds on their solution. Through the rigorous propositions presented, the additional approaches for tightening further the upper bounds, the connection to the stability of the system, and the discussion on different applications of the approach, the potentially wide applicability and effectiveness of the presented methodology have been established. Although high performance computing machines can solve nowadays large dimensionality systems of linear equations, there are at least two reasons for which the presented methodology for coping with infinite dimensional linear equations can still be valuable: for establishing the stability region of a system and for modelling complex systems requiring the formulation of multi-dimensional random variables that lead to an exploding dimensionality of the system equations.

References

1. INFOCOM Network Science Wkshp (NetSciCom 2010), San Diego, USA (2010)
2. Stavrakakis, I.: The Internet of the Future, presentation in session "WP Objective 1.1, WP 2009-2010: Network of the Future". In: ICT 2008, Lyon (2008)
3. Stavrakakis, I.: Random Access Multi-user Communication Networks and Approximation of their Output Processes, Ph.D. dissertation, University of Virginia, UMI order No. 8913912, Ann Arbor Michigan (1988)
4. Stavrakakis, I., Kazakos, D.: A Limited Sensing Protocol for Multiuser Packet Radio Systems. IEEE Transactions on Communications 37(4), 353–359 (1989)
5. Tsybakov, B.S., Vvedenskaya, N.D.: Random Multiple Access Stack Algorithm. Translated from Problemy Peredachi Informatsii 16(3), 80–94 (1980)
6. Vvedenskaya, N.D., Tsybakov, B.S.: Random Multiple Access of Packets to a Channel with Errors. Translated from Problemy Peredachi Informatsii 19(2), 52–68 (1983)
7. Georgiadis, L., Papantoni-Kazakos, P.: Limited Feedback Sensing Algorithms for the Packet Broadcast Channel, Special Issue on Random Access Communications. IEEE Trans. Inform. Theory IT.31, 280–294 (1985)
8. Capetanakis, J.: Tree Algorithm for Packet Broadcast Channel. IEEE Trans. Inform. Theory IT.25, 505–515 (1979)
9. Chung, G.L.: A Course in Probability Theory. Academic, New York (1974)
10. Kantorovich, L.V., Krylov, V.I.: Approximate Methods of Higher Analysis. Interscience, New York (1958)
11. Stavrakakis, I., Kazakos, D.: A Multi-user Random Access Communication System for Users with Different Priorities. IEE Transactions on Communications 39(11) (1991)
12. Stavrakakis, I., Kazakos, D.: On the Approximation of the Output Process of Multiuser Random Access Communication Networks. IEEE Transactions on Communications 38(2) (1990)
13. Stavrakakis, I., Kazakos, D.: Performance Analysis of a Star Topology of Interconnected Networks under 2nd -order Markov Network Output Processes. IEEE Transactions on Communications 38(10) (1990)
14. Stavrakakis, I.: Delay Bounds on a Queuing System with Consistent Priorities. IEEE Transactions on Communications 42(2) (1994)
15. Landry, R., Stavrakakis, I.: Queuing Study of a 3-Priority Policy with Distinct Service Strategies. IEEE/ACM Transactions on Networking 1(5) (1993)
16. http://www.recognition-project.eu/

Modelling Social-Aware Forwarding in Opportunistic Networks

Chiara Boldrini, Marco Conti, and Andrea Passarella

IIT-CNR,
Via G. Moruzzi 1, 56124 Pisa, Italy
{chiara.boldrini,marco.conti,andrea.passarella}@iit.cnr.it

Abstract. Opportunistic networks are one of the most promising evolutions of the traditional Mobile Ad Hoc Networks paradigm. Communications in an opportunistic network rely on the mobility of the users: each message is handed over from node to node, making hop-by-hop decisions to select the node that is better suited for bringing the message closer to its destination. Algorithms exploiting social-awareness are emerging as one of the most efficient categories of forwarding algorithms. However we are currently lacking analytical models able to characterize the performance of social-aware forwarding in opportunistic networks. In this paper we start to fill this gap by proposing an analytical model for the expected number of hops and the expected delay experienced by messages when delivered in an opportunistic social-aware fashion. The model is then used to characterize how the expected delay experienced by messages varies with the different social structures in the network of the users.

Keywords: opportunistic networks, forwarding protocols, social-awareness, analytical model.

1 Introduction

In the broad area of wireless multi-hop networking, Delay Tolerant Networks (DTNs) have recently stood out because of their ability to enable communications even when protocols designed for traditional Mobile Ad Hoc Networks (MANET) cannot do so. In fact, the main requirement of MANET protocols, i.e., the presence of an end-to-end path connecting the source and the destination of a message, can be rarely satisfied in networks, e.g., made up of subnetworks connected only by satellite links [4], or where the nodes are people moving around with their hand-held devices [16]. The latter case is the scenario considered in this paper. In order to differentiate between the different applications of the delay tolerant paradigm, such networks have been named Pocket Switched Networks (PNSs) or *opportunistic networks*, because they opportunistically exploit contacts between users.

Messages in PSNs are routed along a multi-hop path across the nodes of the network. Being PSNs so unstable, source routing is inapplicable as the route

K.A. Hummel et al. (Eds.): PERFORM 2010 (Haring Festschrift), LNCS 6821, pp. 141–152, 2011.

chosen by the source of the message is likely to change within a short time. For this reason, forwarding decisions in opportunistic networks are made hop by hop. The key problem of message forwarding in PSNs is thus the selection of the node to which the message (or a copy of the message, in the case of multi-copy schemes) should be handed over. First and simplest implementations of this new communication paradigm involved a great number of copies of the same message to be spread across the network, in order to maximize the probability that one of them will eventually arrive at the destination [21]. Smarter strategies have been developed later, with the aim of selecting only the *best* relays as next hops for each message. In particular, social-aware strategies have proven [2,11] to be very effective in forwarding messages in an opportunistic network. Their main idea is that, while the connectivity graph of the network might be extremely unstable, the social graph, i.e., the network of relationships between users, is expected to vary on a much larger timescale than that typically of interest for the delivery of messages. This approach is indeed effective because of the correlation between sociality and mobility [17]: knowing social relationships between users enables us to estimate the likelihood of future encounters between nodes, which represent a forwarding opportunity.

Despite being so popular as forwarding strategies, social-aware schemes are typically difficult to model analytically. The main contribution of this paper lies in the definition of an analytical model for the evaluation of social-aware single-copy forwarding schemes. This model, based on Markov Chains, allow us to describe a way for computing significant quantities, such as the expected number of hops or the expected delay, that characterize the forwarding performance.

The paper is structured as follows. In Section 2 we review the state of the art on forwarding modelling for opportunistic networks. In Section 3 we describe our analytical model for social-oblivious and social-aware forwarding. In Section 4 we use the above model for evaluating the performance of four reference forwarding strategies with different underlying social structures for the network of the users. Finally, in Section 5 we conclude the paper.

2   Related Work

As anticipated in the previous section, forwarding protocols can be classified, according to the type of information that they exploit when making forwarding decisions, into social-oblivious and social-aware protocols. Social-oblivious protocols do not use at all information on the way nodes meet or relate with each other. This is the case of the Epidemic protocol [21], whose strategy is to generate and hand over a new copy of the message to each node encountered, and of the Direct Transmission protocol [9], in which messages can only be delivered to the destination when encountered directly. The performance of these protocols is typically poor because either they consume a lot of resources and overload the network (Epidemic) or they are not able to find a path to the destination even when many are available (Direct Transmission). For this reason, they are typically used as a baseline for performance evaluation.

Social-aware protocols, instead, exploit the social structure of the network of users in order to make forwarding decisions. This is because social-awareness enables the prediction of user encounters, which constitute forwarding opportunities. Some social-aware schemes focus only on encounters between nodes. This is the case of PROPHET [13], where the delivery probability of a node for a given destination is estimated based on previous encounters between nodes. Another approach is based on the exploitation of the roles of the nodes in the social graph associated with the network of users. The main idea is that nodes that are more *central* in the social graph are likely to be better forwarders than the other nodes. Bubble Rap [11] and SimBet [5] belong to this category. Social context-aware protocols keep track of a variety of information on the environment – *context* – the users live in (e.g., the people they meet, the friends they have, the places they visit). Context information is then used to quantify the ability of nodes to deliver messages. The HiBOp [2] protocol pertains to this group.

As far as modelling is concerned, quite a few frameworks have been proposed for social-oblivious forwarding schemes [22,10,8,18,19]. Epidemic models, Markov Chains and random walk on graph are the mathematical tools used to model important metrics such as the expected delay. The problem with these model is that they all consider homogeneous networks, i.e., networks where node movements are independent and identically distributed. This is not the case of real networks made up from human users moving with their portable devices: some users may cluster and move together, others may never get in touch with each other. Such heterogeneousness has been so far considered only in [20]. However, authors of [20] focus on multi-copy schemes, while in the following we consider single-copy schemes, i.e., schemes in which there is at any time just one copy of the message to be delivered.

3   A Semi-Markov Model for Message Forwarding

In this section we model the forwarding process as a semi-Markov process, and then we perform a transient study in order to compute the expected number of hops and the expected delay experienced by messages. We start with a general framework, which we then specialize for four forwarding protocols representative of different approaches to forwarding. Let us first introduce in the next section the network model that we consider.

3.1   Network Model

Our model considers a network with N nodes, moving around and meeting with each other. During contacts, nodes can exchange messages. For the sake of simplicity, we hereafter assume that messages can be exchanged only at the beginning of a contact between a pair of nodes (i.e., no periodic probing for new messages to relay during long contact periods), and that the transmission of the relayed messages can be always completed within the duration of a contact. The latter assumption is also justified by the fact that given the high dynamics of an opportunistic networks the file size is expected too be small [14]. In addition, we assume that each message is a bundle [6], an atomic unit that cannot be fragmented.[1] We also assume infinite buffer space on nodes. Given that we are considering single-copy schemes, buffer size is not expected to be critical, at least from low to medium network load. All the above assumptions allow us to isolate, and thus focus on, the effects of node mobility from other effects.

Given that messages are handed over from node to node before reaching their destination, the way nodes move heavily affects the delay experienced by messages. As for the mobility, the main role in the experienced delay is played by the inter-meeting time, which is defined as the time between two consecutive meetings between the same pair of nodes. In this paper we assume that such inter-meeting times can be described with an exponential distribution. Characteristic mobility times have been shown to follow an exponential distribution at least in their tail [12] [7]. Trading accuracy for tractability, here we assume the exponential property for the entire distribution. As a future work, we plan to relax the exponential assumption. In the following we denote as λ_{ij} the rate of the exponential distribution describing the process of encounters between two nodes i and j.

3.2   Reference Forwarding Strategies

We generalize the variety of protocols described in Section 2 into the two main categories of social-oblivious and social-aware forwarding protocols. For these categories, we consider the following policies, which identify important traits of existing forwarding strategies. More specifically, among the social-oblivious schemes we consider the following.

Definition 1 (Direct Transmission). *The source node can only deliver the message to the destination itself.*

Definition 2 (Always Forward). *The source node hands over the message to the first node encountered, and so does each intermediate node. The process stops when the message is delivered to the destination.*

As for the social-aware schemes, a message (be it on the source node or on an intermediate relay) is handed over to another node only if the latter has a higher probability (we call it *fitness*) of bringing the message closer to its destination than the node currently holding the message. Based on how the fitness is computed, we define the following two policies.

Definition 3 (Direct Acquaintance). *The source and each intermediate relay hand over the message to the first encounter having a higher fitness, where the*

---

[1] Fragmentation can indeed add additional delay at the destination or, even worse, impair communication at all when some fragments are lost, due to the high round trip time of opportunistic networks.

fitness F^{DA} is defined as the frequency of a direct meeting with the destination (Equation 1).

$$F_{i,d}^{DA} = \lambda_{i,d}, \forall i \neq d \tag{1}$$

Definition 4 (Social Forwarding). *Messages are delivered through a path with positive gradient of fitness, where the fitness $F_{i,d}^{SF}$ of node i for a message addressed to node d is computed (Equation 2) as the weighted sum of the fitness for a direct acquaintance ($F_{i,d}^{DA}$) and the fitness for an indirect meeting ($F_{i,d}^{I}$).*

$$F_{i,d}^{SF} = \alpha F_{i,d}^{DA} + (1 - \alpha) F_{i,d}^{I}, \quad where\ 0 < \alpha < 1 \tag{2}$$

Component $F_{i,d}^{DA}$ is defined as in Equation 1. The second component is a measure of the likelihood of encountering a node that has high delivery probability for the destination and it is defined according to the following:

$$F_{i,d}^{I} = f(F_{j,d}^{DA}) \quad \forall j \mid \lambda_{ij} \neq 0, j \neq d\,. \tag{3}$$

There is a variety of possible choices for function f in Equation 3. Without loss of generality, in the rest of the paper we use $f \equiv max(\cdot)$.

Differently from the Direct Acquaintance policy, the Social Forwarding strategy is able to detect not only direct meetings with the destination, but also meetings with people that have a high probability of delivering the message to the destination. This strategy enables the exploitation of the delivery skills that are present in the environment surrounding the users, and not only of those of the user itself. In Section 4.2 we will show how important can be this exploitation.

3.3 The Forwarding Process as a Semi-Markov Process

A semi-Markov process is one that changes states in accordance with a Markov chain (called *embedded* or *jump* chain) but where transitions between states can take a random amount of time [15]. As such, it is fully described by the transition matrix associated with its embedded chain and by $T_i^{exit}, \forall i = 0, \cdots, n$, where T_i^{exit} denotes the distribution of the time that the semi-Markov process spends in state i before making a transition.

We express our semi-Markov process in terms of the embedded Markov chain in Figure 1. Assuming that node i is currently holding a message whose

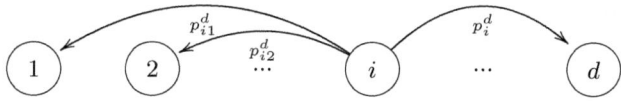

Fig. 1. Fragment of the embedded Markov Chain (valid for all $i \neq d$)

destination[2] is d, the probability p_{ij}^d that node i will delegate the forwarding of the message to another node j is a function of both the likelihood of meeting node j and the probability that node i will hand over the message to node j according to the forwarding policy in use.

Theorem 1 proves that, under the exponential assumption for inter-meeting times (see Section 3.1), the semi-Markov process that describes the forwarding evolution becomes a Continuous Time Markov process, in which T_i^{exit} follows an exponential distribution.

Theorem 1 (Exit time). *T_i^{exit}, the time before the semi-Markov process exits state i, follows an exponential distribution with rate $\sum_{\substack{j=1 \\ j \neq i}}^{N} \lambda_{ij} p_{ij}^{forw}$, where p_{ij}^{forw} represents the probability that node i hands over the message to node j according to the forwarding scheme in use. T_i^{exit}'s expected value is thus given by the following:*

$$E[T_i^{exit}] = \frac{1}{\sum_{\substack{j=1 \\ j \neq i}}^{N} \lambda_{ij} p_{ij}^{forw}} . \tag{4}$$

Proof. See [3].   □

Below we derive the transition probabilities associated with the embedded chain in Figure 1 for each of the forwarding schemes described in Section 3.2.

Proposition 1 (General form of the transition matrix for the forwarding process). *The transition matrix associated with the process of forwarding a message from a source node i to the destination node d is given in Equation 5, where, as an example, $d = N$.*

$$\mathbf{P} = \begin{pmatrix} 0 & p_{12} & \cdots & p_{1,N-1} & p_{1,N} \\ p_{21} & 0 & \cdots & p_{2,N-1} & p_{2,N} \\ \vdots & \vdots & \ddots & \vdots & \vdots \\ \vdots & \vdots & \vdots & \ddots & \vdots \\ 0 & 0 & \cdots & 0 & 1 \end{pmatrix} \tag{5}$$

The state associated with the destination node d is absorbing, because in state d the forwarding process is completed.

Theorem 2 (Transition probabilities p_{ij}). *Probabilities p_{ij} in Equation 5 are given by:*

$$p_{ij} = \frac{\lambda_{ij} p_{ij}^{forw}}{\sum_z \lambda_{iz} p_{iz}^{forw}} , \tag{6}$$

where λ_{ij} denotes the rate of encounters between node i and node j, and p_{ij}^{forw} represents the probability that node i hands over the message to node j according to the forwarding scheme in use.

---

[2] The chain is different for different destinations, because the convenient relays are generally not the same. However, for the sake of readability, in the following we drop superscript d.

Proof. See [3].   □

Both exit time T_i^{exit} and transition probabilities p_{ij} are dependent on p_{ij}^{forw}. Thus, in the following we derive p_{ij}^{forw} for each of the reference forwarding policies in Section 3.2.

Lemma 1 (p_{ij}^{forw} for Direct Transmission). *The probability p_{ij}^{forw} that node i hands over the message to node j when the Direct Transmission policy is in use is given by the following:*

$$p_{ij}^{forw} = \begin{cases} 1 & j = D \\ 0 & otherwise \end{cases} \tag{7}$$

Proof. See [3].   □

Lemma 2 (p_{ij}^{forw} for Always Forward). *The probability p_{ij}^{forw} that node i hands over the message to node j when the Always Forward policy is in use is given by the following:*

$$p_{ij}^{forw} = 1, \quad \forall i, j \tag{8}$$

Proof. See [3].   □

Lemma 3 (p_{ij}^{forw} for Direct Acquaintance and Social Forwarding). *Under the Direct Acquaintance strategy, the probability p_{ij}^{forw} that node i hands over the message to node j can be computed as:*

$$p_{ij}^{forw} = \begin{cases} 1 & F_{i,d}^{DA} < F_{j,d}^{DA} \\ 0 & otherwise \end{cases} \tag{9}$$

Analogously, for the Social Forwarding scheme we have for p_{ij}^{forw}:

$$p_{ij}^{forw} = \begin{cases} 1 & F_{i,d}^{SF} < F_{j,d}^{SF} \\ 0 & otherwise \end{cases} \tag{10}$$

Fitness $F_{i,j}^{DA}$ and $F_{i,j}^{SF}$ are defined in Equations 1 and 2.

Proof. See [3].   □

Theorems 1 and 2 completely define the forwarding Markov process. Thus, we can exploit well known algorithms for Markov chain transient analysis in order to compute significant properties of the forwarding process. In the following, we describe how to compute the expected delay and the expected number of hops travelled by messages.

Theorem 3 (Expected delay). *The expected delay $E[D_i^d]$ for a message generated by node i and addressed to node d can be obtained from the minimal non-negative solution to the following system:*

$$\begin{cases} E[D_i^d] = 0 & i = d \\ E[D_i^d] = E[T_i^{exit}] + \sum_{j \neq d} p_{ij} E[D_j^d] & \forall i \neq d \end{cases} \tag{11}$$

Proof. See [3].   □

Theorem 4 (Expected number of hops). *The expected number of hops $E[H_i^d]$ travelled by a message generated by node i and addressed to node d can be obtained from the minimal non-negative solution to the following system:*

$$\begin{cases} E[H_i^d] = 0 & i = d \\ E[H_i^d] = 1 + \sum_{j \neq d} p_{ij} E[H_j^d] \ \forall i \neq d \end{cases} \tag{12}$$

Proof. See [3]. $\qquad\qquad\qquad\qquad\qquad\qquad\qquad\qquad\qquad\qquad\qquad\qquad\qquad$ □

4   Performance Evaluation of Social-Aware Forwarding

In this section we provide a detailed analysis of the performance of the Direct Transmission, Always Forward, Direct Acquaintance, and Social Forwarding schemes using the analytical model that we have described above. Under the assumptions in Section 3.1, this model is exact (for a comparison between analytical and simulation results please refer to [3]).

In the following we consider 15 nodes, which move around in the network and exchange messages according to the policies defined in Section 3.2. We assume that node movements are triggered by their social relationships with the other nodes of the network. Each scenario we consider is characterized by a different social structure connecting the nodes of the network. Based on this structure, we define node mobility according to the following algorithm. We assume that the default meeting rate is λ for each pair of nodes connected by a social link. For those scenarios in which nodes are grouped into communities, however, assuming the user is in touch with n communities, the rate of contact with users in each of those communities is λ/n.

Solving the systems of equations in Theorems 3 and 4 provides us with a $N \times N$ matrix for the expected delay and a $N \times N$ matrix for the expected number of hops. Thus, the entry at position (i, j) in the matrix gives the expected delay (or number of hops) value for the $i - j$ node pair. For ease of visualization, we rely on a histogram of the expected delay and of the expected number of hops computed for the $N(N-1)$ pairs of interests. The bin width is set to 2 for the histograms of the expected delay and to 1 for the expected number of hops. Finally, please note that, in all the cases analyzed below, the resulting expected delay between any pair of nodes is a function of λ. In order to be able to plot such results we set λ to 1. This choice has absolutely no effect on our performance comparison, because λ appears only as a multiplying factor.

4.1   Homogeneous Network

Let us start our performance evaluation with the case of a complete social graph, i.e., a graph in which an edge connecting any pair of nodes exists. With this configuration all nodes are homogeneous from a mobility standpoint, i.e., every pair of nodes meets at the same rate λ. As a consequence, the concept of community does not apply here.

From Theorem 3, we obtain that the expected delay experienced by messages is the same for all the four policies and equal to $\frac{1}{\lambda}$. This result is not surprising, since all nodes are equivalent in this configuration, and choosing the one or the other does not make any difference. However, the different forwarding strategies may drastically differ in the number of hops needed to bring the message to its destination. Indeed, Table 1 shows that the Direct Transmission, Direct Acquaintance, and Social Forwarding schemes are all able to detect the fact that, as all nodes are equally good as relays, the most convenient strategy is to appoint the source of the message as its unique forwarder. Instead, the Always Forward scheme, which continuously delegates the forwarding of the message to any new encounter, needs much more relays (from which the high number of hops), which in turn imply many (unneeded) transmissions, with the consequence of poor resource utilization.

Table 1. Expected number of hops

	DT	AF	DA	SF
# hops	1	14	1	1

This homogeneous scenario is the one commonly used to evaluate the Epidemic forwarding strategy [21], which under ideal conditions (i.e., infinite bandwidth, infinite buffer space on devices, infinite battery lifetime, no contention, etc.) is the optimal forward policy as far as the expected delay is concerned. Being a multi-copy strategy, the Epidemic protocol does not fit into our model. However, we can exploit results presented in [22] in order to compare our single-copy strategies with Epidemic routing. The expected delay $E[D^{epi}]$ under Epidemic routing converges to $\frac{\ln N}{\beta(N-1)}$ as $N \to \infty$. This value is thus generally much smaller than $\frac{1}{\lambda}$, and it decreases as N increases. However, the price to pay for this quick delivery is in terms of the number of copies disseminated into the network. According to [22], the expected number of copies $E[C^{epi}]$ injected into the network by Epidemic routing is $\frac{N-1}{2}$. As N increases, $E[C^{epi}]$ also increases, thus flooding the network with many copies of the same data. When ideal conditions assumption is released, this will drastically affect the performance of Epidemic routing, and the delay provided will be much smaller than the optimal value, as shown in [1].

4.2 Connected Communities

While in the homogeneous case all nodes were equal as far as their meetings were concerned, here we consider the case of a heterogeneous network. We equally distribute our 15 nodes into 3 communities, namely, $C1$, $C2$, and $C3$. Each community is a complete subgraph, meaning that all nodes within each community are connected with each other. We also add links between communities in the social graph. These links are edges connecting a node in one community to another node in another community. We hereafter refer to the nodes having inter-community links as *travellers*. We assign travellers only to community $C1$, which makes the network still connected (i.e., it exists at least one multi-hop path between every pair of nodes). However, with this configuration, community $C2$ and $C3$ cannot communicate directly, and they have to exploit the forwarding capabilities of the visiting travellers from $C1$. In the following, for ease of reading, we denote with indices from 1 to 5 the nodes in the first community ($C1$), with indices from 6 to 10 the nodes in the community ($C2$), and so on.

Figure 2 shows the expected delay experienced by messages in this scenario. Both the Direct Acquaintance and the Direct Transmission schemes are not able to deliver a subset of messages. The Direct Transmission scheme suffers when the source and the destination of the message do not get in touch with each other directly, thus producing in this case infinite delays. This is because with Direct Transmission nodes can only deliver their messages directly to the destination, thus missing all the opportunities offered by relaying: when the destination is never met, the message cannot be delivered. With Direct Acquaintance a node hands over a message to a node that has a higher probability of meeting the destination, measured in terms of direct encounters (Equation 1). The traveler that visits $C1$ does not meet any nodes of $C3$ directly, thus it is not considered a good relay by the Direct Acquaintance scheme. A more efficient strategy should also consider the transitivity of opportunities (e.g., node a meets b, which in turn meets c, thus a can be considered a good relay for destination c). This transitivity of encounters is detected by the Social Forwarding strategy, which indeed is able to deliver all messages to their destinations. The Always Forward strategy is also able to deliver all messages, but using many relays (Figure 3). The reason is that, being the forwarding opportunities so limited, with the Always Forward strategy the destination is typically found by chance after many (bad) relays have been used.

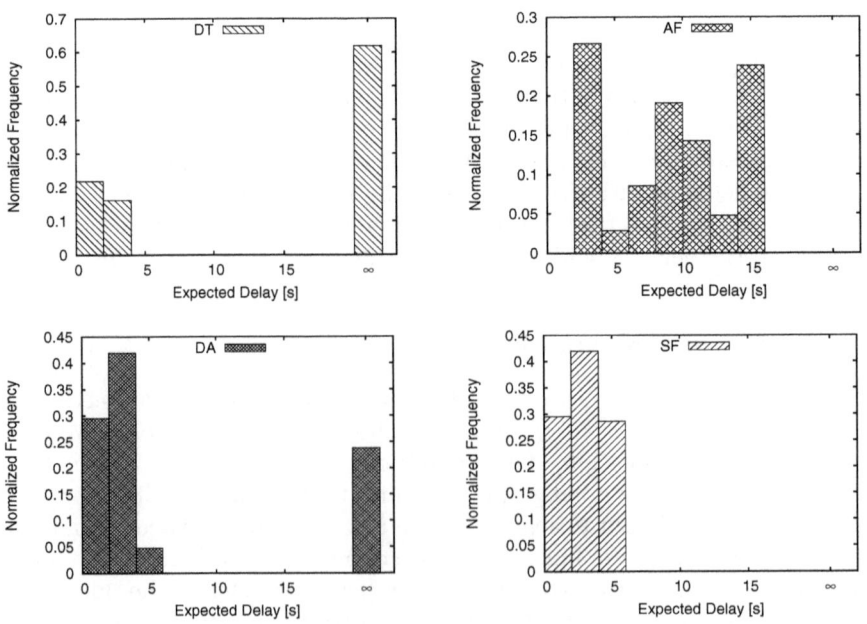

Fig. 2. Expected delay with connected communities (Sec. 4.2)

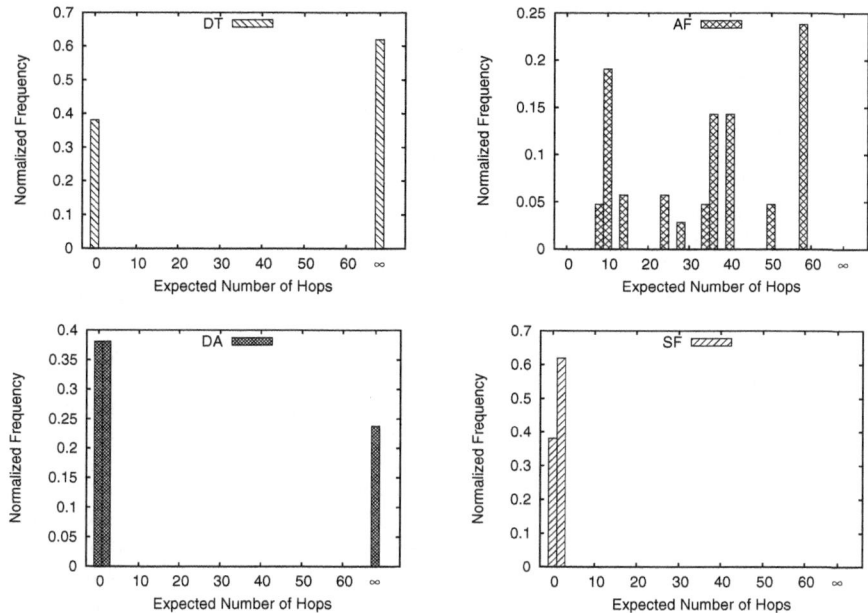

Fig. 3. Expected number of hops with connected communities (Sec. 4.2)

5   Conclusions

In this paper we have proposed an analytical model based on Markov processes for social-aware forwarding in opportunistic networks. Using this model, we have discussed how to compute the expected delay and the expected number of hops of messages delivered according to four reference forwarding schemes, of which two are able to exploit social information when making forwarding decisions. In the second part of the paper, we have used the model to compare the forwarding performance of social-oblivious and social-aware strategies in terms of expected delay and expected number of hops. In general, social-aware policies turn out to provide lower delays while at the same time keeping the number of hops down, thus improving the efficiency of the network. We have also shown how the ability of exploiting indirect connections between nodes may be a key strategy when forwarding opportunities are limited, and for this reason we have identified the Social Forwarding strategy as the most promising social-aware approach.

Acknowledgments. This work was partially funded by the European Commission under the SOCIALNETS (217141) FET-PERADA project.

References

1. Boldrini, C., Conti, M., Passarella, A.: Impact of social mobility on routing protocols for opportunistic networks. In: Proceedings of IEEE WoWMoM 2007. pp. 1–6 (2007)

2. Boldrini, C., Conti, M., Passarella, A.: Exploiting users' social relations to forward data in opportunistic networks: The HiBOp solution. Pervasive and Mobile Computing 4(5), 633–657 (2008)
3. Boldrini, C., Conti, M., Passarella, A.: Modelling social-aware forwarding in opportunistic networks. Tech. rep., IIT-CNR (2010), `http://bruno1.iit.cnr.it/~chiara/perform10_tr.pdf`
4. Burleigh, S., Hooke, A., Torgerson, L., Fall, K., Cerf, V., Durst, B., Scott, K., Weiss, H.: Delay-tolerant networking: an approach to interplanetary internet. IEEE Communications Magazine 41(6), 128–136 (2003)
5. Daly, E., Haahr, M.: Social network analysis for information flow in disconnected Delay-Tolerant MANETs. IEEE Transactions on Mobile Computing pp. 606–621 (2008)
6. Fall, K., Farrell, S.: DTN: an architectural retrospective. IEEE Journal on Selected Areas in Communications 26(5), 828 (2008)
7. González, M., Hidalgo, C., Barabási, A.: Understanding individual human mobility patterns. Nature 453(7196), 779–782 (2008)
8. Groenevelt, R., Nain, P., Koole, G.: The message delay in mobile ad hoc networks. Performance Evaluation 62(1-4), 210–228 (2005)
9. Grossglauser, M., Tse, D.: Mobility increases the capacity of ad hoc wireless networks. IEEE/ACM Transactions on Networking 10(4), 477–486 (2002)
10. Haas, Z., Small, T.: A new networking model for biological applications of ad hoc sensor networks. IEEE/ACM Transactions on Networking 14(1), 27–40 (2006)
11. Hui, P., Crowcroft, J., Yoneki, E.: Bubble rap: social-based forwarding in delay tolerant networks. In: Proceedings of the 9th ACM MobiHoc. pp. 241–250. ACM (2008)
12. Karagiannis, T., Le Boudec, J.Y., Vojnović, M.: Power law and exponential decay of inter contact times between mobile devices. In: Proceedings of the 13th ACM MobiCom '07. pp. 183–194 (2007)
13. Lindgren, A., Doria, A., Schelén, O.: Probabilistic routing in intermittently connected networks. LNCS pp. 239–254 (2004)
14. Ott, J.: Application protocol design considerations for a mobile internet. In: Proceedings of first ACM/IEEE international workshop on Mobility in the evolving internet architecture. p. 80. ACM (2006)
15. Ross, S.: Introduction to probability models. Academic Press (2007)
16. Scott, J., Hui, P., Crowcroft, J., Diot, C.: Haggle: A networking architecture designed around mobile users. In: Proceedings of IFIP WONS (2006)
17. Silvis, J., Niemeier, D., D'Souza, R.: Social networks and travel behavior: Report from an integrated travel diary. In: 11th International Conference on Travel Behaviour Reserach, Kyoto (2006)
18. Spyropoulos, T., Psounis, K., Raghavendra, C.: Efficient routing in intermittently connected mobile networks: The multiple-copy case. IEEE/ACM Transactions on Networking 16(1), 77–90 (2008)
19. Spyropoulos, T., Psounis, K., Raghavendra, C.: Efficient routing in intermittently connected mobile networks: The single case. IEEE/ACM Transactions on Networking 16(1), 63–76 (2008)
20. Spyropoulos, T., Turletti, T., Obraczka, K.: Routing in Delay-Tolerant Networks Comprising Heterogeneous Node Populations. IEEE Transactions on Mobile Computing pp. 1132–1147 (2009)
21. Vahdat, A., Becker, D.: Epidemic routing for partially connected ad hoc networks. Tech. Rep. CS-2000-06, Duke University (2000)
22. Zhang, X., Neglia, G., Kurose, J., Towsley, D.: Performance modeling of epidemic routing. Computer Networks 51(10), 2867–2891 (2007)

On Lookahead Strategy for Movement-Based Location Update: A General Formulation

Vicente Casares-Giner

GIRBA Group - Instituto ITACA, Universidad Politécnica de Valencia,
Camino de Vera s/n. 46022 Valencia, Spain
vcasares@upvnet.upv.es

Abstract. Location management deals with the procedure to update the current location of a mobile terminal (MT) and with the procedure to deliver incoming calls to that called MT. Basically, the performance evaluation of location management procedures are dependent on the MT's mobility behavior, on the MT's cell residence time and on the call arrival process to the MT. In the open literature, the typical analysis for location management has been addressed under the assumptions of a random walk mobility model, the exponential cell sojourn time of the MT -some times relaxed to a general probability distribution function- and the exponential inter-arrival time distribution for incoming calls. However, the random walk model seems not to be valid as many mobile users follow some daily trajectories, such as from home to the working place, from the working place to the shopping center, etc. To reflect a more realistic movement pattern, we propose a directional oriented mobility model. And as a consequence of that, we also propose a lookahead procedure combined with the movement-based location update scheme, with the main idea of saving signaling traffic through the air interface. In the lookahead strategy we analytically derive closed form expressions for the mean number of location update (LU) messages triggered by the MT between two consecutive call arrivals and the paging probabilities to evaluate the paging cost under some selective paging strategies. The analysis has been carried out assuming a general cell residence time and a renewal point process for call arrivals to the MT.

Keywords: Location update, selective paging.

1 Introduction

Mobility management is a key functionality that aims in providing ubiquitous telecommunication services in wireless mobile access networks. Mobility management procedures can be classified as handoff management and location management. The first procedure aims to guarantee the continuity of an on going call when the mobile terminal (MT) visits a new cell. Location management is defined as a set of procedures that allow a MT being located at anytime, anywhere, so that incoming calls may be delivered to that MT. This set consists of location update (LU) and call delivery (CD). The LU process consists of maintaining the

K.A. Hummel et al. (Eds.): PERFORM 2010 (Haring Festschrift), LNCS 6821, pp. 153–166, 2011.
© IFIP International Federation for Information Processing 2011

MT location information updated in the system database. The database entry of an MT is updated whenever the MT triggers an LU message or an incoming call is delivered. The CD procedure is decomposed into two steps: interrogation and terminal paging (PG). Firstly, in the interrogation step, the system database is queried to obtain the Registration Area (RA) where the MT is supposed to be in. Afterwards, the PG procedure follows, i.e., the MT is searched by polling the set of cells of the RA.

There is a trade off between LU and PG procedures. As the number of LU messages increases, the LU cost becomes higher but the PG cost decreases as the MT position is known more accurately. On the other hand, the lower is the number of LUs, the lower is the LU cost; but the uncertainty of the MT position is higher and the PG cost increases.

The performance of a location management procedure is dependent on the mobility patterns of the MT. One of the most extended mobility models used in the evaluation of location tracking is the random walk mobility model, [1] [2]. However, the random walk model is rather unrealistic since many MTs follow daily trajectories such as, as we have said before, from home to the working place, from home to the recreation center, etc. Therefore, in order to evaluate in a proper way the signaling load associated to mobility tracking, more realistic mobility models are required [3].

In this work we consider a mobility model with a certain directionality, and as a consequence of that information, an enhanced movement-based mobility tracking procedure is proposed. The proposal is based on the lookahead tracking principle and it is illustrated and evaluated in a 2-D scenario.

2   Movement-Based Location Update

2.1   Scenario and Mobility Model

Although hexagonal and square cell layouts in a 2-D scenario can be considered, here we analyze our proposal in the hexagonal cell layout [4]. Cells are located in a plane and they are enumerated according to Fig. 1. The mobility model is as follows. After a certain sojourn or cell dwell time in cell (i, j), characterized by a probability density function (pdf) $f_{m;i,j}(t)$, our MT will travel in the direction of angles $\{i\pi/3\}$ with probabilities $\{p_i\}$, counterclockwise, for $i = 0, 1, 2, 3, 4, 5$, respectively. We will assume that all cells have the same size, and that $f_{m;i,j}(t) = f_m(t)$. The following probabilities $\{p_i\}$ are considered for our single parameter mobility model

$$p_0 = \frac{\beta^3}{D(\beta)}; \quad p_1 = p_5 = \frac{\beta^2}{D(\beta)}; \quad p_2 = p_4 = \frac{\beta}{D(\beta)}; \quad p_3 = \frac{1}{D(\beta)}$$

$$\text{with } D(\beta) = (\beta^3 + 2\beta^2 + 2\beta + 1) = (\beta + 1)(\beta^2 + \beta + 1);$$

where β is positive number, $\beta \in (0, 1)$ or $\beta \in (1, \infty)$, and $D(\beta)$ is a normalization factor. Clearly, the parameter β provides a wide range of the directionality of the MT's movement. For instance, if $\beta = 1$, $p_i = 1/6, \forall i$, we have the standard random walk model. For $\beta > 1$ we have $p_o > p_1 = p_5 > p_2 = p_4 > p_3$. Clearly when $\beta \to \infty$ $p_0 \to 1$ the MT will always travel in the same direction, i.e. according to Fig. 1, towards the right hand side. The higher is β the higher the directionality of our MT will be.

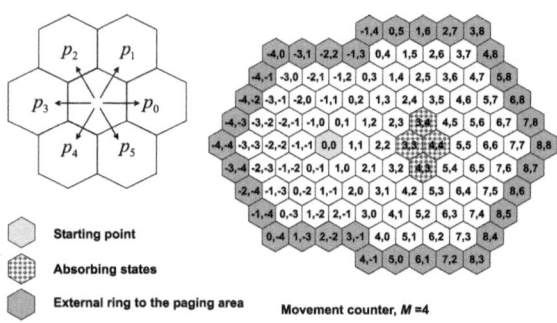

Fig. 1. Illustration of the lookahead strategy for movement based location update, with movement-counter threshold $M = 4$ and $\Delta = \{(3,3),(3,4),(4,3),(4,4)\}$

2.2   Lookahead Strategy

Let us consider that at time $t = 0^+$ our MT is located at cell $(i,j) = (0,0)$, see Fig. 1. We choose a set of target cells, the Δ set, that our MT will likely find along its trajectory. For example, if $\beta > 1$ cells $(1,1)$, $(2,2)$, $(3,3)$, ... can be reached by our MT with higher probability than cells $(-1,-1)$, $(-2,-2)$, $(-3,-3)$, ... Then, each time the MT visits a cell that does not belong to the Δ set, its movement counter is increased in one unit. When the movement counter reaches a certain threshold, say M, it will send a LU message to the network. But, if the MT visits any cell of the Δ set before the threshold M is reached, it will reset its movement counter. Fig. 1 depicts the case of $M = 4$ and $\Delta = \{(3,3),(3,4),(4,3),(4,4)\}$. Obviously the distance d_Δ between the starting point, the cell $(0,0)$ and the nearest target cell, measured in terms of cells, must be such that $0 \le d_\Delta \le M$; $d_\Delta = 3$ in Fig. 1. The case $d_\Delta > M$ is equivalent to having no target cells since, in this case, our MT will trigger LU messages according to the standard movement-based scheme, i.e. no absorption into the Δ set will happen. Notice that, each time the MT triggers a LU message the procedure starts anew at cell $(0,0)$, i.e. the scenario depicted in Fig. 1 is repeated.

2.3   Modeling and Analysis

In this paper we generalize the work [5] in two aspects; first by allowing more than one cell in the Δ set, secondly by considering a general renewal point process for modeling call arrivals to the MT. Then, for a predetermined number of movements m we are interested in the number of LU messages triggered by our MT between two consecutive call arrivals. We say that our MT is in state S_i if it is roaming in cell i. Although a natural cell identification in a 2-D space is the use of two indexes, for the sake of simplicity in the mathematic notation we will use a single index. Since the RA where the MT is roaming is configured by a finite number of cells, a correspondence between the identification with two indexes and the identification with a single index can be established. Then, starting from state S_0, cell $(0,0)$ in Fig. 1, if an absorption into the Δ set happens at movement M or before, the MT will reset its movement counter. If the MT visits a cell that does not belong to the Δ set it will increase in one unit its movement counter. When the movement counter reaches the threshold M without being absorbed into the Δ set, the MT will trigger a LU message and the process starts anew at state S_0.

The analysis can be formulated by building a suitable Markov chain and using some useful taboo probabilities [6]. We call ${}_H f_{i,j}^{(n)}$ the conditional probability that the Markov chain enters state S_j for the first time at the nth step, having initially started from state S_i and avoiding the set of states H.

With no loss of generality, in the sequel we assume that the Δ set of absorbing states is configured by three cells, j, k and l, that is, $\Delta = \{j, k, l\}$. Also, we denote by $\overline{\Delta}_s$ the complementary set of cells in Δ with respect to cell s. In our case $\overline{\Delta}_j = \{k, l\}$, $\overline{\Delta}_k = \{j, l\}$ and $\overline{\Delta}_l = \{j, k\}$.

Then, if initially the MT is in state S_0 the probability that in M movements no absorption into the Δ set occurs, $P_{nab(0,\Delta)}(M)$, is given by

$$P_{nab(0,\Delta)}(M) = 1 - \sum_{n=1}^{M} f_{0,\Delta}^{(n)} = 1 - \sum_{n=1}^{M} \left({}_{\overline{\Delta}_j} f_{0,j}^{(n)} + {}_{\overline{\Delta}_k} f_{0,k}^{(n)} + {}_{\overline{\Delta}_l} f_{0,l}^{(n)} \right) \quad (1)$$

Also, if initially the MT is in state S_s, for $(s = j, k, l)$ the probability that in M movements no absorption into the Δ set occurs, $P_{nab(s,\Delta)}(M)$, is given by

$$\begin{aligned} P_{nab(s,\Delta)} &= 1 - \sum_{n=1}^{M} f_{s,\Delta}^{(n)} = \\ 1 - \sum_{n=1}^{M} &\left({}_{\overline{\Delta}_j} f_{s,j}^{(n)} + {}_{\overline{\Delta}_k} f_{s,k}^{(n)} + {}_{\overline{\Delta}_l} f_{s,l}^{(n)} \right); \quad \text{for } s = j, k, l \end{aligned} \quad (2)$$

NOTE: Details on how the taboo probabilities $\{{}_H f_{i,j}^{(n)}\}$ are evaluated are given in the internal report [7].

3   The Number of Location Update Messages

3.1   Mean Number of LU Messages for a Fixed Number of Movements

Let $P_0^{(m)}(i)$ $(P_s^{(m)}(i)$, for $s = j, k, l)$ denote the probability of having i LU messages triggered in m movements, under the assumption that the MT starts its trajectory in cell 0 $(s = j, k, l)$. Let $M_0^{(m)}$ $(M_s^{(m)}$, for $s = j, k, l)$ denote the mean number of LU messages. Then, we have

$$M_0^{(m)} = \begin{cases} 0; & \text{for } m < M \\ \sum_{n=1}^{M} \mathbf{f}_{0,\Delta}^{(n)} \mathbf{M}_{\Delta}^{(m-n)} + P_{nab(0,\Delta)}(M)[1 + M_0^{(m-M)}]; & \text{for } m \geq M \end{cases} \quad (3)$$

and

$$\mathbf{M}_{\Delta}^{(m)} = \begin{cases} \mathbf{0}; & \text{(a column vector of zeros)}; \quad \text{for } m < M \\ \sum_{n=1}^{M} \mathbf{f}_{\Delta,\Delta}^{(n)} \mathbf{M}_{\Delta}^{(m-n)} + \mathbf{P}_{nab(\Delta,\Delta)}(M)[1 + M_0^{(m-M)}]; & \text{for } m \geq M \end{cases} \quad (4)$$

where (more details are given in [7])

$$\mathbf{M}_{\Delta}^{(m)} = \left(M_j^{(m)}, M_k^{(m)}, M_l^{(m)} \right)^T = \sum_{i=0}^{\infty} i \mathbf{P}_{\Delta}^{(m)}(i);$$

$$\text{with} \quad \mathbf{P}_{\Delta}^{(m)}(i) = \left(P_j^{(m)}(i), P_k^{(m)}(i), P_l^{(m)}(i) \right)^T;$$

$$\mathbf{P}_{nab(\Delta,\Delta)}(M) = \left(P_{nab(j,\Delta)}(M), P_{nab(k,\Delta)}(M), P_{nab(l,\Delta)}(M) \right)^T;$$

$$\mathbf{f}_{0,\Delta}^{(n)} = \left(\overline{\Delta_j} f_{0,j}^{(n)}, \overline{\Delta_k} f_{0,k}^{(n)}, \overline{\Delta_l} f_{0,l}^{(n)} \right); \quad \mathbf{f}_{\Delta,\Delta}^{(n)} = \begin{pmatrix} \overline{\Delta_j} f_{j,j}^{(n)} & \overline{\Delta_k} f_{j,k}^{(n)} & \overline{\Delta_l} f_{j,l}^{(n)} \\ \overline{\Delta_j} f_{k,j}^{(n)} & \overline{\Delta_k} f_{k,k}^{(n)} & \overline{\Delta_l} f_{k,l}^{(n)} \\ \overline{\Delta_j} f_{l,j}^{(n)} & \overline{\Delta_k} f_{l,k}^{(n)} & \overline{\Delta_l} f_{l,l}^{(n)} \end{pmatrix}$$

Therefore, the mean number of LU messages triggered by the MT between two consecutive call arrivals, $\overline{\#_{LU}}$, can be expressed as

$$\overline{\#_{LU}} = \sum_{m=0}^{\infty} M_0^{(m)} \alpha(m) = \sum_{m=M}^{\infty} M_0^{(m)} \alpha(m) \quad (5)$$

where $\alpha(m)$ is the probability of m cell boundary crossings between two consecutive call arrivals. $\alpha(m)$ is derived in the next section.

3.2 On the Number of Cell Boundary Crossing between Two Call Arrivals

We assume that calls arrive to the MT following a renewal point process with inter-arrival time *pdf* given by $f_c(t)$. We denote $f_c^*(s)$ the Laplace transform (LT) of $f_c(t)$. Then, $\alpha(k)$ is given by (see [7]),

$$\alpha(k) = \begin{cases} \dfrac{1}{2\pi i} \displaystyle\int_{c-i\infty}^{c+i\infty} \dfrac{[1 - f_{mr}^*(s)]}{s} f_c^*(-s) ds; & \text{for } k = 0 \\ \dfrac{1}{2\pi i} \displaystyle\int_{c-i\infty}^{c+i\infty} \dfrac{f_{mr}^*(s)[1 - f_m^*(s)][f_m^*(s)]^{k-1}}{s} f_c^*(-s) ds; & \text{for } k > 0 \end{cases} \quad (6)$$

In (6), $f_m^*(s)$ is the LT of $f_m(t)$; the pdf of the MT dwell time or residence time in a cell, with mean value $1/\lambda_m = -f_m^{*\prime}(0)$. $f_{mr}^*(s)$ is the LT of the residual cell residence time, $f_{mr}^*(t)$. Following the residual life theorem [8] we have

$$f_{mr}(t) = \lambda_m \int_t^\infty f_m(t)\mathrm{d}t = \lambda_m[1 - F_m(t)]; \qquad f_{mr}^*(s) = \frac{\lambda_m}{s}[1 - f_m^*(s)] \quad (7)$$

3.3   Mean Number of LU between Two Consecutive Call Arrivals

In order to obtain a closed form expression for (5) first we define the following generating functions

$$M_0^*(z) = \sum_{m=0}^\infty M_0^{(m)} z^m; \quad \mathbf{M}_\Delta^*(z) = \sum_{k=0}^\infty \mathbf{M}_\Delta^{(k)} z^{\cdot} \qquad (8)$$

$$\mathbf{F}_{0,\Delta}^*(z) = \sum_{n=1}^M \mathbf{f}_{0,\Delta}^{(n)} z^n = \left(F_{0,j}^*(z), F_{0,k}^*(z), F_{0,l}^*(z) \right);$$

$$\text{with} \quad F_{0,s}^*(z) = \sum_{n=1}^M {}_{\overline{\Delta}_s} f_{0,s}^{(n)} z^n, \quad \text{for } s \in \Delta \equiv \{j,k,l\}$$

$$\mathbf{F}_{\Delta,\Delta}^*(z) = \sum_{n=1}^M \mathbf{f}_{\Delta,\Delta}^{(n)} z^n = \begin{pmatrix} F_{j,j}^*(z) & F_{j,k}^*(z) & F_{j,l}^*(z) \\ F_{k,j}^*(z) & F_{k,k}^*(z) & F_{k,l}^*(z) \\ F_{l,j}^*(z) & F_{l,k}^*(z) & F_{l,l}^*(z) \end{pmatrix};$$

$$\text{with} \quad F_{r,s}^*(z) = \sum_{n=1}^M {}_{\overline{\Delta}_s} f_{r,s}^{(n)} z^n, \quad \text{for } r,s \in \Delta \equiv \{j,k,l\}$$

Then, from (3) and (4) a linear relationship between $M_0^*(z)$ and $\mathbf{M}_\Delta^*(z)$ is obtained. That is

$$M_0^*(z) = \mathbf{F}_{0,\Delta}^*(z)\mathbf{M}_\Delta^*(z) + P_{nab(0,\Delta)}(M)z^M \left(\frac{1}{1-z} + M_0^*(z) \right)$$

$$\mathbf{M}_\Delta^*(z) = \mathbf{F}_{\Delta,\Delta}^*(z)\mathbf{M}_\Delta^*(z) + \mathbf{P}_{nab(\Delta,\Delta)}(M)z^M \left(\frac{1}{1-z} + M_0^*(z) \right)$$

$$(9)$$

Therefore, solving (9) we get the expression for $M_0^*(z)$

$$M_0^*(z) =$$

$$= \frac{1}{1-z} \frac{\left\{ \mathbf{F}_{0,\Delta}^*(z)[\mathbf{I} - \mathbf{F}_{\Delta,\Delta}^*(z)]^{-1}\mathbf{P}_{nab(\Delta,\Delta)}(M) + P_{nab(0,\Delta)}(M) \right\}z^M}{1 - \left\{ \mathbf{F}_{0,\Delta}^*(z)[\mathbf{I} - \mathbf{F}_{\Delta,\Delta}^*(z)]^{-1}\mathbf{P}_{nab(\Delta,\Delta)}(M) + P_{nab(0,\Delta)}(M) \right\}z^M} \qquad (10)$$

Secondly, from (3)-(6) we can write, after some algebra, [7],

$$\overline{\#_{\mathrm{LU}}} = \sum_{m=0}^{\infty} M_0^{(m)} \alpha(m) = \sum_{k=M}^{\infty} M_0^{(k)} \alpha(k) =$$

$$= \frac{1}{2\pi i} \int_{c-i\infty}^{c+i\infty} \frac{f_{mr}^*(s)[1-f_m^*(s)]}{s f_m^*(s)} M_0^*(f_m^*(s)) f_c^*(-s) ds \qquad (11)$$

$$= -\sum_{p \in \sigma_c} Res_{s=p} \frac{f_{mr}^*(s)[1-f_m^*(s)]}{s f_m^*(s)]} M_0^*(f_m^*(s)) f_c^*(-s)$$

The last equality in (11) is derived applying the Cauchy's Residue Theorem, i.e., σ_c denotes the set of poles of $f_c^*(-s)$ in the right half complex plane, and $Res_{s=p}$ denotes the residue at poles $s = p \in \sigma_c$. $M_0^*(f_m^*(s))$ is the expression (10) evaluated at $z = f_m^*(s)$.

4   Terminal Paging Procedure

Blanket or non selective and selective or multi-step PG procedures are considered. For PG evaluation, in the first case we only need the number of cells that conforms the RA. Once the parameter M and the Δ set are fixed, the total number of cells to poll can easily be obtained. For instance, when $M = 4$ and $\Delta = \{(3,3),(3,4),(4,3),(4,4)\}$ Fig. 1 accounts for the set of cells, the Ω set, in which the MT can be in. The cardinality of Ω is $C(\Omega) = 73$. Obviously, for selective PG we need the PG probabilities. They are derived in the next lines.

4.1   Paging Probabilities

Let $\pi_{i,t}^{(m)}$ denotes the probability that starting at cell i the MT is located at cell t after m movements. Then, following the arguments used in section 3.1 we can write, for $i = 0$

$$\pi_{0,t}^{(m)} = \begin{cases} p_{0,t}^{(m)}; & \text{for } m < M \\ \displaystyle\sum_{n=1}^{M} \mathbf{f}_{0,\Delta}^{(n)} \boldsymbol{\pi}_{\Delta,t}^{(m-n)} + P_{nab(0,\Delta)}(M) \pi_{0,t}^{(m-M)}; & \text{for } m \geq M \end{cases} \qquad (12)$$

and for the Δ set, cells $i = s \in \Delta \equiv \{j,k,l\}$

$$\boldsymbol{\pi}_{\Delta,t}^{(m)} = \begin{cases} \mathbf{p}_{\Delta,t}^{(m)}; & \text{for } m < M \\ \displaystyle\sum_{n=1}^{M} \mathbf{f}_{\Delta,\Delta}^{(n)} \boldsymbol{\pi}_{\Delta,t}^{(m-n)} + \mathbf{P}_{nab(\Delta,\Delta)}(M) \pi_{0,t}^{(m-M)}; & \text{for } m \geq M \end{cases} \qquad (13)$$

with

$$\boldsymbol{\pi}_{\Delta,t}^{(m)} = \left(\pi_{j,t}^{(m)}, \pi_{k,t}^{(m)}, \pi_{l,t}^{(m)} \right)^T; \quad \mathbf{p}_{\Delta,t}^{(m)} = \left(p_{j,t}^{(m)}, p_{k,t}^{(m)}, p_{l,t}^{(m)} \right)^T;$$

Therefore the set of probabilities $\{\pi_{i,t}\}$ can be expressed as

$$\pi_{i,t} = \sum_{m=0}^{\infty} \pi_{i,t}^{(m)} \alpha(m) \tag{14}$$

For $i = 0$, (14) gives the probability that the MT is located at cell j when a call arrival occurs. To evaluate the set $\{\pi_{0,t}\}$ we follow a similar procedure to that in section 3. To that end, first, let us define the following generating functions, $\pi_{0,t}^*(z)$ and $\boldsymbol{\pi}_{\Delta,t}^*(z)$

$$\pi_{0,t}^*(z) = \sum_{m=0}^{\infty} \pi_{0,t}^{(m)} z^m; \quad \boldsymbol{\pi}_{\Delta,t}^*(z) = \sum_{m=0}^{\infty} \boldsymbol{\pi}_{\Delta,t}^{(m)} z^m; \tag{15}$$

Then, from equations (12) and (13) a linear relationships between $\pi_{0,j}^*(z)$ and $\boldsymbol{\pi}_{\Delta,t}^*(z)$ is obtained. That is

$$[1 - z^M P_{nab(0,\Delta)}(M)]\pi_{0,t}^*(z) - \mathbf{F}_{0,\Delta}^*(z)\boldsymbol{\pi}_{\Delta,t}^*(z) =$$
$$= \sum_{n=0}^{M-1} p_{0,t}^{(n)} z^n - \sum_{n=1}^{M-1} \mathbf{f}_{0,\Delta}^{(n)} \Big[\sum_{k=0}^{M-n-1} \mathbf{p}_{\Delta,t}^{(k)} z^k \Big] z^n = \Upsilon(0, \Delta, z, t);$$

$$-z^M \mathbf{P}_{nab(\Delta,\Delta)}(M)\pi_{0,t}^*(z) + [\mathbf{I} - \mathbf{F}_{\Delta,\Delta}^*(z)]\boldsymbol{\pi}_{\Delta,t}^*(z) =$$
$$= \sum_{n=0}^{M-1} \mathbf{p}_{\Delta,t}^{(n)} z^n - \sum_{n=1}^{M-1} \mathbf{f}_{\Delta,\Delta}^{(n)} \Big[\sum_{k=0}^{M-n-1} \mathbf{p}_{\Delta,t}^{(k)} z^k \Big] z^n = \boldsymbol{\Gamma}(\Delta, \Delta, z, t);$$
$$\tag{16}$$

Therefore, solving (16) for $\pi_{0,t}^*(z)$ we have

$$\pi_{0,t}^*(z) =$$
$$= \frac{\mathbf{F}_{0,\Delta}^*(z)[\mathbf{I} - \mathbf{F}_{\Delta,\Delta}^*(z)]^{-1}\boldsymbol{\Gamma}(\Delta, \Delta, z, t) + \Upsilon(0, \Delta, z, t)}{1 - \{\mathbf{F}_{0,\Delta}^*(z)[\mathbf{I} - \mathbf{F}_{\Delta,\Delta}^*(z)]^{-1}\mathbf{P}_{nab(\Delta,\Delta)}(M) + P_{nab(0,\Delta)}(M)\}z^M} \tag{17}$$

Secondly, from (12)-(13) and (6) the paging probabilities, (14) for $i = 0$, can be written as

$$\pi_{0,t} = \sum_{k=0}^{\infty} \pi_{0,t}^{(k)} \alpha(k) = \pi_{0,t}^{(0)} \alpha(0) + \sum_{k=1}^{\infty} \pi_{0,t}^{(k)} \alpha(k) =$$

$$= \frac{1}{2\pi i} \int_{c-i\infty}^{c+i\infty} \left[\frac{[f_m^*(s) - f_{mr}^*(s)]}{s f_m^*(s)} p_{0,t}^{(0)} + \frac{f_{mr}^*(s)[1 - f_m^*(s)]}{s f_m^*(s)} \pi_{0,t}^*(f_m^*(s)) \right] f_c^*(-s) ds$$

$$= -\sum_{p \in \sigma_c} Res_{s=p} \left[\frac{[f_m^*(s) - f_{mr}^*(s)]}{s f_m^*(s)} p_{0,t}^{(0)} + \frac{f_{mr}^*(s)[1 - f_m^*(s)]}{s f_m^*(s)} \pi_{0,t}^*(f_m^*(s)) \right] f_c^*(-s)$$

$$(18)$$

As in (11), Cauchy's Residue Theorem is used to reach the last equality in (18).

4.2   Paging Procedure

We have evaluated the PG cost for the conventional blanket or single step paging, $\eta = 1$ and for two selective paging schemes with delays $\eta = 2$ and $\eta = 3$. In the illustrative example we have used a single absorbing state for the Δ set. Also, we have implemented the line-paging philosophy suggested in [9]. To that end we have defined three sets of disjoint cells, A, B and C, that is, $A \cap B = \varnothing$, $A \cap C = \varnothing$, $B \cap C = \varnothing$, and $A \cup B \cup C = \Omega$, being $\varnothing$ the empty set and Ω the full set of cells where the MT can be in. Denoting by $C\{X\}$ the cardinality of X, we have chosen

$$
\begin{aligned}
A &= \{(0,0), (1,1), ... \quad , (\min\{\Delta + M - 1, 2\Delta\}, \min\{\Delta + M - 1, 2\Delta\})\}, \\
&\quad C\{A\} = \min(\Delta + M - 1, 2\Delta) + 1; \\
B &= \{\text{The first ring of cells around the set A}\}, \\
&\quad C\{B\} = \begin{cases} 2C\{A\} + 4; \text{if } \Delta < M - 1 \\ 2C\{A\} + 1; \text{if } M - 1 \le \Delta \le M \end{cases}, \\
C &= \{\text{The remaining set of cells: } \Omega \text{ - } A \cup B\}, \\
&\quad C\{C\} = C\{\Omega\} - C\{A \cup B\} = C\{\Omega\} - C\{A\} - C\{B\}
\end{aligned}
\tag{19}
$$

and $C\{\Omega\} = 3(M-1)^2 + 3(M-1) + 1 + \Delta(2M - 1)$.

5   Numerical Analysis

The following assumptions have been established. A Gamma distributed for the cell residence time, with pdf and LT given by

$$f_m(t) = \frac{(\gamma \lambda_m)^\gamma t^{\gamma - 1}}{\Gamma(\gamma)} e^{-\gamma \lambda_m t}, \quad f_m^*(s) = \left(\frac{\gamma \lambda_m}{s + \gamma \lambda_m} \right)^\gamma \tag{20}$$

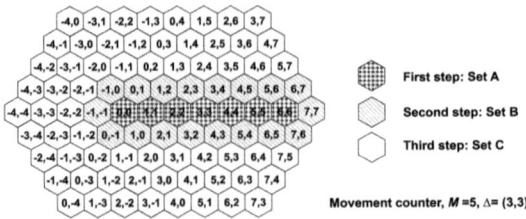

Fig. 2. Paging schemes for $M = 5$ and $\Delta = \{(3,3)\}$. Cells are grouped into three sets and they are sorted for paging procedure. Blanket polling: set $\Omega = A \cup B \cup C$. Two step paging: first A, secondly $B \cup C$. Three step paging: first A, secondly B, thirdly C.

with mean value $1/\lambda_m$ and with the squared coefficient of variation $C_m^2 = 1/\gamma$, [8]. For the call inter-arrival time, on one hand, the Erlangian-Geo-type $f_{c,e}(t)$, with LT

$$f_{c,e}^*(s) = \prod_{i=0}^{n-1} \frac{\lambda_i}{s + \lambda_i}, \quad \lambda_i = \frac{\lambda_c}{a^i} \frac{1 - a^n}{1 - a} \tag{21}$$

with mean value $1/\lambda_c$ and $C_{c,e}^2 = \dfrac{(1 + a^n)(1 - a)}{(1 - a^n)(1 + a)} \leq 1$ for $0 < a < 1$ and n any natural number. On the other hand, the hyper-exponential distribution H_2, $f_{c,h}(t)$, with LT

$$f_{c,h}^*(s) = a\frac{\lambda_1}{s + \lambda_1} + (1 - a)\frac{\lambda_2}{s + \lambda_2}, \quad \lambda_2 = n\lambda_1, \quad \frac{1}{\lambda_c} = \frac{a}{\lambda_1} + \frac{(1 - a)}{n\lambda_1} \tag{22}$$

with mean value equal to $1/\lambda_c$ and $C_{c,h}^2 = \dfrac{1 + 2an(n - 1) - a^2(n - 1)^2}{1 + 2a(n - 1) + a^2(n - 1)^2} \geq 1$ for $0 < a < 1$, equivalently $\frac{1}{n} \leq \frac{\lambda_1}{\lambda_c} \leq 1$, and n any natural number. It is easy to verify that the maximum C_h^2 is achieved for $a = 1/(n + 1)$.

All plots have been obtained for a Call-to-Mobility ratio (CMR) equals to $\lambda_c/\lambda_m = 0.10$. From Fig. 3 we observe that for $C_m^2 \leq 1$ the mean number of LU messages, $\overline{\#_{LU}}$ in (5), reduces when C_c^2 increases. However, when $C_m^2 > 1$ the mean number of LU messages, $\overline{\#_{LU}}$ is rather insensitive to C_c^2.

A significant saving is achieved when selective paging is implemented as it can be seen in Fig. 4 and Fig. 5. The saving is quite noticeable from blanket polling (not shown in the paper, since it is quite trivial) to the two step PG or polling scheme. Also a noticeable saving is obtained from two step polling to three step polling. As additional observation we realize that, for a fixed number of PG steps, and for a fixed C_m^2, the paging cost is no sensitive to the inter-arrival process. Also, for a fixed number of paging steps, for a fixed value of C_c^2 and for $C_m^2 \leq 1$ ($C_m^2 > 1$) the PG cost is rather insensitive (quite sensitive) to the inter-arrival time distribution.

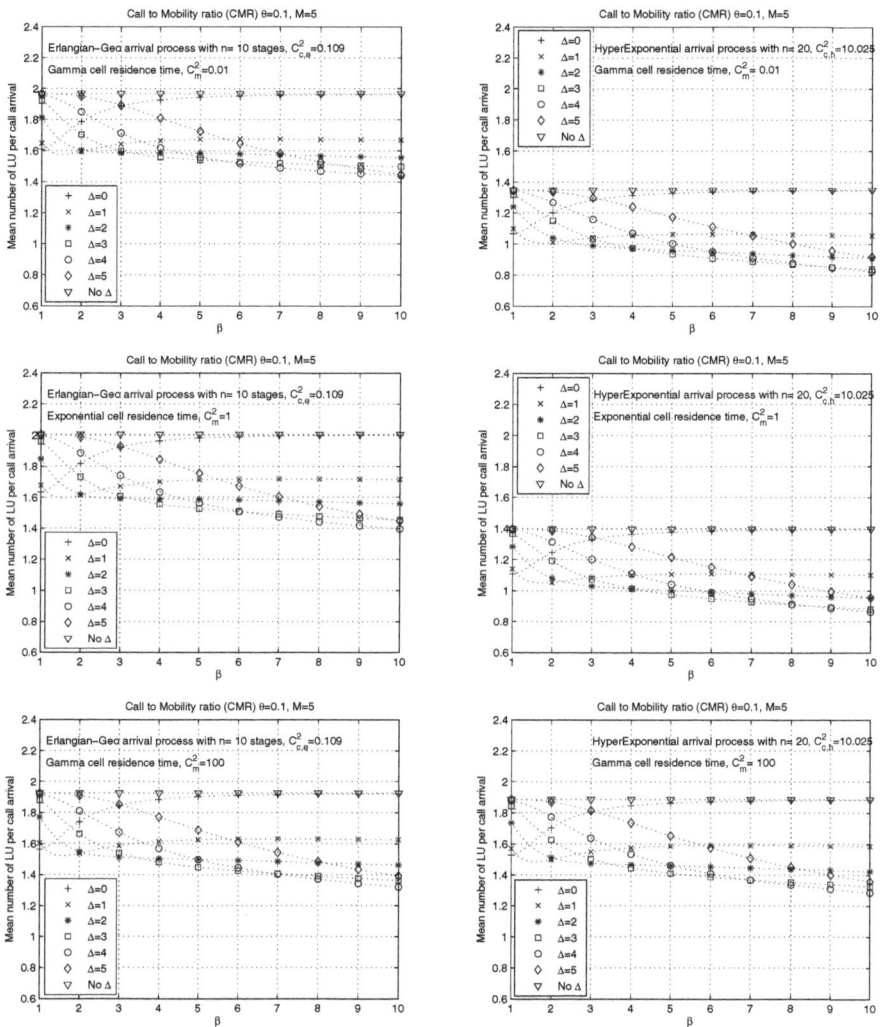

Fig. 3. Average number of LU messages between two consecutive call arrivals in terms of the mobility model parameter β with the target Δ cell as varying parameter ($\Delta = i$ means the target cell (i, i) in Fig. 2). Inter-arrival call process: Erlangian-Geo with n= 10 stages and $C^2_{c,e} = 0.109$ (low-variance) on the left and hyper-exponential with n=20 and $C^2_{c,h} = 10.025$ (high-variance) on the right. Cell residence time, from top to bottom: Gamma distribution with $C^2_m = 0.01$ (low-variance) , $C^2_m = 1$ (exponential) and $C^2_m = 100$ (hig-variance), respectively.

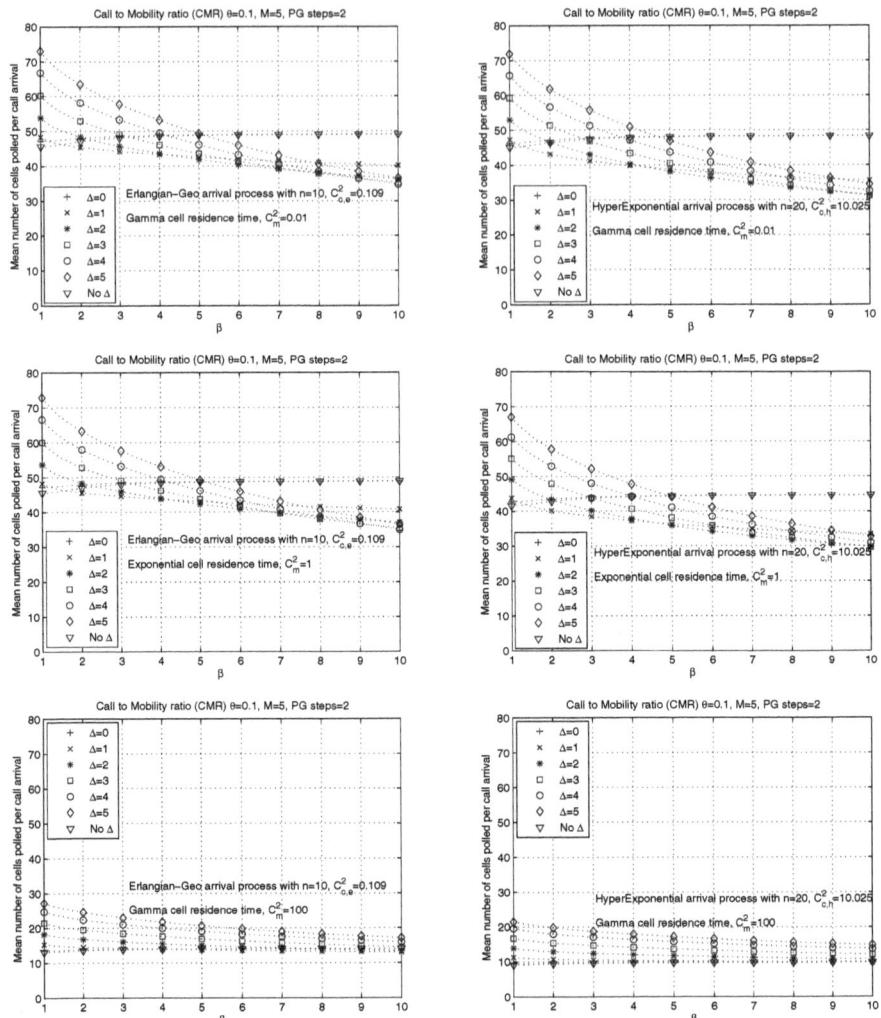

Fig. 4. Two steps paging. Average number of cells polled per call arrival in terms of the mobility model parameter β with the target Δ cell as varying parameter ($\Delta = i$ means the target cell (i, i) in Fig. 2). Inter-arrival call process: Erlangian-Geo with n= 10 stages and $C_{c,e}^2 = 0.109$ (low-variance) on the left and hyper-exponential with n=20 and $C_{c,h}^2 = 10.025$ (high-variance) on the right. Cell residence time, from top to bottom: Gamma distribution with $C_m^2 = 0.01$ (low-variance) , $C_m^2 = 1$ (exponential) and $C_m^2 = 100$ (hig-variance), respectively.

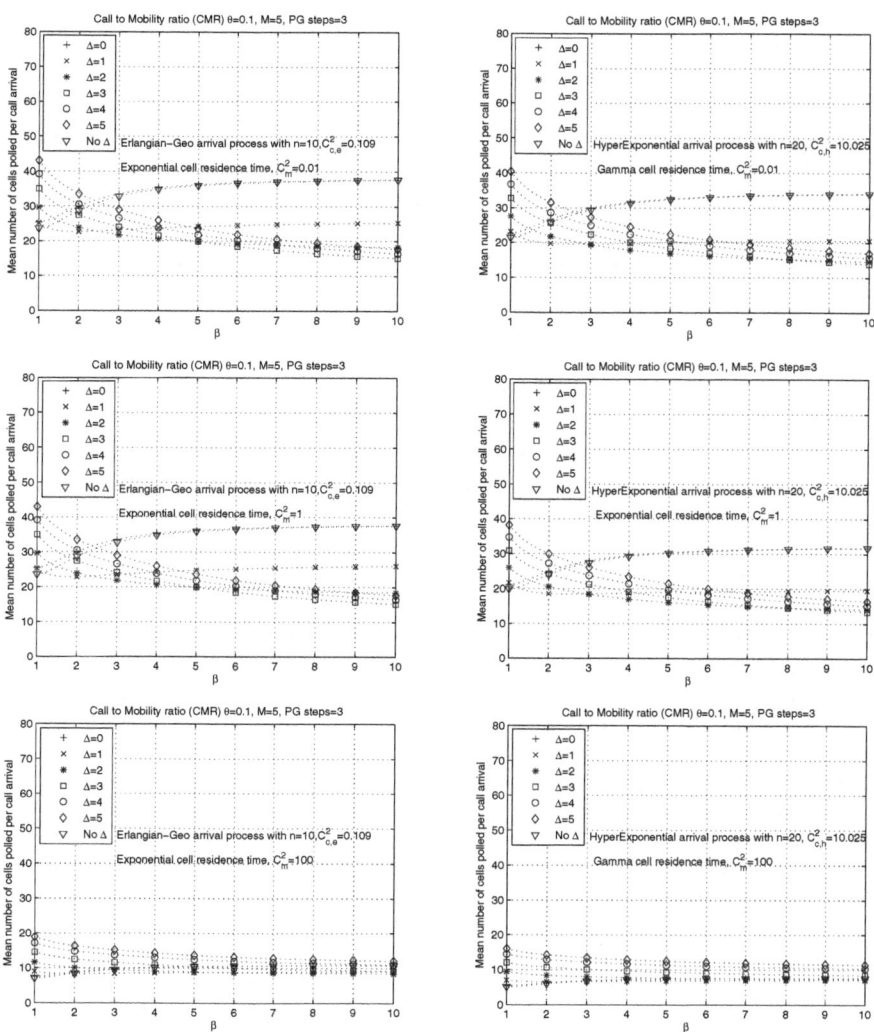

Fig. 5. Three steps paging. Average number of cells polled per call arrival in terms of the mobility model parameter β with the target Δ cell as varying parameter ($\Delta = i$ means the target cell (i, i) in Fig. 2). Inter-arrival call process: Erlangian-Geo with n= 10 stages and $C_{c,e}^2 = 0.109$ (low-variance) on the left and hyper-exponential with n=20 and $C_{c,h}^2 = 10.025$ (high-variance) on the right. Cell residence time, from top to bottom: Gamma distribution with $C_m^2 = 0.01$ (low-variance) , $C_m^2 = 1$ (exponential) and $C_m^2 = 100$ (hig-variance), respectively.

6   Conclusions

In this paper we have proposed and studied within a general framework, a lookahead location update procedure with selective paging. We have seen that the information about the directionality of the mobile terminal's movement can be used to save signaling cost associated to movement-based location management procedures. With a lookahead scheme the size of the registration area increases and a selective paging scheme is recommended to avoid a noticeable increment in the paging cost. The net effect is a significant saving in the total signaling traffic on the common air interface. We believe that this scheme could be used by mobile operators where mobile terminals follow certain daily or customary trajectory.

Acknowledgments. This research has been financed by the Spanish Ministerio de Ciencia e Innovación project TSI2007-66869-C02-02 and by the Universidad Politécnica de Valencia under the grant PAID-05-08. Also, thanks for the support received from the European Network of Excellence Euro NF and to its research community for many scientific and fruitful interactions with its partners.

References

1. Camp, T., Boleng, T., Davies, J.: A survey of mobility models for ad hoc network research. Wireless Communications and Mobile Computing 2(5), 483–502 (2002)
2. Bettstetter, C., Hartenstein, H., Perez-Costa, X.: Stochastic properties of the random waypoint mobility model. Wireless Networks 10(5), 555–567 (2004)
3. Martinez-Arrue, I., Garcia-Escalle, P., Casares-Giner, V.: Location Management Based on the Mobility Patterns of Mobile Users. In: Cerdà-Alabern, L. (ed.) EuroNGI/EuroFGI 2008. LNCS, vol. 5122, pp. 185–200. Springer, Heidelberg (2008)
4. Escalle-García, P., Casares-Giner, V., Mataix-Oltra, J.: Reducing location update and paging costs in a Personal Communications Services Network. IEEE Trans. on Wireless Communications 1(2), 200–209 (2002)
5. Casares-Giner, V., Escalle-García, P.: On Movement-Based Location Update. A Lookahead Strategy. In: Proceedings of the 5th Euro-NGI Conference on Next Generation Internet Networks, pp. 213–220. IEEExplore (2009)
6. Karlin, S., Taylor, H.M.: A first course in stochastic processes, 2nd edn. Academic Press (1975)
7. Casares-Giner, V.: On lookahead strategy for movement-based location update. A general formulation. Internal Report Universidad Politecnica de Valencia (2010)
8. Kleinrock, L.: Queuing Theory. Theory, vol. 1. John Wiley (1975)
9. Hwang, H.W., Chang, M.F., Tseng, C.C.: A direction-based location update scheme with a line-paging strategy for PCS networks. IEEE Communications Letters 4(5), 149–151 (2000)

Bayesian Estimation of Network-Wide Mean Failure Probability in 3G Cellular Networks

Angelo Coluccia[1], Fabio Ricciato[1,2], and Peter Romirer-Maierhofer[2]

[1] University of Salento, Lecce, Italy
[2] FTW Forschungszentrum Telekommunikation Wien, Vienna, Austria
{ricciato,coluccia,romirer}@ftw.at

Abstract. Mobile users in cellular networks produce calls, initiate connections and send packets. Such events have a binary outcome — success or failure. The term "failure" is used here in a broad sense: it can take different meanings depending on the type of event, from packet loss or late delivery to call rejection. The Mean Failure Probability (MFP) provides a simple summary indicator of network-wide performance — i.e., a Key Performance Indicator (KPI) — that is an important input for the network operation process. However, the robust estimation of the MFP is not trivial. The most common approach is to take the ratio of the total number of failures to the total number of requests. Such simplistic approach suffers from the presence of heavy-users, and therefore does not work well when the distribution of traffic (i.e., requests) across users is heavy-tailed — a typical case in real networks. This motivates the exploration of more robust methods for MFP estimation. In a previous work [1] we derived a simple but robust sub-optimal estimator, called EPWR, based on the weighted average of individual (per-user) failure probabilities. In this follow-up work we tackle the problem from a different angle and formalize the problem following a Bayesian approach, deriving two variants of non-parametric optimal estimators. We apply these estimators to a real dataset collected from a real 3G network. Our results confirm the goodness of the proposed estimators and show that EPWR, despite its simplicity, yields near-optimum performance.

1 Introduction

The users of a third-generation (3G) mobile network generate various types of activity. Consider the following events: transmission of IP packets, opening of Transport- and Application-layer connections (i.e., envoy of TCP SYN packets and HTTP GET commands), activation of phone call, SMS envoy, data connection and signaling procedures (Attach Request, Location Area Update, Authentication Request, Paging etc.). All such events have in common two characteristics. First, each event is naturally *associated with an individual user* in the mobile network: the caller (for outgoing calls) or the callee (for incoming ones), the sender (for uplink packets) or the receiver (for downlink packets). Second, each event has a *binary outcome: success or failure*. The term "failure" can take on different meanings depending on the type of event: for packets, failure can represent the missed delivery (e.g. due to queue loss or corruption by link-level errors) or late delivery after a delay threshold.

K.A. Hummel et al. (Eds.): PERFORM 2010 (Haring Festschrift), LNCS 6821, pp. 167–178, 2011.

Generally speaking, failures can be caused by the *unavailability* of some resource — e.g. due to exhaustion by too many concurrent requests — or by its *unreachability*. The involved resources can be shared — e.g., the GGSN or a Core Network link — or dedicated to individual users — e.g. a dedicated radio channel, or the receive buffer inside the terminal itself. The *failure probability* experienced by each individual user will be affected by the availability and reachability of both the shared and dedicated resource components.

Even if the status of shared resources is good, user-specific conditions can severely impact the individual failure probability: for example, a user with poor radio link will experience a high rate of packet losses, while a terminal configured with a wrong Access Point Name (APN, ref. [2]) will have all its PDP-context activations rejected. On the other hand, failures affecting multiple users at the same time often indicate a problem and/or overload in the shared section. Therefore, monitoring the incidence of failures across users is of great importance for the operation and troubleshooting of the network infrastructure.

Network equipments typically maintain logs and/or counters of attempts and failures, from which synthetic indicators are derived — often called Key Performance Indicators (KPI). Failures (and successes) can be also measured by a passive monitor either from the observation of an explicit failure indication (e.g. a Negative ACK or reject message) or by the absence of an explicit success indication (e.g. positive ACK). One of the most common KPI is trivially the ratio between the total number of failures and the total number of attempts (or requests). Here we show that such simplistic approach has some fundamental limitations and does not work well in scenarios of practical interest. The problem arises when the individual rate of requests (calls, procedures, packets) varies wildly across users — a case that is typically encountered in real networks when the distribution of user activity is often heavy-tailed. In this work we formulate the problem in terms of Bayesian estimation and provide (two variants of) an optimal estimator. Furthermore, we compare its performance to a simpler estimator developed in a previous work [1], discussing the trade-offs between optimality and simplicity.

2   Problem Formulation

2.1   System Model

To preserve generality, we will adopt the term "REQUEST" to refer to a general resource access attempt — to avoid specific terms like packet, call, connection. Unsuccessful requests are referred to as "FAILURES". Each request is associated with one mobile "USER".

For a generic measurement timebin (e.g. 1 minute), let I denote the total number of *active users* for which at least one request was observed. In operational networks I is often quite large, from thousands to millions depending on the timebin duration. For every user i ($i = 1 \ldots I$) we introduce the following variables:

- n_i is the total number of requests associated to i ($n_i \geq 1$);
- m_i is the number of failures ($0 \leq m_i \leq n_i$);
- $r_i \overset{\text{def}}{=} \frac{m_i}{n_i}$ denotes the empirical failure ratio for i;
- a_i the (unknown) failure probability for i ($a_i \in [0, 1]$).

We denote by $N \stackrel{\text{def}}{=} \sum_i n_i$ and $M \stackrel{\text{def}}{=} \sum_i m_i$ respectively the total number of requests and failures across all terminals. To simplify the notation we will occasionally use the vectors $\boldsymbol{n} \stackrel{\text{def}}{=} [n_1 n_2 \cdots n_I]^T$, $\boldsymbol{m} \stackrel{\text{def}}{=} [m_1 m_2 \cdots m_I]^T$ and $\boldsymbol{a} \stackrel{\text{def}}{=} [a_1 a_2 \cdots a_I]^T$. For the sake of mathematical tractability we assume independence between failures: each request for user i fails with probability a_i independently from any other request of the same or other users. Therefore m_i is the sum of n_i Bernoulli trials with probability a_i, and all m_i's are independent Binomial random variables:

$$m_i \sim \mathcal{B}(n_i, a_i) \Rightarrow \begin{cases} \mathrm{E}[m_i] = n_i a_i \\ \mathrm{VAR}[m_i] = n_i a_i (1 - a_i). \end{cases} \tag{1}$$

Throughout the paper we assume that $\boldsymbol{m}$ and $\boldsymbol{n}$ have been measured in some way and serve as input for the problem at hand.

2.2 Goal Definition

A central component of the model is that the (unknown) failure probabilities a_i's are regarded as i.i.d. random variables generated from a common underlying distribution $p(a)$ with mean value $\bar{a} \stackrel{\text{def}}{=} \mathrm{E}[a]$. Therefore it is natural to take $\bar{a}$ as the summary indicator representative of "network-wide failure probability". In other words, we are interested in obtaining an estimator of the Mean Failure Probability $\bar{a}$ from the measured vectors $\boldsymbol{n}$ and $\boldsymbol{m}$.

The goodness of such an estimator can be evaluated against the following criteria:

– **Optimality.** We consider only unbiased estimators and take minimum variance as the optimality criterion. In fact, when the estimate of $\bar{a}$ is used for performance monitoring, lower variance (i.e., smaller statistical fluctuations) allows better discrimination of change-points and/or trends.
– **Generality.** In passive monitoring the vector $\boldsymbol{n}$ is given and can not be controlled. Typically, the traffic volume is distributed very unevenly across users, often with long-tails. Moreover, the traffic distribution change across time, following daily or weekly cycles and long-term trends in user activity (see e.g. [4, §VI-A]). Therefore we seek a robust estimator that does not rely on specific assumptions about (the distribution of) $\boldsymbol{n}$ nor requires manual re-tuning when the distribution changes.
– **Simplicity.** The ideal estimator should be easy to implement, fast to compute and conceptually simple — as a matter of fact, methods which can be understood straightforwardly by practitioners are more likely to be adopted in practice.

2.3 Resolution Approach

In a previous contribution [1] we focused on a particular class of weighted estimators and casted the problem in terms of constrained optimization. We derived a very simple sub-optimal estimator, hereafter called Empirical Piecewise-linear Weighted Ratio (EPWR), which showed excellent performance in simulations. The EPWR involves a cut-off parameter θ to be set heuristically — we showed that it is not too sensitive to the exact value of θ as far as extreme settings (very small or very large) are avoided.

In this work we tackle the problem from a more theoretically-grounded perspective: following a formal Bayesian approach, we identify two estimators that are well suited for our purposes. Moreover, we compare the performance of the Bayesian estimators against the EPWR estimator derived in [1] plus two other simplistic estimators. The comparison is carried out on a sample dataset from a real operational network, based on data obtained with the METAWIN system [7].

3   Simple Estimators

Hereafter we present two common estimators and highlight their limitations. In §3.3 we recall the EPWR estimator which was derived earlier in [1].

3.1   Empirical Global Ratio (EGR)

The Key Performance Indicator (KPI) most widely adopted by practitioners is simply the ratio between the total number of failures and the total number of requests across all users, formally:

$$S_{\text{EGR}} \stackrel{\text{def}}{=} \frac{\sum_{i=1}^{I} m_i}{\sum_{i=1}^{I} n_i} = \frac{M}{N}. \tag{2}$$

We refer to such quantity as the *Empirical Global Ratio* (EGR). It can be seen that S_{EGR} is indeed an unbiased estimator for the mean failure probability:

$$\text{E}[S_{\text{EGR}}] = \frac{\sum_{i=1}^{I} \text{E}[\text{E}[m_i|a_i]]}{\sum_{i=1}^{I} n_i} = \frac{\sum_{i=1}^{I} n_i \text{E}[a_i]}{\sum_{i=1}^{I} n_i} = \bar{a}$$

It is worth remarking that S_{EGR} is the simplest estimator to implement in practice. In fact, it does not require knowledge of the full vectors n, m (each composed of I elements) but only of two global counters N and M. Therefore, it does not require the measurement platform to associate requests and failures to individual users.

Intuitively, the problem with S_{EGR} is the presence of few users with very high traffic (large n_i) that occasionally inflate the value of eq. (2), thus increasing the estimator variance. One possible approach to "correct" S_{EGR} is to pick users with large n_i as "outliers" and filter them out before computing the ratio in eq. (2). Such strategy has two drawbacks: for one, it requires an heuristically-tuned method to classify outliers. Second, filtering out "fat" users with large n_i means discarding their sample estimates for $\bar{a}$ — actually the most reliable ones — and for long-tailed n_i's the loss of information might not be negligible. In other words, the variance increase due to a lower number of reliable measurement samples will partially offset the reduction due to filtering.

3.2   Empirical Mean Ratio (EMR)

Note that the empirical loss ratio r_i for each user i is an unbiased estimator of the mean failure probability $\bar{a}$, formally $\text{E}[r_i] = \text{E}[\text{E}[r_i|a_i]] = \text{E}[a_i] = \bar{a}$. Therefore an intuitive summary indicator can be obtained as the arithmetic mean across all users:

$$S_{\text{EMR}} \overset{\text{def}}{=} \frac{1}{I} \sum_{i=1}^{I} r_i = \frac{1}{I} \sum_{i=1}^{I} \frac{m_i}{n_i}. \tag{3}$$

We refer to such indicator as the *Empirical Mean Ratio* (EMR). From a system-level perspective, the implementation of S_{EMR} is more costly than S_{EGR}: in fact, it requires the measurement platform to count requests and failures separately for each user. The additional resource consumption (memory, processing) should be compensated by better accuracy of the estimation. However, S_{EMR} is a sub-optimal estimator for $\bar{a}$. The problem lies in the fact that the variance of r_i (conditioned to a_i) is inversely proportional to the number of packets n_i, i.e. $\text{VAR}[r_i|a_i] = a_i(1 - a_i)/n_i$: intuitively, the larger the sample size, the better the accuracy of the estimate. Therefore, large variability of n_i's maps to large variability of the uncertainty (variance) of the individual estimates — a case of *heteroscedasticity*. In the simple arithmetic mean as in eq. (3), more accurate estimates (for large n_i) weight the same as poor ones (for small n_i), and if the number of low-traffic users is high they will drive up the overall variance.

Similarly to S_{EGR}, one possible "correction" for S_{EMR} is to filter out (discard) the observations associated to low-traffic users and compute the ratio in eq. (3) only on the samples above a minimum sample size $n_i \geq \gamma$. Again, the drawback of such strategy is that a certain amount of information available in the data is simply discarded, which is a grossly suboptimal approach to the problem.

3.3 Empirical Piecewise-Linear Weighted Ratio (EPWR)

We have seen that both S_{EMR} and S_{EGR} suffer respectively from small and large users — the user size refers to its traffic volume n_i — and that a simplistic workaround would be to just discard the measurements associated with the smallest or biggest users for S_{EMR} and S_{EGR} respectively. This approach involves a certain loss of information. Moreover, the decimation of samples works *against* the goal of reducing the uncertainty of the estimate. Therefore we are set to find a more clever strategy: instead of selectively *discarding* measurement samples based on their size, one can simply *weight* them differently. Following this idea, in a previous work [1] we have derived a simple sub-optimal estimator that takes the form of a weighted average of the r_i's:

$$S_{\text{EPWR}} \overset{\text{def}}{=} \sum_{i=1}^{I} w_i r_i \tag{4}$$

with piecewise-linear weigths given by:

$$w_i = \frac{x_i}{\sum_{i=1}^{I} x_i}, \quad x_i = \min(n_i, \theta) \ \forall i \tag{5}$$

Such definition involves a single parameter θ that represents a sort of "cut-off" point dividing the users into two regions: those with $n_i < \theta$ are weighted proportionally to their size n_i, while those with $n_i > \theta$ are weigthed equally — proportionally to θ. The analysis reveals that the optimal value of the cut-off parameter is given by $\dot{\theta} = \frac{\bar{a} - \bar{a} - \sigma_a^2}{\sigma_a^2}$, i.e.

it depends on the first two moments of the distribution of a. Notably the optimal setting does not depend on n, which is an advantage in applications where n varies in time. Unfortunately, in practice the optimal value of $\dot\theta$ cannot be identified since $\bar{a}$ and σ_a^2 are unknown. However it can be shown that the performances of S_{EPWR} depend only weakly on the exact value of θ, provided that it falls in a "reasonable" intermediate range away from extreme values (very small or very large), indicating that an heuristically fixed value for θ would be sufficient to achieve near-optimal performance in most practical cases. This claim, supported by simulation results in [1], will be further confirmed by the numerical results presented below in §5.

4 Bayesian Estimators

Since a is regarded as a random vector, we can apply Bayesian techniques, which provide a theoretically well grounded approach to the estimation problem at the cost of somewhat higher complexity of the resolution procedure. In this section we discuss different possible approaches under the common framework of Bayesian inference.

The structure of our problem is that of a Bayesian hierarchical model [5]: the data m are described by a probability distribution whose parameters a are random variables themselves. The Bayesian approach requires to provide the *a priori* distribution $p(a)$, which is unknown in our case. If $p(a)$ were known, the optimal estimator for a would be obtained by following the classical Bayes' procedure, i.e. minimizing a Bayes risk. Two of the most common choices for the Bayes risk are the MAP (*Maximum A Posteriori*) criterion and the MMSE (*Minimum Mean Square Error*) criterion. In both cases the key element is the posterior distribution $p(a|m)$, expressed by the Bayes' Theorem as:

$$p(a|m) = \frac{p(m, a)}{p(m)} = \frac{p(m, a)}{\int p(m, a)\,\mathrm{d}a}$$

where $p(m, a) = p(m|a)p(a)$.

When $p(a)$ is unknown, as in our case, it is possible to choose a parametric distribution family with unknown parameters for the prior, and to resort to a procedure called *Empirical Parametric Bayes* [5]. The basic idea is quite simple: since the parameters of the prior distribution — called *hyperparameters* — are unknown, their estimates are used instead. It is then possible to derive a Bayesian estimator (MAP or MMSE) in the usual way. The estimation of the hyperparameters is preferably obtained via Maximum Likelihood (ML), although other techniques are sometimes used — e.g. Method of Moments [5].

In some applications it is not clear which family of distributions should be adopted. In such cases, a well-established practice is to use the so-called *conjugate prior*, i.e. the prior distribution $p(a)$ that results in a posterior distribution $p(a|m)$ belonging to the same family. In our case, the conjugate prior corresponding to the Binomial distribution is the Beta distribution, with support in $[0, 1]$, defined as:

$$p(a_i) = \frac{1}{\mathrm{B}(\alpha, \beta)} a_i^{\alpha-1}(1 - a_i)^{\beta-1} \qquad i = 1, \dots, I \tag{6}$$

where

$$B(x, y) = \int_0^1 t^{x-1}(1 - t)^{y-1} \, dt \tag{7}$$

is the Beta function [10]. The Beta distribution is quite general and flexible: it has two positive shape parameters α, β which allow for a great variety of shapes, ranging from uniform to U-shape, convex or concave, symmetric or skewed. Moreover, consider that Bayesian estimation is usually found to be robust against deviations from the ideal choice of the prior distribution family. The mean of a Beta distribution is given by

$$X \sim \text{Beta}(\alpha, \beta) \Rightarrow E[X] = \frac{\alpha}{\alpha + \beta} \tag{8}$$

The ML estimates of the hyperparameters are given by:

$$\hat{\alpha}, \hat{\beta} \overset{\text{def}}{=} \arg\max_{\alpha,\beta} p(\boldsymbol{m}|\alpha, \beta) \tag{9}$$

where the marginal distribution $p(\boldsymbol{m}|\alpha, \beta)$ is obtained by marginalization:

$$p(\boldsymbol{m}|\alpha, \beta) = \int_0^1 \cdots \int_0^1 p(\boldsymbol{m}, \boldsymbol{a}|\alpha, \beta) \, da_1 \cdots da_I. \tag{10}$$

Due to independence, we can factorize the joint density function as follows:

$$p(\boldsymbol{m}, \boldsymbol{a}|\alpha, \beta) = \prod_{i=1}^I p(m_i, a_i|\alpha, \beta) = \prod_{i=1}^I p(m_i|a_i, \alpha, \beta)p(a_i)$$

$$= \prod_{i=1}^I \binom{n_i}{m_i} a_i^{m_i}(1 - a_i)^{n_i - m_i} \cdot \frac{1}{B(\alpha, \beta)} a_i^{\alpha-1}(1 - a_i)^{\beta-1}$$

and recalling eq. (7) we can rewrite eq. (10) as follows:

$$p(\boldsymbol{m}|\alpha, \beta) = \int_0^1 \cdots \int_0^1 \prod_{i=1}^I p(m_i, a_i|\alpha, \beta) \, da_i = \prod_{i=1}^I \binom{n_i}{m_i} \frac{B(\alpha + m_i, \beta + n_i - m_i)}{B(\alpha, \beta)} \tag{11}$$

i.e., the marginal distribution of $\boldsymbol{m}$ is the product of I Beta-Binomial distributions $\text{BetaBin}(n_i, \alpha, \beta)$. This expression can be used in eq. (9) to obtain the ML estimates of the hyperparameters. The solution cannot be expressed in closed-form and must be obtained numerically. Notably, the function (11) is unimodal (see [13] for a proof) and its negative logarithm is convex — a consequence of the log-concavity of the likelihood function for Beta [12] [8] — therefore standard numerical methods can be applied (e.g. the simplex method) which are likely to locate the optimum quickly and accurately.

Given the estimates of the hyperparameters, the posterior distribution

$$p(a_i|m_i, \hat{\alpha}, \hat{\beta}) = \frac{p(m_i, a_i|\hat{\alpha}, \hat{\beta})}{p(m_i|\hat{\alpha}, \hat{\beta})}$$

is obtained in a similar way, and as expected is a Beta distribution (conjugate prior):

$$a_i | m_i, \hat{\alpha}, \hat{\beta} \sim \text{Beta}(\hat{\alpha} + m_i, \hat{\beta} + n_i - m_i) \tag{12}$$

The (empirical) Bayes estimator for the generic a_i is then derived in the classical way by minimizing a Bayes risk. Two of the most common criteria are MAP and MMSE.

The MAP estimator maximizes the *a posteriori* probability:

$$\hat{a}_i^{\text{MAP}} \overset{\text{def}}{=} \arg\max_{a_i} p(a_i | m_i, \hat{\alpha}, \hat{\beta}) = \frac{\hat{\alpha} + m_i - 1}{\hat{\alpha} + \hat{\beta} + n_i - 2} \tag{13}$$

for $i = 1 \ldots I$, as can be easily verified by taking the derivatives of $\log p(a_i | m_i, \hat{\alpha}, \hat{\beta})$ (ref. eq. (6) and (12)). From the denominator of eq. (13) we observe that the MAP estimator may lead to inconsistent values for small n_i. This is a problem in network applications, where typically a considerable fraction of users generate only very few requests. Therefore the MAP estimator is not well suited for our purposes.

The MMSE estimator minimizes the mean squared error, and coincides with the conditional mean [5]. Recalling eq. (8) we can write:

$$\hat{a}_i^{\text{MMSE}} \overset{\text{def}}{=} \text{E}\left[a_i \Big| m_i, \hat{\alpha}, \hat{\beta} \right] = \frac{\hat{\alpha} + m_i}{\hat{\alpha} + \hat{\beta} + n_i}$$

for $i = 1 \ldots I$, and by taking the arithmetic mean we obtain the following estimator:

$$Q_{\text{MMSE}} \overset{\text{def}}{=} \frac{1}{I} \sum_{i=1}^{I} \hat{a}_i^{\text{MMSE}} = \frac{1}{I} \sum_{i=1}^{I} \frac{\hat{\alpha} + m_i}{\hat{\alpha} + \hat{\beta} + n_i}. \tag{14}$$

Otherwise, an alternative approach is to estimate directly $\bar{a}$ from eq. (8) with the ML estimates of the hyperparameters:

$$Q_{\text{HYP}} \overset{\text{def}}{=} \frac{\hat{\alpha}}{\hat{\alpha} + \hat{\beta}} \tag{15}$$

where the subscript "HYP" indicates that the estimator is obtained solely from the hyperparameters' estimates, without the posterior information m.

Note that in both cases the $\hat{\alpha}, \hat{\beta}$ are estimated from the data vector m, n: the difference between the two estimators lies in the fact that Q_{MMSE} is suboptimal — because it heuristically adopts an arithmetic mean — but uses the posterior information m for both parameters and hyperparameters, while Q_{HYP} is optimal in the ML sense but uses the posterior information only for estimating the hyperparameters and not $\bar{a}$. Despite such difference, we found that Q_{MMSE} and Q_{HYP} always lead to extremely similar values when applied to our real datasets, as discussed later in §5.

Finally we remark that both estimators require a numerical procedure to be computed. However the shape of the objective function (unimodal, convex) allows the use of fast and accurate numerical methods — for the sake of space we do not provide here further details of the implemented numerical procedure. Nonetheless, the computational gap with S_{EPWR} remains large: in a MATLAB simulation with $I = 10^5$ users, the computation time of Q_{MMSE} and Q_{HYP} is about 10 seconds on a standard computer (Core2 Duo 2Ghz), against a few milliseconds of S_{EPWR}.

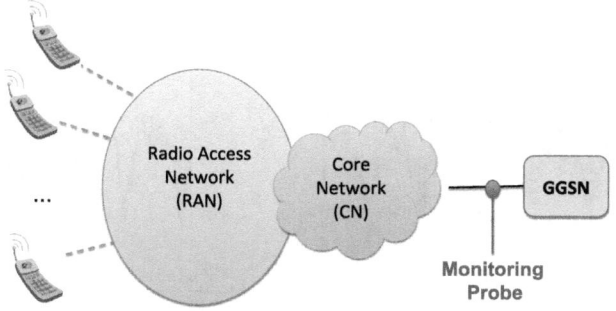

Fig. 1. Reference measurement scenario

5 Numerical Results from Real Datasets

In the previous sections we have presented three sub-optimal estimators, very simple to implement and to compute — S_{EGR}, S_{EMR} and S_{EPWR} — and derived two Bayesian estimators — Q_{HYP} and Q_{MMSE} — which are more complex but optimal given the problem model at hand. Hereafter we compare their performance on a real dataset.

Our dataset is based on measurements collected from an operational 3G cellular network by the METAWIN system [7]. The measurement setting is sketched in Fig. 1: a passive monitor located on the Gn interface near the GGSN observes the TCP traffic in both directions (for more details on 3GPP network architecture refer e.g. to [2]). We aim at revealing congestion and/or other performance glitches in the network section between the monitoring point and the mobile terminals from the observation of TCP handshaking packets between mobile clients and Internet-side servers. To this purpose we collect two datasets: DATA:INV and DATA:RTT.

In DATA:INV we count for each mobile station i all SYNACK packets flowing in downlink (variable n_i) as well as the number of them which *failed to be unambiguously associated* to a corresponding uplink ACK (variable m_i). Such definition includes those cases where either the SYNACK or the corresponding ACK were lost in the network section from the monitoring point to the mobile terminal, but also other cases not necessarily related to loss events. For example, when two or more identical SYNACKs are observed but only one ACK (between the same end-points) the SYNACK-to-ACK association remains ambiguous, i.e. it is not possible to decide which one of the SYNACK triggered the ACK[1]. In other words, the SYNACK packet in downlink denotes a "request" event, and the presence [resp. lack] of an *unambiguously corresponding* ACK packet in uplink denotes the "success" [resp. "failure"] event.

A second dataset DATA:RTT was obtained from the (semi-)RTT measurements: for each correctly (and unambiguously) acknowledged SYNACKs, we measure the client-side RTT, i.e. the elapsed time between the timestamps of the SYNACK and the corresponding ACK. For this dataset, we mark a "failure" event when the RTT exceeds a

---

[1] In [3] we showed that such cases are not infrequent in GPRS/UMTS networks due to the presence of "early retransmitter" servers, with initial retransmission timeout set to sub-second values in the same order of the Round Trip Time (RTT) in these networks.

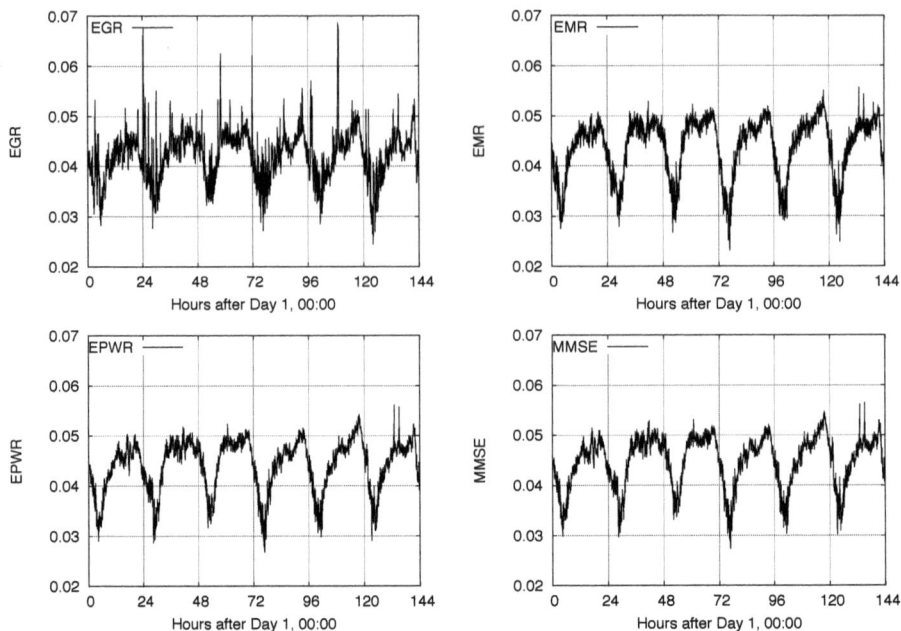

Fig. 2. Estimated mean failure probability for DATA:INV dataset (missing or ambiguous SYNACK/ACK associations)

fixed threshold T — we set $T = 0.5$ sec and $T = 1.5$ sec respectively for UMTS/HSPA and GPRS/EDGE. For more details about the measurement setting see [3].

The rationale for extracting such measurements is that congestion in the downlink path towards the terminals (e.g. at some SGSN or RNC) would expectedly result in an increase of downlink packet loss and/or delay, and therefore should be reflected in a increase of the "failure" rate in DATA:RTT and/or DATA:INV. Note that serious connectivity problems in the Radio Access Network might impede even the transmission of uplink SYN packets, which would translate into a reduction of SYNACK — i.e. we would observe *missing* rather than *unacknowledged* SYNACKs. While in principle such kind of events could be revealed by monitoring the absolute number of SYNs and corresponding SYNACKs, this aspect is left outside the scope of this work.

The measurements are binned in 5 minutes intervals. For this work we consider a measurement period of one week collected in August 2010. The analysis was conducted separately for UMTS/HSPA and GPRS/EDGE users, but for the sake of space we report only results for UMTS/HSPA. The time-series computed with different estimators are reported in Fig. 2 and Fig. 3 respectively for DATA:INV and DATA:RTT datasets. We found that Q_{HYP} and Q_{MMSE} always lead to almost identical results, with only negligible differences in a few timebins — for this reason and for the sake of space we report only the time-series of Q_{MMSE} and skip the graphs for Q_{HYP}.

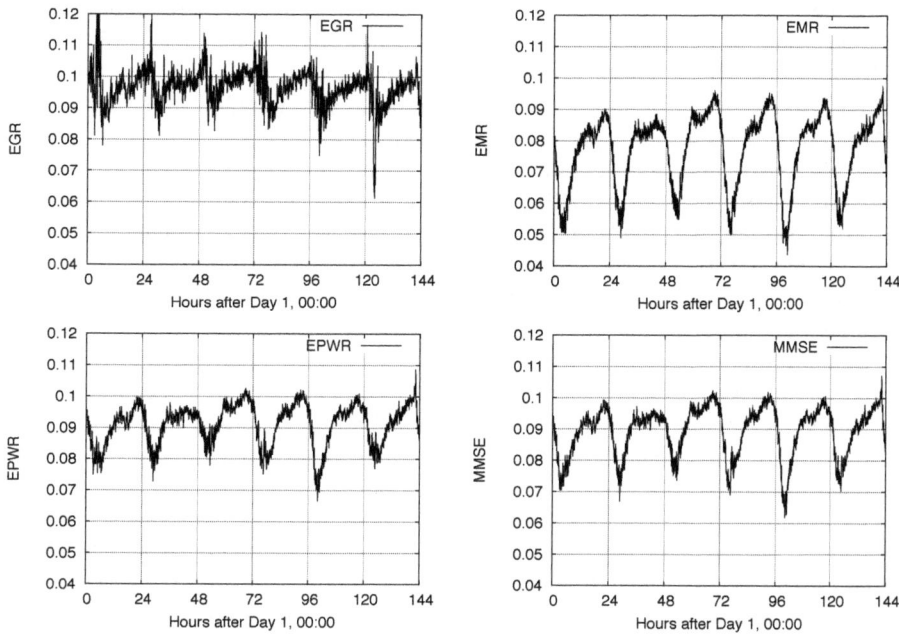

Fig. 3. Estimated mean failure probability for DATA:RTT dataset (unambiguous SYNACK/ACK pairs with semi-RTT exceeding 500 ms)

From Fig. 2 and Fig. 3 we first observe that S_{EGR} exhibits larger fluctuations (higher variance) and occasionally large spikes due to the sporadic presence of heavy users (high n_i) with many failures[2] (high m_i). Second, both S_{EMR} and S_{EPWR} perform quite close to the optimal reference Q_{MMSE}. To dig further, we resorted to the inspection of the scatterplots $\langle S_{EMR}, Q_{MMSE}\rangle$ and $\langle S_{EPWR}, Q_{MMSE}\rangle$ (not shown here) which revealed that S_{EPWR} correlates slightly better than S_{EMR} to the reference Q_{MMSE} values: in DATA:INV the Pearson the correlation coefficient is 0.997 and 0.954 respectively for $\langle S_{EMR}, Q_{MMSE}\rangle$ and $\langle S_{EPWR}, Q_{MMSE}\rangle$, while in DATA:RTT S_{EMR} exhibits a larger bias.

6   Conclusions

So called Key Performance Indicators (KPI) play an important role in the operation of real mobile networks, as they provide a synthetic view of the network-wide status and quality. A large class of network events can be modeled as binary REQUESTS associated to USERS (e.g. origin or destination entity) and having one of two possible outcomes, i.e. SUCCESS or FAILURE. For these, a very popular KPI is the ratio between the total number of failures to the total number of requests, regardless of per-user associations. We have shown that such simplistic KPI — referred to as EGR in this work — suffers from the presence heavy-users. The problem is of practical relevance

---

[2] The "spikes" in DATA:INV are due to mobile terminals receiving very high rate of SYNACKs to which they do not respond: they are likely involved in TCP scanning or SYN flooding.

in real networks, where the distribution of requests across users is often heavy-tailed. In some cases, the variability of EGR makes it useless for any practical exploitation.

To overcome the limitations of EGR, network operators should adopt more robust KPIs based on separate counts of success and failures per individual users. The problem is then how to make the best possible use of such data.

We have introduced a system model that motivates the adoption of the Mean Failure Probability (MFP) as a natural KPI. Since MFP is unknown, it must be estimated from the observed data. We have shown that EGR can be considered an unbiased estimator for MFP, but not the optimal one. In a previous work we had derived a more robust estimator, namely the EPWR, very simple to implement, that involves a free parameter to be tuned heuristically. In this paper we have formalized the problem in terms of Bayesian estimation, deriving two Bayesian estimators which are provably optimal (in two different senses) given the system model. The analysis of two real datasets from an operational 3G mobile network has confirmed that all the proposed estimators are considerably more stable than EGR, and that EPWR performs very closely to the optimal reference provided by the Bayesian solution.

Our estimators can be adopted by network operators and/or equipment vendors as robust KPI. Owing to the generality of the system model, and to the abstract definition of the notions of "request", "failure" and "user" therein, the concepts and estimators proposed in this work can be applied to a wide range of different measurements, in communication networks and other application domains.

References

1. Coluccia, A., Ricciato, F., Romirer-Maierhofer, P.: On Robust Estimation of Network-wide Packet Loss in 3G Cellular Networks. In: 5th IEEE Broadband Wireless Access Workshop (BWA 2009), Honolulu (November 2009)
2. Kaaranen, H., et al.: UMTS Networks — Architecture, Mobility and Services, 2nd edn. Wiley (2005)
3. Romirer-Maierhofer, P., Ricciato, F., D'Alconzo, A., Franzan, R., Karner, W.: Network-Wide Measurements of TCP RTT in 3G. In: Papadopouli, M., Owezarski, P., Pras, A. (eds.) TMA 2009. LNCS, vol. 5537, pp. 17–25. Springer, Heidelberg (2009)
4. D'Alconzo, A., Coluccia, A., Ricciato, F., Romirer-Maierhofer, P.: A Distribution-Based Approach to Anomaly Detection for 3G Mobile Network. In: IEEE GLOBECOM 2009 (2009)
5. Lehmann, E.L., Casella, G.: Theory of Point Estimation. Springer, Heidelberg (1998)
6. Wolter, K.M.: Introduction to Variance Estimation. Springer Series in Statistics (2007)
7. Metawin and Darwin projects, http://userver.ftw.at/~ricciato/darwin/
8. Minka, T.P.: Estimating a Dirichlet distribution, Microsoft Technical Report (2003)
9. Robbins, H.: An Empirical Bayes Approach to Statistics. In: Proc. Third Berkeley Symposium on Mathematical Statistics and Probability, vol. 1, pp. 157–163. Univ. of California Press (1956)
10. Abramowitz, M., Stegun, I.A.: Handbook of Mathematical Functions with Formulas, Graphs, and Mathematical Tables. Dover Publications, New York (1972)
11. Yeredor, A.: The Joint MAP-ML Criterion and its Relation to ML and to Extended Least-Squares. IEEE Trans. on Signal Processing 48(12) (December 2000)
12. George, E.I., Makov, U.E., Smith, A.F.M.: Conjugate Likelihood Distributions. Scandinavian Journal of Statistics 20(2), 147–156 (1993)
13. Kevin, B., Reeds, J.: Compound Multinomial Likelihood Functions are Unimodal: Proof of a Conjecture of I. J. Good. The Annals of Statistics 5(1), 79–87 (1977)

Time Is Perception Is Money – Web Response Times in Mobile Networks with Application to Quality of Experience

Markus Fiedler[1], Patrik Arlos[1], Timothy A. Gonsalves[2],
Anuraag Bhardwaj[1,3], and Hans Nottehed[4]

[1] Blekinge Institute of Technology, Karlskrona, Sweden
{mfi,pal}@bth.se
[2] Indian Institute of Technology Mandi, India
tag@iitmandi.ac.in
[3] Indian Institute of Technology Madras, Chennai, India
anuraagbady@gmail.com
[4] info24, Kista, Sweden
hans.nottehed@info24.se

Abstract. The number of mobile operators providing Internet access to end users is growing. However, irrespective of the access network, we observe a distinct sensitivity of user perception to response and download times, in particular for interactive services on the web. In order to facilitate the choice of the right network for a given task, this paper presents a systematic study of web download time and corresponding throughput as a function of the file size. Based on measurement data from three Swedish mobile operators and a particular strategy of choosing file sizes, we find surprisingly simple, yet sufficiently accurate approximations of download times. These approximations are based on simple-to-measure parameters and provide valuable quantitative insights into the acceleration of HTTP/TCP/IP-based data delivery. The paper discusses the emergence of these approximations and related errors. Furthermore, it correlates the findings with Quality of Experience, thus building bridges between performance, user perception and provisioning issues.

Keywords: Download time, interactive service, web service, throughput, user perception, file size, measurements, Quality of Experience.

1 Introduction

The success of interactive services depends on their responsiveness, as perceived by their users. Increasingly many such services are built from web services. These are using the HyperText Transfer Protocol (HTTP), running on top of the Transmission Control Protocol (TCP) and the Internet Protocol (IP), for exchanging data and configuration information in form of HyperText Markup Language (HTML) or eXtensible Markup Language (XML) files. This implies that the response time of a service is typically dominated by the *download time* of a

K.A. Hummel et al. (Eds.): PERFORM 2010 (Haring Festschrift), LNCS 6821, pp. 179–190, 2011.
© IFIP International Federation for Information Processing 2011

corresponding HTML/XML document. The users find themselves at the end of service chains and have to wait until all the involved sub-services have executed and delivered their results, e.g. via an InforMation eXchange (IMX) [1] that composes the information to be sent to the user.

We also observe that the number of service chains that include one or multiple mobile links is increasing. The simple exchangeability of mobile operators by replacing the SIM card in a USB dongle (or to switch between different dongles) has opened up new possibilities of choosing different operators if download times are perceived too long and/or throughput is felt to be insufficient. As waiting times increase, so does the risk that the user churns, i.e. leaves one operator for another that offers better performance, eventually entailing loss of revenue for the original provider. On the other hand, assigning more resources to a user increases cost for the provider. Thus, the latter needs to find a reasonable balance between user-perceived performance and cost.

This background motivates us to take a closer look at response times of HTTP-based downloads via different (in our case Swedish) mobile operators, and to relate them to user-perceived *Quality of Experience* (QoE) [2, 3]. The typical procedure to benchmark (mobile) networks is as follows: Large files are downloaded, the corresponding *quasi-stationary throughput* is calculated as amount of downloaded data divided by the download time, and the result is reported together with other performance measures such as round-trip times or jitter values [4]. However, quasi-stationarity is preceded by a *transient phase*. TCPs startup behaviour [5, 6] causes the transient throughput to approach the quasi-stationary throughput from below, which points at an acceleration of the download process. While being of minor interest for large files, the transient phase has a pronounced impact on the first few seconds of a file transfer [6], which is just in the order of magnitude of user patience [7] and thus deserves specific attention. The use of the quasi-stationary throughput to estimate download times during the transient phase will lead to over-optimistic and undesirable underestimations of the download time, and should thus be avoided.

This paper presents an original study of the download time and corresponding throughput as functions of the size of the downloaded file, as seen by the end user that does not have any insight into specific parameters and conditions of the mobile network used. The corresponding approximations are simple to parameterise from online measurements on application level and simple to deploy. They provide a useful basis for deciding which operator to choose for particular communication situations, here transfers of files upto some Megabytes). Vice versa, the formulae can be applied by operators and providers to adjust the trade-off between user perception, provisioning and cost, respectively.

The remainder of the paper is structured as follows. Section 2 discusses related work. Section 3 introduces notations and definitions to be used throughout the paper and the setup of the measurements. Section 4 presents measurement results for three different operators, followed by subsequent analysis and classification. Section 5 proposes and evaluates three- and two-parameter approximation formulae for download times as function of the file sizes. Section 6 presents and discusses closed formulae relating QoE with file sizes through the presented approximation models. Section 7 concludes the paper and takes an outlook on future work.

2   Related Work

It has been recognised that users get the more distracted, the longer they have to wait [7, 8]. Obviously, this implies a "negative impact on attitudes toward delay" [9]. In particular, it has been recognised that user perception and rating quickly decrease during the first seconds of waiting [3, 10]. This effect can amongst others be modelled by differential equations [11, 12], where the derivative of QoE with respect to the delay is negative. The resulting QoE–QoS relationships capture the effect of delays on QoE, typically described by negative-exponential [11–13] or logarithmic functions [10, 12].

Much work has been done on the performance of TCP, the underlying protocol for web transfers. However, the steady-state loss and delay performance and throughput are targeted in the majority of contributions, see e.g. [3, 14–17]. For example, [16] studies the impact of variable rate and variable delay on long-lived TCP performance in a Third Generation (3G) mobile network. The study was based on traces from a 3G 1 X network taken around 2002, and simulation results based on the network simulator ns-2. Also, [17] finds that the download time is proportional to the file size, which implies a constant throughput. The study employs a variable file size from some ten KB[1] on. Reference [18] defines web objects as between 1 and 100 KB, and [5] states that average and median of the flow sizes are smaller than 10 KB.

Motivated by the latter observation, [5] addresses explicity the performance problems faced by short flows due to the start-up behaviour of TCP, and proposes an analytic model that explicity covers such short flows. The model is validated by simulations. No measurements in real environments are taken into account. The latter provide the ground for reference [6], according to our knowledge the only reference to study measurements of the ramping-up of TCP in a real mobile network. For GPRS, the duration of the slow start phase was found to last at least six seconds before the congestion window is expanded sufficiently in order to allow for approaching the quasi-stationary throughput. No further mobile technologies (3G etc.) were analysed. We conclude that a study of the impact of the file size on the download time and thus on QoE, based on measurements on contemporary mobile networks and compiled into easy-to-deploy formuale, is missing. This paper closes the gap.

3   Notations, Definitions and Setup

We will now introduce the variables to be used throughout the paper; the *size of the file to be downloaded* X; the *download time* T; and the *perceived throughput*

---

[1] 1 KB = 1000 B, 1 MB = 1000 KB; 1 KiB = 1024 B, 1 MiB = 1024 KiB; we will use the appropriate factor depending on the context.

on application level R, with the corresponding binary logarithms interrelated through:

$$\mathrm{lb}(R/\mathrm{bps}) = \mathrm{lb}(8\,X/\mathrm{B}) - \mathrm{lb}(T/\mathrm{s}) = 3 + \mathrm{lb}(X/\mathrm{B}) - \mathrm{lb}(T/\mathrm{s})\,. \tag{1}$$

In order to cover the whole range from small to large files, we increase the file sizes by factor two, starting from the smallest possible file size of one byte (just the end marker) in order to be able to see any particular behaviour for one-packet transmissions. We have chosen a maximal file size of 4 MiB, which comes closest to the file size of 3.7 MB used by the tool described in [4]. Table 1 contains the classification of file sizes that will be used throughout the paper. Furthermore, we denote the *quasi-stationary throughput* that can be observed for very large files as R_∞, amongst others shown by the tool [4].

Table 1. Classification of file sizes

Label	Size $X(i) = 2^i$ B	Range of i
S	1 B ... 1 KiB	0 ... 10
M	2 KiB ... 256 KiB	11 ... 18
L	512 KiB ... 4 MiB	19 ... 22

The setup used for the measurements is as follows: Host P, a PC with Windows XP connected to the mobile operators using a Huawei E220 USB modem, requests a file via HTTP (not FTP) from the Linux-based host S that is running an Apache web server. Host S is not publicly known, so all HTTP traffic destined to S originates from P, and disturbances caused by competing traffic are avoided. For each file size as specified in Table 1, fourty subsequent downloads are performed ($j = 1 \ldots 40$) without caching the file, before moving to the next file size ($i = i + 1$). The download time for file size i and replication j, defined as $T_j(i)$, is obtained through time stamp $T_j^s(i)$ just before retrieving the file using the `get` function found in the Perl module `LWP::Simpe`. Completion of the transfer triggered a new time stamp $T_j^e(i)$. From this, we calculate $T_j(i) = T_j^e(i) - T_j^s(i)$ and the throughput $R_j(i) = X(i)/T_j(i)$ according to (1).

Three Swedish mobile operators are considered, denoted by A, B and C, each of them offering HSDPA (High-Speed Downlink Packet Access). Taking the role of an end user, no specific insights into configurations and load conditions of the mobile networks were available. The experiments were conducted during December 2009 during business days. All units were stationary during the tests, and no obvious network congestion was observed during the series of tests.

4   Results and Analysis

4.1   Median Values of Download Time and Throughput

Typically, the spread of the distribution of $R_j(i)$ over j is quite small. In 20 out of a total of 2640 measurements, we observed download times exceeding their neighboring values by up to ten seconds. This effect is rather common and discussed a.o. in [19]. As these extraordinary long download times have a significant impact on the average, we use the median (and skip that notion frequently for sake of brevity).

The quasi-stationary throughput per operator was approximated by the median of the throughput obtained for the largest investigated file size of 4 MiB, i.e. $R_\infty = R_{med}(22)$. The corresponding values are given in Table 2, accompanied by average m_R, standard deviation s_R and cofficient of variation c_V. Furthermore, the median of a set of round-trip time measurements RTT_{med} performed with the tool [4] is shown. Obviously, operator A provides the highest quasi-stationary throughput in combination with the smallest relative variation, while the reverse holds for operator C. Operator B, which is actually sharing the 3G network infrastructure with operator A, yields a considerably lower throughput than operator A with comparable relative variation, despite of quite similar RTT values.

Table 2. Quasi-stationary throughput estimations and round-trip measurements for different operators

Op.	R_∞/bps	lb(R_∞/bps)	$m_R(22)$/bps	$s_R(22)$/bps	$c_V(22)$	RTT_{med}/ms
A	949584	19.9	935374	38594	0.04	125
B	530114	19.0	530987	28409	0.05	130
C	311365	18.3	323128	43934	0.13	336

Figure 1 provides a first impression on the median of the download time obtained from the three different operators. For small files, the download time is quite independent of the file size, with fastest delivery provided by operator C, followed by B and A. A discontinuity is observed for operator A: For very small sizes ($X \leq 32$ B), the download time is found between 1 s and 1.2 s, while it then drops to values between 0.25 s and 0.38 s for 64 B $\leq X \leq 1$ KB, which means that small packets may take more time than large packets to get delivered, an observation also reported and discussed in [19]. For medium-sized files, the download time is growing with the file size for all operators, however with different gradients. Operator B initially displays the longest download times. From around 20 KB on, operator C takes most time to deliver the file. Operator A provides fastest delivery for files larger than 30 KB. For large files, the order of delivery is operator A, followed by operators B and C.

As particular examples, we are taking a closer look at the medians of download times T_{med} and throughputs R_{med} provided by operator B, accompanied by error bars indicating the asymmetrical span between maximum and minimum of the 40 contributing values. From figures 2 and 3, we can see three areas in which T_{med} and R_{med} are almost linear in log-log representation. This is illustrated piecewise by linear regressions and points at *power-type relationships* between download time, throughput and file size, respectively. Quite similar behaviours are obtained for operator A and C, cf. Figure 1. For small files, cf. Table 1,

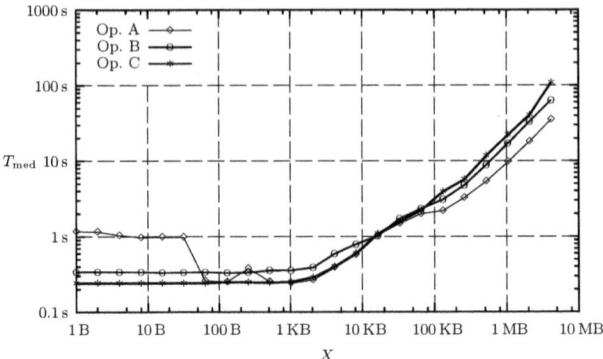

Fig. 1. Median of the download time T_{med} via different operators versus file size X

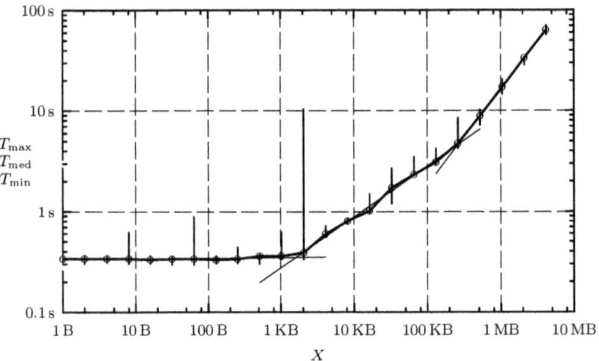

Fig. 2. Median of the download time T_{med} via operator B versus file size X, with error bars and potential regressions

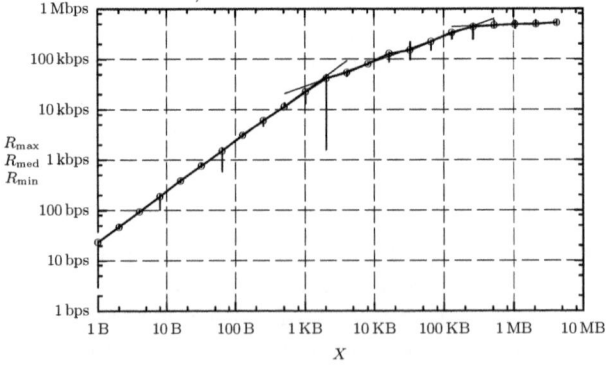

Fig. 3. Median of the perceived throughput R_{med} via operator B versus file size X, with error bars and potential regressions

the download time is practically constant, and consequently, the throughput grows proportionally with the file size. For large files, we observe saturation of R_{med}. Its logarithm shows a quite weak, yet linear dependence on $lb(X/B)$, while $lb(T_{med})$ grows in almost the same manner as $lb(X/B)$, with a gradient around 1. For medium-sized files, both $lb(T_{med})$ and $lb(R_{med})$ raise in a similar fashion, with a gradient around $1/2$. Obviously, we are facing a transision area in which both download time and throughput grow as the file size grows, which indicates acceleration while downloading due to TCP start-up. The throughput has not reached its quasi-stationary value yet, which implies that the real download time is higher than any estimation that builds upon the quasi-stationary value R_∞.

4.2   Regressions

We now construct linear least-square regressions on the logarithms of X, T and R for the areas S, M and L for all operators of the types

$$lb(\hat{T}/s) = a_T lb(X/B) + b_T ; \tag{2}$$

$$lb(\hat{R}/s) = a_R lb(X/B) + b_R ; \tag{3}$$

and evaluate the coefficient of determination $\mathcal{R}^2$ [3], which mostly signals good matches ($\mathcal{R}^2 \to 1$) with exception of some weak trends ($a \to 0$).

Table 3. Regressions for $\hat{T}$ and $\hat{R}$

Size	Op.	a_T	b_T	$\mathcal{R}^2$	a_R	b_R	$\mathcal{R}^2$
	A	-0.276	0.566	0.785	1.276	2.434	0.987
S	B	0.006	-1.582	0.211	0.994	4.582	0.999
	C	0.006	-2.072	0.602	0.994	5.072	1.000
	A	0.520	-7.486	0.976	0.480	10.486	0.971
M	B	0.506	-6.907	0.995	0.494	9.907	0.953
	C	0.627	-8.769	0.997	0.373	11.768	0.991
	A	0.906	-14.83	0.999	0.094	17.83	0.933
L	B	0.949	-14.89	0.999	0.051	17.89	0.953
	C	1.054	-16.57	0.987	-0.054	19.57	0.164

Table 3 shows the obtained regressions. As expected from (1), we observe for each operator and file size class $a_T + a_R = 1$ and $b_T + b_R \simeq 3$. A closer look at the results provides the following insights. For small files, the download time regression for operator A has a coefficient $a_T < 0$, which is due to the above-described discontinuity. For the other operators B and C, there is hardly any dependence of the download time on the size for small files ($a_T \simeq 0$). For medium-sizes files, the coefficients a_T and a_R are found close to 0.5, which has shown to be the key for the approximation presented in the next section. For large files, operator C shows an unexpected negative trend in the throughput, seen from $a_R < 0$. The other two operators show slightly increasing trends ($0 < a_R < 0.1$), i.e. the throughput still rises sligthly as the file size increases.

5   Download Time Approximations

The regressions shown in Figure 2 suggest the use of maximum of the three approximations

$$\hat{T} = \max\{\hat{T}^{\mathrm{S}}, \hat{T}^{\mathrm{M}}, \hat{T}^{\mathrm{L}}\} \tag{4}$$

to estimate the download time. The components of (4) belong to the different regimes of file sizes as defined in Table 1 and are given as follows:

- For small files

$$\hat{T}^{\mathrm{S}} = \mathrm{const.} \tag{5}$$

  The time $\hat{T}^{\mathrm{S}}$ needed to send a one-packet file depends on the operator.
- For medium-sized files

$$\hat{T}^{\mathrm{M}}/\mathrm{s} = 8 \left(\frac{X/\mathrm{B}}{R_\infty/\mathrm{bps}}\right)^{a_T}. \tag{6}$$

  This part builds upon the observation that $b_R \simeq a_T \mathrm{lb}(R_\infty/\mathrm{bps})$, cf. tables 2 and 3. For $a_T = 0.5$, the general power-type relationship (6) reduces to a simple square-root formula, which reminds of an accelerated movement with constant acceleration a in which the time to reach a distance s is given by $t = \sqrt{2s/a}$. Translated to our case ($s = X$), the acceleration of the file transfer would amount to $a = R_\infty/32$ bps.
- For large files

$$\hat{T}^{\mathrm{L}}/\mathrm{s} = 8 \frac{X/\mathrm{B}}{R_\infty/\mathrm{bps}}. \tag{7}$$

  The motivation for this part of the approximation is found in the observation $a_T \to 1$ and $a_R \to 0$ in Table 3, i.e. the throughput is almost constant and can thus be approximated by R_∞.

Obviously, the approximation depends on *three parameters*:

1. the *one-packet download time* $\hat{T}^{\mathrm{S}}$ to be measured from downloading files of a typical size (e.g. 1 KB);
2. the quasi-stationary throughput R_∞ obtained from downloading large files of several MB and taking the median of the measured values; and
3. the acceleration factor a_T obtained from downloading a sequence of medium-sized files, followed by the regression (2).

The three-parameter approximation can be simplified by letting $a_T = 0.5$, which fits operators A and B almost exactly and operator C approximately ($a_T \simeq 0.625$). This leaves us with *two parameters* $\hat{T}^{\mathrm{S}}$ and R_∞.

We will now investigate the approximation errors, given as the difference of approximated and measured median of the download time

$$e_T(X(i)) = \hat{T}(X(i)) - T_{\mathrm{med}}(X(i)) \tag{8}$$

and seen from Figure 4. Positive values of $e_T(X(i))$ mean that the approximation overestimates the download time for the given operator and file size $X(i)$.

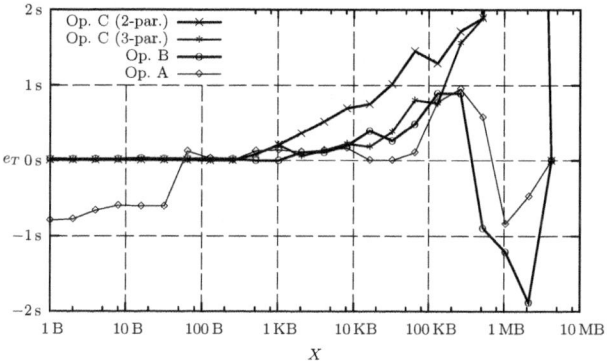

Fig. 4. Estimation errors for different operators

For small files, the only significant deviation is seen for operator A due to the discontinuity mentioned above. For medium-sized files, the approximation estimates on the safe side. The deviations are less than one second for operators A and B, while the two-parameter approximation for operator C displays a steady drift towards two seconds. The performance of the three-parameter approximation for operator C is comparable to that of A and B until a file size of 128 KiB thanks to the better capture of the gradient in the log scale by the third parameter $a_T = 0.625$. Upon reaching the area of large file sizes, a trend towards underestimations of no more than two seconds for operator A and one second for operator B is observed. The approximation for operator C is very conservative, overestimating the median by up to 13 s for a file size of 2 MiB, which indicates the need to use a refined value of R_∞ to better account for this domain. All errors vanish for a file size of 4 MiB, as this was the anchor point for R_∞.

6   Links to Quality of Experience

In the following, we assume that we can quantify the QoE by employing a numerical linear Mean Opinion Score (MOS)-type scale from 5 (= excellent) to 1 (= unacceptable). We employ a logarithmic relationship between QoE (user rating) and QoS (response time) found in standard [10] and recently re-confirmed *a.o.* by [3, 20]. In our example, the numerical values stem from the case of a time scale of reference of 6 s and a mix of two user groups. They can (be) change(d) according to context and user expectations, as illustrated in [10]. Starting from

$$\text{QoE} = \max\{\min\{4.38 - 0.9\text{lb}(T/\text{s}), 5\}, 1\} \tag{9}$$

and inserting the above results (5)–(7), we arrive at

$$\text{QoE} = \begin{cases} \min\{4.38 - 0.9\text{lb}(\hat{T}^S/\text{s}), 5\} & \text{for S} \\ \max\{\min\{1.68 + 0.9a_T(\text{lb}(R_\infty/\text{bps}) - \text{lb}(X/\text{B})), 5\}, 1\} & \text{for M} \\ \max\{\min\{1.68 + 0.9(\text{lb}(R_\infty/\text{bps}) - \text{lb}(X/\text{B})), 5\}, 1\} & \text{for L} \end{cases} \tag{10}$$

As expected, the QoE decreases as file size and download time grow, and increases as throughput increases. However, the sensitivity of QoE to the parameters depend on the file size itself: Due to the acceleration behaviour, the sensitivity is smaller for medium-sized files than for large files. Figure 5 illustrates the QoE estimations from (10), based on the three-parameter approximation and Table 2. The discontinuity for operator A is due to the change of domain between medium-sized and large files. Best QoE is reached for operator A, followed by operator C for files up to 30 KB. From then on, operator B provides the second-best perception.

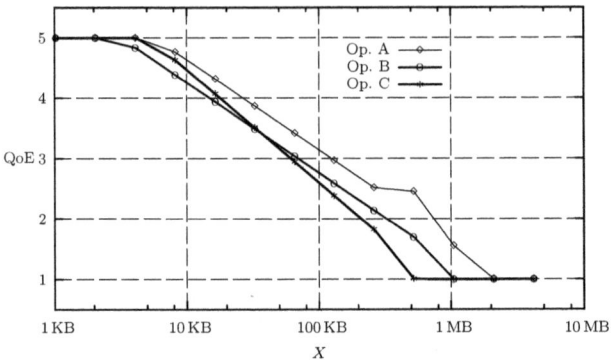

Fig. 5. QoE estimations versus file size X for different operators

Figure 5 also shows that, in order not to challenge user patience beyond feasibility (i.e. keep the QoE above 3), downloaded files should not be larger than approximately 120 KB for operator A, 70 KB for operator B and 60 KB for operator C, respectively.

7   Conclusions and Outlook

This paper presented a measurement-based study of download performance and approximations of download times and corresponding throughputs as functions of the size of the downloaded file. Through a specific strategy of choosing file sizes, employing a factor of two, we were able to see linear relationships between the logarithms of download time, throughput and file size for small, medium-sized and large files. From a subsequent analysis of the parameters of the approximations, we found surprisingly simple approximation formulae with reasonably tight error bounds for the download times. For the small- and medium-sized files that represent the majority of the web downloads and that yield download times in the order of user patience, we obtain an error bound of roughly one second in most cases. The formulae will help users and application providers to "choose the right network for the right task", alternatively to limit data sizes in order to keep download times of an acceptable level. Furthermore, it will help providers to choose the level of provisioning they would like to offer in order to yield a reasonable trade-off between quality and cost.

Besides of refinements of the proposed approaches and formulae, future work will include validation and parameterisation of the download time and throughput approximations for other access network technologies such as WLAN, ADSL and Ethernet. At this point, a deeper analysis of TCP traces and a quantitative comparison to the results obtained from the model proposed in [5] would be of interest. We will also study the impact of data loss and delay variation on the proposed approximations, as they will affect the quasi-stationary throughput. Finally, the practical applicability of the formulae to assess and implement network selection policies for seamless communications will be studied.

Acknowledgements. The authors would like to thank the Swedish Agency for Innovation Systems, VINNOVA, for sponsoring this work through the EViMonA project (d-nr 2007/ 02505). They are also grateful to the Indian-Swedish project that supported the collaboration, and in particular Prof. Ashok Jhunjhunvala and Prof. Sara Eriksén for their great effort and commitment.

References

1. Info24, Homepage, `http://www.info24.se/` (last seen July 31, 2010)
2. ITU-T Recommendation P.10/G.100 (incl. Amendment 2), Vocabulary for performance and quality of service (July 2006) (2008)
3. Shaikh, J., Fiedler, M., Collange, D.: Quality of Experience from user and network perspectives. Annals of Telecommunications, Special Issue on Quality of Experience: 1/Metrics and Performance Evaluation 65(1-2) (January-February 2010), electronically available at `http://www.springerlink.com`, doi:10.1007/s12243-009-0142-x
4. Gonsalves, T., Bhardwaj, A.: Comparison of AT-Tester with other popular testers for Quality of Service Experience (QoSE) of an internet connection (August 2009), `http://www.broadbandasia.info/`
5. Mellia, M., Stoica, I., Zhang, H.: TCP model for short lived flows. IEEE Comm. Letters 6(2), 85–87 (2002)
6. Chakravorty, R., Pratt, I.: WWW performance over GPRS. In: Proc. 4th Int. Workshop on Mobile and Wireless Communication Networks (MWCN 2002), pp. 527–531 (September 2002)
7. Zona Research Inc., "The economic impacts of unacceptable web-site download speeds," Report (1999)
8. Nielsen, J.: Usability Engineering. Morgan Kaufman (1994)
9. Rose, G.M., Evaristo, R., Straub, D.: Culture and consumer responses to web download time: a four-continent study of mono and polochronism. IEEE Trans. on Engineering Management 50(1), 31–44 (2003)
10. ITU-T Recommendation G.1030, Estimating end-to-end performance in IP networks for data applications (November 2005)
11. Fiedler, M., Hoßfeld, T., Tran-Gia, P.: A generic quantitative relationship between Quality of Experience and Quality of Service. IEEE Network, Special Issue on Improving Quality of Experience for Network Services (2), 36–41 (2010)

12. Fiedler, M., Hoßfeld, T.: Quality of Experience-related differential equations and provisioning-delivery hysteresis. In: Proc. 21st ITC Specialist Seminar on Multimedia Applications, Miyazaki, Japan (March 2010),
http://www.ieice.org/proceedings/ITC-SS21/ITC-SS21-Proceedings.pdf
13. Hoßfeld, T., Tran-Gia, P., Fiedler, M.: Quantification of Quality of Experience for Edge-Based Applications. In: Mason, L.G., Drwiega, T., Yan, J. (eds.) ITC 2007. LNCS, vol. 4516, pp. 361–373. Springer, Heidelberg (2007)
14. Chen, X., Wang, W., Nie, J.: Analysis of web response time in asymmetrical wireless network. In: Proc. 11th IEEE Singapore Int. Conf. on Communication Systems (ICCS 2008), pp. 1427–1430 (November 2008)
15. Miyagi, M., Ohkubo, K., Kataoka, M., Yoshizawa, S.: Performance prediction method for web-access response time distribution using formula. In: Proc. IEEE/IFIP Network Operations and Management Symposium (NOMS 2004), vol. 1, pp. 905–906 (April 2004)
16. Chan, M.C., Ramjee, R.: TCP/IP performance over 3G wireless links with rate and delay variation. In: MobiCom 2002: Proc. of the 8th Annual International Conference on Mobile Computing and Networking, pp. 71–82. ACM, New York (2002)
17. Voskarides, S., et al.: Practical evaluation of GPRS use in telemedicine system in Cyprus. In: Proc. 4th Int. IEEE EMBS Special Topic Conference on Information Technology Applications in Biomedicine, pp. 39–42 (April 2003)
18. Baccarelli, E., Biagi, M., Cordeschi, N., Pelizzoni, C.: Minimization of download times of large files over wireless channels. IEEE Trans. on Mobile Computing 6(10), 1105–1115 (2007)
19. Arlos, P., Fiedler, M.: Influence of the packet size on the one-way delay on the down-link in 3G networks. In: Proc. ISWPC 2010, Modena, Italy (May 2010)
20. Reichl, P., Egger, S., Schatz, R., d'Alconzo, A.: The logarithmic nature of QoE and the role of the Weber-Fechner Law in QoE assessment. In: Proc. IEEE ICC 2010, Cape Town, South Africa (May 2010)

On Traffic Domination
in Communication Networks

Walid Ben-Ameur[1], Pablo Pavon-Marino[2], and Michał Pióro[3,4]

[1] TELECOM SudParis, 9, rue Charles Fourier, 91011 Évry, France
[2] Technical University of Cartagena, Pza. Hospital 1, 30202 Cartagena, Spain
[3] Warsaw University of Technology, Nowowiejska 15/19, 00-665 Warszawa, Poland
[4] Lund University, Box 118, 221 00 Lund, Sweden
walid.benameur@int-evry.fr, pablo.pavon@upct.es, mpp@tele.pw.edu.pl

Abstract. Input data for communication network design/optimization
problems involving multi-hour or uncertain traffic can consist of a large
set of traffic matrices. These matrices are explicitly considered in problem formulations for link dimensioning. However, many of these matrices
are usually dominated by others so only a relatively small subset of matrices would be sufficient to obtain proper link capacity reservations, supporting all original traffic matrices. Thus, elimination of the dominated
matrices leads to substantially smaller optimization problems, making
them treatable by contemporary solvers. In the paper we discuss the
issues behind detecting domination of one traffic matrix over another.
We consider two basic cases of domination: (i) total domination when
the same traffic routing must be used for both matrices, and (ii) ordinary domination when traffic dependent routing can be used. The paper
is based on our original results and generalizes the domination results
known for fully connected networks.

Keywords: network optimization, traffic matrices domination, multi-
hour optimization, uncertain traffic, graph theory.

1 Introduction

Input data for optimization problems related to communication network design
and planning can consist of a large set of traffic matrices. This is the case for
example when multi-hour traffic or uncertain traffic is considered [1,9,11]. By
using a set of matrices rather than a single traffic matrix, it is possible to take
into account the non-coincidence of the peak load hours in different parts of the
network, or to use a large set of measured matrices when the traffic is hardly
predictable. In consequence, a set of traffic matrices is explicitly used in problem
formulations in order to dimension the network links. However, many of the input
matrices are usually dominated by others, and in effect only a relatively small
subset of them can be sufficient to obtain proper link capacity reservations, supporting all original traffic matrices. Thus, elimination of the dominated matrices
leads to substantially smaller optimization problems, making them treatable by

K.A. Hummel et al. (Eds.): PERFORM 2010 (Haring Festschrift), LNCS 6821, pp. 191–202, 2011.
© IFIP International Federation for Information Processing 2011

the optimization solvers. Moreover, the domination relation between traffic matrices is of interest not only for identifying and then removing the dominated traffic matrices. In [7] it was shown that the domination can also be exploited to devise more efficient planning algorithms under multi-hour or uncertain traffic demand, introducing a small "upper bound" set of artificial traffic matrices dominating the entire set of the original traffic matrices and yet yielding a very good approximation of the true optimal solution.

In the paper we discuss the issues behind detecting domination of one traffic matrix over another. We consider two basic cases: total domination (when the same traffic routing must be used for both matrices – in this case we in fact consider domination of a set of traffic matrices over one additional traffic matrix), and ordinary domination (when traffic-dependent routing can be used). The paper is based on our original results presented in [2,8].

Given a network graph $G(V, E)$, and two traffic matrices $\hat{h}$ and h, $\hat{h}$ *totally dominates* h ($\hat{h} \succeq h$ in short) if for each link capacity reservation $u : E \to \mathcal{R}_+$ and for each flow pattern $f : P \to \mathcal{R}_+$ (where P is the set of routing paths) such that (u, f) supports matrix $\hat{h}$, the solution (u, f) does also support matrix h.

The definition of ordinary traffic domination is slightly different. We say that $\hat{h}$ *ordinarily dominates* h ($\hat{h} \models h$) if for each link capacity reservation $u : E \to \mathcal{R}_+$ for which there exists a flow pattern $\hat{f} : P \to \mathcal{R}_+$ such that $(u, \hat{f})$ supports matrix $\hat{h}$, there exists a (in general different) flow pattern $f : P \to \mathcal{R}_+$ such that (u, f) supports matrix h. Obviously, total domination implies ordinary domination, i.e., $\hat{h} \succeq h$ implies $\hat{h} \models h$.

The known result (due to Gianpaolo Oriolo [6]) on total domination is as follows: in a complete graph $\hat{h} \succeq h$ if, and only if $\hat{h} \geq h$ component-wise. In [8] we have given a complete characterization of total domination which we describe in Section 2 below. In fact, total domination is discussed in a more general setting that examines when a set of traffic matrices dominates one additional traffic matrix.

For ordinary domination the known result (also due to [6]) states that in a complete graph $\hat{h} \models h$ if, and only if, matrix h can be routed in the network with link capacities equal to the elements of $\hat{h}$. This result, however, cannot be extended to arbitrary graphs simply because in an incomplete graph there may no direct links between the end nodes of some demands, making the Oriolo result not applicable. Consequently, in Section 3, we present a different kind of general necessary and sufficient condition for ordinary domination in an arbitrary graph, using findings of [2].

2 Total Domination

2.1 Definition

Consider an undirected network graph $G = G(V, E)$ with the set of nodes (vertices) V and the set of undirected links (edges) E. A path p between nodes s and node t in graph G is given by a sequence of nodes $v_1 = s, v_2, v_3 \ldots, v_{n-1}, v_n =$

t), $v_i \in V, i = 1, \ldots, n$, such that $v_i v_{i+1} \in E, i = 1, 2, \ldots, n-1$. Path p is called elementary if all the nodes are different, and in effect an elementary path can be treated as a subset of links: $p \subseteq E$. Link capacity reservation $u : E \to \mathcal{R}_+$ will be identified with vector $(u_e, e \in E)$ where $u_e = u(e)$.

Let D denote the set of (undirected) traffic demands in the considered network. Each demand $d \in D$ is characterized by its end nodes $a(d)$ and $b(d), (a(d) \neq b(d))$, and its traffic volume h_d. The vector $h = (h_d, d \in D)$ is referred to as traffic vector. (We prefer to use the notion of the traffic vector instead of the traffic matrix used in the introduction.) For each demand $d \in D$ we specify a set of its candidate paths P_d – a subset of all elementary paths between nodes $a(d)$ and $b(d)$ that are selected for carrying traffic of demand d. We assume that there is at most one demand between any two nodes so that the path sets $P_d, d \in D$ are mutually disjoint (in the sequel a demand d between nodes s and t will sometimes be denoted, somewhat informally, by $d = st \in D$). Finally, we put $P = \bigcup_{d \in D} P_d$ – the set of all admissible paths. Having the predefined path set P, for each link e and demand d we can define $Q_{ed} \subseteq P_d$ – the set of all candidate paths for demand d that contain link e: $Q_{ed} = \{p \in P_d : p \ni e\}$. The length of path $p \in P$ with respect to a given vector of link metrics (weights) $\pi = (\pi_e, \ e \in E)$ will be denoted by $|p|_\pi$.

Given graph $G(V, E)$, set of demands D, and set P of candidate paths, we define a static (fixed) flow allocation pattern as a vector $f = (f_p, \ p \in P)$, where for each $d \in D$ and $p \in P_d$, the entity f_p represents a fraction $(0 \leq f_p \leq 1)$ of traffic volume of demand $d \in D$ assigned to path p.

Consider a finite set of traffic vectors $H = \{h^t : t \in T\}$ to be supported by the network, and one additional traffic vector $h = (h_d, d \in D)$ outside set H. Set H can represent $|T|$ different traffic hours, or a set of observed realizations of a random traffic vector. For every $h^t, t \in T$, its traffic volume for demand $d \in D$ is denoted by h_d^t so that $h^t = (h_d^t, d \in D)$.

We say that a capacity reservation vector u and a flow allocation pattern f support H with respect to the given path set P if (u, f) satisfies the following linear constraints:

$$\sum_{p \in P_d} f_p = 1 \qquad\qquad d \in D \qquad\qquad (1a)$$

$$\sum_{d \in D} \sum_{p \in Q_{ed}} h_d^t f_p \leq u_e \qquad\qquad e \in E, \ t \in T \qquad\qquad (1b)$$

$$f_p \geq 0 \qquad\qquad p \in P. \qquad\qquad (1c)$$

Constraint (1a) assures that each demand is satisfied, while constraint (1b) – that each link must support the load induced by any vector from H. Constraint (1c) ensures non-negativity of flows (and hence of capacity reservations, provided each link is in at least one path).

In the sequel we shall assume that the network graph $G(V, E)$ and the set of demands D are fixed. Then, for a given path set P and a set of traffic vectors H, the feasible set (1) will be abbreviated by $\mathcal{P}_P(H)$. Note that $\mathcal{P}_P(H)$ is a polyhedron in $\mathcal{R}_+^{|E|+|P|}$.

Now, we are ready to introduce a formal definition of total domination, extending the one from [6] to a set of traffic vectors H and a traffic vector h defined for a demand set D.

Definition 1. *Let P be a given set of candidate paths for D in a network with arbitrary graph G. We say that H totally dominates h with respect to P ($H \succeq_P h$ in short) if, and only if, each feasible solution in $\mathcal{P}_P(H)$ is also a feasible solution in $\mathcal{P}_P(h)$, i.e., $\mathcal{P}_P(H) \subseteq \mathcal{P}_P(h)$.*

(In the sequel we will skip subscript P in $\mathcal{P}_P$ and in $\succeq_P$ when the path set P is fixed.)

2.2 A Sufficient Condition and a Necessary Condition for Total Domination

Consider a network with an arbitrary graph G, arbitrary demand set D and the corresponding traffic vectors H and h.

Proposition 1. $\exists \, \hat{h} \in conv(H), \; \hat{h} \geq h \Rightarrow H \succeq_P h$.

Proof. Consider a set of scalar coefficients $\alpha_t \geq 0, t \in T, \sum_{t \in T} \alpha_t = 1$ defining the convex combination $\hat{h} = \sum_{t \in T} \alpha_t h^t \in conv(H)$. Let $(u, f) \in \mathcal{P}_P(H)$. Then (u, f) is also a feasible solution of (1) for the convex combination $\hat{h}$ ($(u, f) \in \mathcal{P}_P(\hat{h})$), because the convex combination of constraints (1b) for $\mathcal{P}_P(H)$ with the above defined coefficients $\alpha_t, t \in T$ yields constraint (1b) defining polyhedron $\mathcal{P}_P(\hat{h})$.

Now suppose that $\hat{h} \geq h$. This implies that for the given $(u, f) \in \mathcal{P}_P(\hat{h})$, constraint (1a) is satisfied also for $\mathcal{P}_P(h)$ which means that (u, f) supports h. ∎

Fig. 1. A two-link network

The sufficient condition for total domination formulated above is in general not necessary, as illustrated in Fig.1. The figure shows a 3-node, 2-link graph with three demands $13, 12, 23$, and with the set of admissible paths $P = \{\{123\}, \{12\}, \{23\}\}$. Assume $H = \{\hat{h}\}, \hat{h} = (1, 0, 0)$ ($\hat{h}_{13} = 1, \hat{h}_{12} = \hat{h}_{23} = 0$), and $h = (0, 1, 1)$ ($h_{13} = 0, h_{12} = h_{23} = 1$). Certainly, since the network is a tree, only one flow allocation pattern exists. It is obvious that every capacity vector u supporting traffic vector $\hat{h}$ (i.e., $u \geq (1, 1)$), also supports h (and, in fact, vice versa). Then, $\hat{h}$ totally dominates h with respect to P (and vice versa). Still, it is not true neither that $\hat{h} \geq h$ nor that $h \geq \hat{h}$.

Below we give a (technical) property of the path set P under which the condition $\exists \, \hat{h} \in conv(H), \; \hat{h} \geq h$ holds is also necessary for the total domination.

Proposition 2. *Let G, H, h be as in Proposition 1 and let $P = \bigcup_{d \in D} P_d$. Suppose that there exists a link $\hat{e} \in E$ (the so called* enabling link*) with the following property:*

$$\forall \, d \in D \; (\exists \, p \in P_d, \; \hat{e} \in p) \wedge (\exists \, p \in P_d, \; \hat{e} \notin p), \tag{2}$$

Then, if H totally dominates h wrt P, then there exists a $\hat{h} \in conv(H)$ such that $\hat{h} \geq h$:

$$H \succeq_P h \Rightarrow \exists \, \hat{h} \in conv(H), \; \hat{h} \geq h. \tag{3}$$

Proposition 2 is proved in [8] using the means of the LP dual theory. In the next two subsections we will use property (2) to find a general characterization of total domination.

2.3   Two-Connected Networks

Graph $G(V, E)$ is called *2-connected* if it has at least three nodes and does not contain any *cut vertex*, i.e., any vertex $v \in V$ such that $G \setminus v$ has more connected components than G. (Below, we follow definitions and results given in [3].)

Proposition 3. *Suppose that a network graph is 2-connected (or composed of one link) and that for each demand $st \in D$ its set of candidate paths is composed of all elementary st-paths. Then*

$$H \succeq h \; \Leftrightarrow \; \exists \, \hat{h} \in conv(H), \; \hat{h} \geq h. \tag{4}$$

Proof. 2-connected graphs enjoy the two following properties: (i) G is 2-connected if, and only if, any two nodes $v, w \in V$ are connected by two node disjoint paths [3], and (ii) G is 2-connected if, and only if, for any two nodes $v, w \in V$ and link $e \in E$ there exists an elementary path between v and w containing link e [3].

These properties imply that in a 2-connected graph each link $\hat{e} \in E$ is enabling for any demand set D (provided P_d contains all elementary paths between its end nodes). Consider an arbitrary demand $st \in D$. There must be a path $p \in P_d$ such that $\hat{e} \notin p$. Otherwise, all elementary st-paths would contain $\hat{e}$, contradicting property (i). Also, by property (ii), there must be a path $p' \in P_d$ such that $\hat{e} \in p'$. Trivially, property (4) is valid also for graphs composed of just one link (and its end nodes). ∎

2.4   General Characterization of Total Domination

In this section we will present the main result of Section 2 – a general necessary and sufficient condition for total domination in a network with an arbitrary connected graph and path sets P_d containing all elementary paths for each demand $d \in D$. For simplifying the considerations, we assume (without loosing generality – we can assign the zero traffic volume to non-existing demands) that the network contains demands corresponding to all node pairs, i.e., $D = V^{|2|}$ ($A^{|2|}$ denotes the family of all two-element subsets of set A).

A connected graph $G(V, E)$ containing cut vertices is called *separable*. The maximal induced subgraphs of G that are not separable are called *blocks*. A block is either 2-connected or is formed by just one link. The blocks of a graph are unique. Any two blocks intersect in at most one node and this node must be a cut vertex. Two nodes s and t are in the same block if, and only if, they are connected by a path not traversing a cut vertex. Finally, any cycle must be contained in a block (see [3]).

Consider a network with a connected separable graph $G = (V, E)$ and a fixed traffic vector h. Suppose that graph G is composed of B blocks $G^b = (V^b, E^b), b \in \mathcal{B}$, where $\mathcal{B} = \{1, 2, ..., B\}$, and C is the set of cut vertices of G. Clearly, each elementary path between a pair of nodes st traverses the same sequence of blocks (the set of indices of these blocks will be denoted by $B(st)$, and the same sequence of cut vertices (denoted by $C(st)$). (Otherwise, there would be a cycle not contained in a single block.)

We treat the blocks as separate networks with their own sets of demands $D^b = (V^b)^{|2|}$ and the corresponding sets of candidate paths $P^b_d, d \in D^b$ composed of all elementary paths in G^b ($P^b = \bigcup_{d \in D^b} P^b_d$). The traffic vector h^b for each block $b \in \mathcal{B}$ is induced by the traffic vector h. The volume h^b_{st} of each demand $st \in D^b$ is either left unchanged (when both s and t are not cut vertices) or adjusted to account for the demands that have only one end node in the considered block b (i.e., when $s \in V^b, t \notin V^b$ and $b \in B(st)$) or transit the block (i.e., when $s \notin V^b, t \notin V^b$ and $b \in B(st)$). More precisely:

$$h^b_{st} = h_{st} \qquad\qquad\qquad\qquad\qquad s, t \in V^b \setminus C, s \neq t \qquad (5a)$$

$$h^b_{st} = h_{st} + \sum \{h_{sw} : w \notin V^b, t \in C(sw)\} \qquad s \in V^b \setminus C, t \in V^b \cap C \quad (5b)$$

$$h^b_{st} = h_{st} + \sum \{h_{vw} : v, w \notin V^b, s, t \in C(vw)\} \quad s, t \in V^b \cap C, s \neq t. \quad (5c)$$

Proposition 4. *Consider a network with a connected graph $G = (V, E)$ split into blocks $G^b, b \in \mathcal{B}$. Let H be a set of traffic matrices, and let h be an additional traffic vector. For each $b \in \mathcal{B}$, let H^b, h^b denote the set of traffic vectors defined for H and h in the way described by (5). Then,*

$$H \succeq h \Leftrightarrow \forall b \in \mathcal{B} \, \exists \, \hat{h}^b \in conv(H^b), \, \hat{h}^b \geq h^b. \qquad (6)$$

Although Proposition 4 is quite intuitive, its detailed proof is a bit lengthy and is omitted in this survey paper. The proof is given in [8].

In the example from Fig.1, the graph is split into two blocks with $V^1 = \{1, 2\}$ and $V^2 = \{2, 3\}$, and the adjusted traffic vectors are: $\hat{h}^1 = (1), h^1 = (1), \hat{h}^2 = (1), h^2 = (1)$. Hence, $\hat{h} \succeq h$ because $\hat{h}^1 \succeq h^1$ on G^1 and $\hat{h}^2 \succeq h^2$ on G^2 (and vice versa, $h \succeq \hat{h}$ because $h^1 \succeq \hat{h}^1$ on G^1 and $h^2 \succeq \hat{h}^2$ on G^2).

3   Ordinary Domination

3.1   Definition

Given graph $G(V, E)$, set of demands D, and set P of candidate elementary paths, we define a flow allocation pattern as a vector $f = (f_p, \ p \in P)$, where for each $d \in D$ and $p \in P_d$, the entity f_p represents a flow of demand $d \in D$ assigned to its path p. Note that this definition of a flow allocation pattern is a bit different than the analogous definition in Subsection 2.1 which has involved fractional flows while the current one involves absolute flows.

Consider a traffic vector h. We say that a capacity reservation vector u supports h if there exists a flow allocation pattern f such that (u, f) satisfies the following linear constraints:

$$\sum_{p \in P_d} f_p = h_d \qquad\qquad d \in D \qquad\qquad (7a)$$

$$\sum_{d \in D} \sum_{p \in Q_{ed}} f_p \leq u_e \qquad\qquad e \in E \qquad\qquad (7b)$$

$$f_p \geq 0 \qquad\qquad p \in P. \qquad\qquad (7c)$$

Constraint (7a) assures that each demand is satisfied, while constraint (7b) – that each link must support the load induced by the traffic vector h and flow f. Constraint (7c) ensures non-negativity of flows (and hence of capacity reservations, provided each link is in at least one path).

In the sequel we shall assume that the network graph $G(V, E)$, the set of demands D, and the set of candidate paths P are fixed. A capacity reservation vector $u = (u_e, e \in \mathcal{E})$ is said to support a demand vector h if, and only if, there exists a flow pattern f such that (7) is satisfied. Then, for a given traffic vector h, the set of all supporting capacity reservation vectors will be abbreviated by $\mathcal{U}_P(h)$. Note that $\mathcal{U}_P(h)$ is a polyhedron in $\mathcal{R}_+^{|E|}$.

A formal definition of ordinary domination, extending the one from [6] to networks with arbitrary graphs and demand sets D is as follows.

Definition 2. *Let P be the set of all elementary candidate paths for D in a network with arbitrary graph G. We say that $\hat{h}$ ordinarily dominates h ($\hat{h} \models_P h$ in short) if, and only if, each element u of the set $\mathcal{U}_P(\hat{h})$ (i.e., each capacity vector u supporting $\hat{h}$) is also an element of the set $\mathcal{U}_P(h)$ (i.e., u supports h), i.e., $\mathcal{U}_P(\hat{h}) \subseteq \mathcal{U}_P(h)$.*

(In the sequel we will skip subscript P in $\mathcal{U}_P(\hat{h}), \mathcal{U}_P(h)$ and in $\models_P$ when the path set P is fixed. Also, we will sometimes call ordinary domination just domination.)

As already mentioned in the introduction, in an incomplete graph there may no direct links between the end nodes of some demands, making the Oriolo result for ordinary domination not applicable. This can be seen in the example of Fig.1 where $\hat{h} \models h$, still h cannot be routed in $\hat{h}$. On the other hand, note that $\hat{h}$ can be routed in h and hence $h \models \hat{h}$.

3.2   Necessary and Sufficient Condition for Ordinary Domination

Consider a network with an arbitrary graph G, arbitrary demand set D, a given fixed set P of candidate paths (for example the set of all elementary paths for all demands in D), and two traffic vectors $\hat{h}$ and h defined for D. Let Π denote the set of all vectors $\pi = (\pi_e, \ e \in E)$ such that $\pi_e \geq 0, e \in E$ and $\sum_{e \in E} \pi_e = 1$. Further, let $\lambda(\pi) = (\lambda_d(\pi), \ d \in D)$ denote the vector of the lengths of the shortest paths for the demands calculated for the link weights given by π.

Proposition 5. $\hat{h} \models h$ if, and only if,

$$\forall \pi \in \Pi, \ \lambda(\pi)(\hat{h} - h) \geq 0. \tag{8}$$

Proof. Consider the following allocation problem related to (7) with a fixed capacity reservation vector u.

$$\text{minimize} \quad z \tag{9a}$$

$$[\lambda_d] \quad \sum_{p \in P_d} f_p = h_d \qquad\qquad d \in D \tag{9b}$$

$$[\pi_e] \quad \sum_{d \in D} \sum_{p \in Q_{ed}} f_p \leq u_e + z \qquad\qquad e \in E \tag{9c}$$

$$f_p \geq 0 \qquad\qquad p \in P. \tag{9d}$$

Certainly, capacity vector u supports h if, and only if, the optimal solution z^* of (9) is non-positive ($z^* \leq 0$). The problem dual to (9) reads:

$$\text{maximize} \quad W(\lambda, \pi) = \sum_{d \in D} \lambda_d h_d - \sum_{e \in E} \pi_e u_e \tag{10a}$$

$$\lambda_d \leq \sum_{e \in p} \pi_e \qquad\qquad d \in D, \ p \in P_d \tag{10b}$$

$$\sum_{e \in E} \pi_e = 1 \tag{10c}$$

$$\pi_e \geq 0 \qquad\qquad e \in E. \tag{10d}$$

Observe, that in any optimal solution (λ^*, π^*) we have that λ_d^* is equal to the length of the shortest path of demand $d \in D$ with respect to link weights π^*, i.e., $\lambda_d^* = \lambda_d(\pi^*)$. Since the optimal value of the dual function W^* ($W^* = W(\lambda^*, \pi^*)$) is equal to z^*, we deduce that capacity vector u supports h if, and only if, the optimal solution W^* of (10) is non-positive ($W^* \leq 0$), i.e., when $\sum_{d \in D} \lambda_d^* h_d \leq \sum_{e \in E} \pi^* u_e$ or, in the vector notation, $\lambda^* h \leq \pi^* u$. Because (10) is a maximization problem and because of (10b), the last condition is equivalent to: $\forall \pi \in \Pi, \ \lambda(\pi)h \leq \pi u$.

Suppose that (8) is satisfied. By the above characterization, for each $u \in \mathcal{U}(\hat{h})$ we have that $\sum_{d \in D} \lambda_d \hat{h}_d \leq \sum_{e \in E} \pi_e u_e$ for each $\pi \in \Pi$. By (8), $\lambda(\pi^*)h \leq \lambda(\pi^*)\hat{h}$, and hence $\lambda(\pi^*)h \leq \pi_e^* u_e$ which means that $u \in \mathcal{U}(h)$. Thus $\hat{h} \models h$.

Now assume that $\hat{h} \models h$ and take some $\pi \in \Pi$. We define a specific flow $f(\pi)$ and the corresponding capacity reservation vector $u(\pi) \in \mathcal{U}(\hat{h})$ as follows. For every demand $d \in D$, put the entire demand volume $\hat{h}_d$ on one selected path in P_d that is shortest with respect to π, and define $u_e(\pi), e \in E$ as the resulting link loads. Consider the primal-dual solution pair $(f(\pi), u(\pi); \pi, \lambda(\pi))$. It follows from the saddle point conditions (see for example [4]) that the primal point $(f(\pi), u(\pi))$ is an optimal solution of (9), and the dual point $(\pi, \lambda(\pi))$ is an optimal solution of (10), so $\lambda(\pi)\hat{h} = \pi u(\pi)$. Since $u(\pi) \in \mathcal{U}(\hat{h})$, by assumption we have that $u(\pi) \in \mathcal{U}(h)$ and hence $\lambda(\pi)h \leq \pi u(\pi) = \lambda(\pi)\hat{h}$. Thus, $\lambda(\pi)(h - \hat{h}) \leq 0$ which means that (8) holds. ∎

It can be shown (we omit a formal proof here) that problem

$$\text{minimize} \quad \lambda(\pi)(\hat{h} - h), \quad \pi \in \Pi \tag{11}$$

is $\mathcal{NP}$-hard. This fact strongly suggests that the problem of determining whether or not $\hat{h} \models h$ is $\mathcal{NP}$-hard as well.

3.3 Two Special Cases

In this section we will use the result of Proposition 5 to characterize domination in two important special cases. For the results discussed below we need an assumption that the path lists $P_d, d \in D$ contain all elementary paths between $a(d)$ and $b(d)$ in the network graph $G(V, E)$.

Case 1: $\hat{h}$ **directly routeable in** $G(V, E)$

Assume that $\hat{h}$ is directly routeable in $G(V, E)$. By this we mean that when $\hat{h}_d > 0$ then the network graph contains link $a(d)b(d)$ (in other words, for any demand $d \in D$ with $\hat{h}_d > 0$, its end nodes $a(d)$ and $b(d)$ are connected by a link in E). Define the capacity reservation vector $\hat{u} = (\hat{u}_e, e \in E)$ as: $\hat{u}_e = \hat{h}_d$ if $e = a(d)b(d)$ and $\hat{h}_d > 0$, and $\hat{u}_e = 0$, otherwise.

Proposition 6. $\hat{h} \models h$ *if, and only if,* $\hat{u}$ *supports* h.

Proof. Suppose that $\hat{h} \models h$. Since (trivially) $\hat{u}$ supports $\hat{h}$, by Definition 2 $\hat{u}$ supports h.

Now suppose that $\hat{u}$ supports h and consider any link metric vector $\pi \in \Pi$. For each pair of nodes $v, w \in V$ define α_{vw} as the length of the shortest elementary path in graph G with respect to link metrics π (we can assume, without loss of generality, that G is connected). Let $\hat{\pi} = (\hat{\pi}_e, e \in E)$ be the vector of link metrics given by $\hat{\pi}_e = \alpha_{vw}$ where $e = vw$. Notice that $\hat{\pi}_e \leq \pi_e, e \in E$ and that $\lambda_d(\pi) = \alpha_{a(d)b(d)}, d \in D$.

We will also show that $\lambda_d(\hat{\pi}) = \alpha_{a(d)b(d)}, d \in D$ which means that $\lambda_d(\hat{\pi}) = \lambda_d(\pi), d \in D$, i.e., the lengths of the shortest paths in G calculated for metrics $\hat{\pi}$ do not change with respect to the lengths calculated for metrics π. To see this observe first that $\lambda_d(\hat{\pi}) \leq \lambda(\pi)$ because $\hat{\pi}_e \leq \pi_e, e \in E$. Moreover, the strict inequality $\lambda_d(\hat{\pi}) < \lambda_d(\pi)$ (for a $d \in D$) would imply (contrary to the definition of $\lambda_d(\pi)$) that there exists an elementary path p between nodes $a(d)$ and $b(d)$ with $|p|_\pi < \lambda_d(\pi)$ (recall that $|p|_\pi$ denotes the length of path p calculated for link metrics π). Suppose that $\hat{p} = \{e_1, e_2, \ldots, e_n\}$ is a shortest path between $a(d)$ and $b(d)$ with respect to metrics $\hat{\pi}$ ($|\hat{p}|_{\hat{\pi}} = \lambda_d(\hat{\pi})$). Let $p^1, p^2, \ldots, p^n$ be a sequence of elementary paths in graph G between the end nodes of links $e_1, e_2, \ldots, e_n$, respectively, with $|p^i|_\pi = \hat{\pi}_{e_i}, i = 1, 2, \ldots, n$. If $\lambda_d(\hat{\pi}) < \lambda_d(\pi)$ then the concatenation of the paths $p^1, p^2, \ldots, p^n$ would contain an elementary path p between $a(d)$ and $b(d)$ with $|p|_\pi \leq \lambda_d(\hat{\pi})$. Hence, the strict inequality $\lambda_d(\hat{\pi}) < \lambda(\pi)$ would lead to a contradiction.

By assumption $\hat{u}$ supports h which means that $\lambda(\hat{\pi})h \leq \hat{\pi}\hat{u}$ (the fact that $\hat{\pi}$ is not normalized, i.e., that $\sum_{e \in E} \hat{\pi}_e$ can be less than 1, does not matter here). Hence, since $\hat{\pi}\hat{u} = \lambda(\pi)\hat{h}$ we finally get $\lambda(\pi)h \leq \lambda(\pi)\hat{h}$, and thus, by Proposition 5, $\hat{h} \models h$. ∎

The result of Proposition 6 was first proven in [6], but only for fully connected network graphs.

Case 2: **Ring networks**

An important class of communication networks are ring networks whose graph $G(V, E)$ forms a cycle, i.e., a graph with n vertices (network nodes) and n edges (network links), and with $V = \{v_0, v_1, \ldots, v_{n-1}\}$ and $E = \{e_0, e_1, \ldots, e_{n-1}\}$ where $e_i = v_i v_{i+1}$, $i = 0, 1, \ldots, n-1$ (in arithmetic modulo n).

We say that demand $d \in D$ crosses an *edge-cut* $\{e_i, e_j\}$ if one node of demand d is in $v_{j+1}, v_{j+2}, \ldots, v_i$ and the other node is in $v_{i+1}, v_{i+2}, \ldots, v_j$. Let $h(e_i, e_j)$ denote the load of cut $\{e_i, e_j\}$ induced by demand vector h, i.e., the sum of volumes h_d for all demands $d \in \mathcal{D}$ that cross the considered cut.

A theorem of Okamura and Seymour [5], when specialized to a cycle G, asserts that u supports h if, and only if, the *edge-cut condition*, that is,

$$\forall\, 0 \leq i, j < n, \quad h(e_i, e_j) \leq u_{e_i} + u_{e_j} \tag{12}$$

holds (see [10]). Let $\hat{h}(e_i, e_j)$ denote the load of cut $\{e_i, e_j\}$ induced by demand vector $\hat{h}$.

Proposition 7. $\hat{h} \models h$ *if, and only if,*

$$\forall\, 0 \leq i, j < n, \quad \hat{h}(e_i, e_j) \geq h(e_i, e_j). \tag{13}$$

Proof. Implication $\Leftarrow$ follows directly form the edge cut condition. Indeed, $u \in \mathcal{U}(\hat{h})$ means that

$$\forall\, 0 \leq i, j < n, \quad \hat{h}(e_i, e_j) \leq u_{e_i} + u_{e_j}. \tag{14}$$

Hence, by assumption (13), property (12) holds, that is, $u \in \mathcal{U}(h)$.

The inverse implication can be proved analogously to the corresponding implication in Proposition 5. Let $\pi(e_i, e_j) \in \Pi$ denote the vector of multipliers corresponding to cut $\{e_i, e_j\}$ defined as follows: $\pi_e(e_i, e_j) = \frac{1}{2}$ if $e = e_i$ or $e = e_j$, and $\pi_e(e_i, e_j) = 0$, otherwise. Observe that for a cut $\{e_i, e_j\}$, inequality in (12) can equivalently be written as

$$\lambda(\pi(e_i, e_j))h \leq \pi(e_i, e_j)u. \tag{15}$$

Now assume that $\hat{h} \models h$ and consider a cut $\{e_i, e_j\}$. We define a capacity reservation vector $u(e_i, e_j) \in \mathcal{U}(\hat{h})$ as follows. For every demand $d \in D$, put the entire demand volume $\hat{h}_d$ on a shortest path with respect to $\pi(e_i, e_j)$ and for each $e \in E$ define $u_e(e_i, e_j)$ as the resulting link load. Note that the volumes $\hat{h}_d$ of the demands that do not cross cut $\{e_i, e_j\}$ do not contribute to $u_{e_i}(e_i, e_j)$ nor $u_{e_j}(e_i, e_j)$ since, by the definition of $\pi(e_i, e_j)$, their shortest paths are of length equal to 0. By the same argument the demands that cross the cut have the shortest path lengths equal to $\frac{1}{2}$, and their demand volumes contribute to $u_{e_i}(e_i, e_j) + u_{e_j}(e_i, e_j)$. This means that $\hat{h}(e_i, e_j) = u_{e_i}(e_i, e_j) + u_{e_j}(e_i, e_j)$ and hence $\lambda(\pi(e_i, e_j))\hat{h} = \pi(e_i, e_j)u(e_i, e_j)$. It follows that $\pi(e_i, e_j)$ is the optimal solution of the dual problem (10) for the capacity vector $u(e_i, e_j)$. By assumption, the capacity reservation vector $u(e_i, e_j)$ supports also h so for any $\pi \in \Pi$, $\lambda(\pi)h \leq \pi u(e_i, e_j)$. Thus, we finally see that

$$\frac{1}{2}h(e_i, e_j) = \lambda(\pi(e_i, e_j))h \leq \pi(e_i, e_j)u(e_i, e_j) = \lambda(\pi(e_i, e_j))\hat{h} = \frac{1}{2}\hat{h}(e_i, e_j) \tag{16}$$

which means that $\hat{h}(e_i, e_j) \geq h(e_i, e_j)$.   ■

4   Concluding Remarks

In the paper we have discussed necessary and sufficient (n-s) conditions for total and ordinary traffic domination. For the first case (total domination) we have presented a general n-s condition which can be easily checked in polynomial time. For the second case we have found an n-s condition that gives a strong evidence that checking for ordinary domination is $\mathcal{NP}$-hard, except for the two special cases we have been able to find. In fact, devising effective approximation methods for checking for ordinary domination seems to be an important and challenging research direction.

It should be noted that in Section 2 the results for total domination are given for undirected network graphs are not generally applicable to directed (even to bi-directed) graphs. On the contrary, the results for ordinary domination presented in Proposition 5 and in Proposition 6 are valid for both undirected and directed network graphs.

Acknowledgement. While working on the paper during his stay as an invited professor at Warsaw University of Technology, W. Ben-Ameur was supported by European Union in the framework of European Social Fund. P. Pavón was supported by the FP7 BONE project, by the MEC project TEC2010-21405-C02/TCM CALM, and by "Programa de Ayudas a Grupos de Excelencia de la R.

de Murcia, F. Séneca" – he had also stayed at Warsaw University of Technology while working on the results presented in the paper. M. Pióro was supported by the Polish Ministry of Science and Higher Education (grants no. 280/N-DFG/2008/0 and N517 397334), and by the Swedish Research Council (grant no. 621-2006-5509).

References

1. Ben-Ameur, W., Kerivin, H.: Routing of uncertain traffic demands. Optimization and Engineering 3, 283–313 (2005)
2. Ben-Ameur, W., Pióro, M.: On traffic domination and total traffic domination in communication networks with arbitrary graphs. Technical report, Institute of Telecommunications, Warsaw University of Technology
3. Jungnickel, D.: Graphs, Networks and Algorithms. Springer, Heidelberg (1999)
4. Lasdon, L.: Optimization Theory for Large Systems. McMillan Publishing Co., Inc. (1970)
5. Okamura, H., Seymour, P.D.: Multicommodity flows in planar graphs. Journal of Combinatorial Theory 31(1), 75–81 (1981)
6. Oriolo, G.: Domination between traffic matrices. Mathematics of Operations Research 33(1), 91–96 (2008)
7. Pavon-Marino, P., Garcia-Manrubia, B., Aparicio-Pardo, R.: Multi-hour network planning based on domination between sets of traffic matrices. To be Published in Computer Networks
8. Pavon-Marino, P., Pióro, M.: On total traffic domination in non-complete graphs. submitted to OR Letters
9. Pióro, M., Medhi, D.: Routing, Flow, and Capacity Design in Communication and Computer Networks. Morgan-Kaufmann (2004)
10. Shepherd, B., Zhang, L.: A cycle augmentation algorithm for minimum cost multicommodity flows on a ring. Discrete Applied Mathematics 110(2-3), 301–315 (2001)
11. Terblanche, S.E.: Contributions towards survivable Network Design with Uncertain Traffic Requirements. PhD thesis, North-West University, South Africa (2008)

Improving Clustering Techniques in Wireless Sensor Networks Using Thinning Process

Monique Becker, Ashish Gupta, Michel Marot, and Harmeet Singh

CNRS–SAMOVAR–UMR 5157 – TELECOM SudParis ; 9 Rue Charles Fourier,
91011 EVRY, France
{Monique.Becker,Ashish.Gupta,Michel.Marot}@telecom-sudparis.eu,
harmeet@iitk.ac.in

Abstract. We propose a rapid cluster formation algorithm using a thinning technique : rC-MHP(rapid Clustering inspired from Matérn Hard-Core Process). In order to prove its performance, it is compared with a well known cluster formation heuristic: Max-Min. Experimental results show that rC-MHP outperforms Max-Min in terms of messages needed to choose the cluster head, cluster head maintenance and memory requirement, comprehensively in sparse as well as in dense networks. We show that rC-MHP has a scalable behavior and it is very easy to implement. rC-MHP can be used as an efficient clustering technique.

1 Introduction

An ideal wireless sensor node consumes very little power, is software programmable, is capable of fast data acquisition and processing and costs little to purchase and install. To preserve energy, clustering is necessary in large networks. To our best knowledge, there are experimental analysis of sensors but not of clustering. Some objectives of clustering are listed below:

- Data aggregation and updates take place in Cluster Heads(CHs).
- Reduce network traffic and the contention for the channel.
- Cluster structure gives the impression of a structured and more stable network.

This paper proposes and validates a new clustering algorithm, rC-MHP (rapid Clustering inspired from Matérn Hard-Core Process). We observed from our experimental work on sensor networks (see [1], [2]) and from literature, e.g. [3], that sensor networks' behavior appeared to be much more complex than the theory led us to expect it to be. So, we validate rC-MHP empirically. Of course, in an empirical approach we can only have a limited number of sensors. Later on, these results can be integrated in simulations for the large networks.

In stochastic geometry, (see [4] page- 145-165), *"Thinning operation uses some definite rule to delete points of a basic process."* The thinning processes can be characterized into two types: independent and dependent thinning processes. In case of independent thinning, the points are independent of each other

K.A. Hummel et al. (Eds.): PERFORM 2010 (Haring Festschrift), LNCS 6821, pp. 203–214, 2011.

(location, in case of sensor nodes) and vice-versa in the other case. The Matérn Hard-Core Process (MHP) [4] is a dependent thinning process which is usually applied to a stationary Poisson point process. Here, we apply it for choosing cluster heads in rC-MHP.

Max-Min d cluster formation [5], proposes a distributed and scalable way of forming clusters in ad hoc networks. In Max-Min, each node has a weight and, based on the weight, a node decides whether it can be a cluster head or not. The *d-dominating*[1] set of CHs is first selected by using nodes identifiers and then clusters are formed. It does not give an optimal solution as the problem is NP-hard but it is a well known and efficient heuristic in selecting nodes having a high criterion value. In [6] the authors further corrected, validated and generalized Max-Min heuristic and showed that the rule 2 of the cluster formation may create loops in the network. The cluster formation seems very distributed once the weights are allocated to the nodes.

This paper shows that if each node wants to know the clustering criteria for its neighbors then there is a serious implementation problem in the real world. We test and compare a well known cluster formation heuristic (Max-Min) with our proposed method(rC-MHP) and show that rC-MHP is far more suited for the sensor networks. Max-Min is theoretically distributed and thus scalable. However, it is unscalable in practice because of the specific features of sensors: their limited memory. In contrast, rC-MHP behaves like a scalable algorithm. This paper shows, due to insignificant memory requirement and clustering overheads, rC-MHP outperforms Max-Min as an efficient way of clustering. These results are based on real experiments conducted on Tmote Sky sensors.

1.1   Description of the Matérn Hard Core Process

If the characteristics of the basic process are known then it is straightforward to calculate the characteristics of the point process produced by an independent thinning (cf. [4], page- 145-165). Thus, if Φ is the result of a $p(x)$-thinning of Φ_b, then its intensity measure $\wedge$ for a Borel set B is given by

$$\wedge (B) = \int_B p(x)\lambda_b \, dx \tag{1}$$

The *Matérn Hard core Process* (MHP)[4] is essentially a dependent thinning applied to a stationary Poisson Point process Φ_b of intensity λ_b. The points of Φ_b are marked independently by random numbers uniformly distributed over $\{0; 1\}$. The dependent thinning retains the point x of Φ_b with mark $m(x)$ if the sphere $b(x, h)$ contains no point of Φ_b with marks smaller than $m(x)$. Formally, the thinning process Φ is given by:

$$\Phi = \{x \in \Phi_b : m(x) < m(y) \forall y \in \Phi_b \cap b(x, h) \backslash \{x\}\} \tag{2}$$

---

[1] A d dominating set of the CHs is a set such that any node is not more than d hops away from its CH.

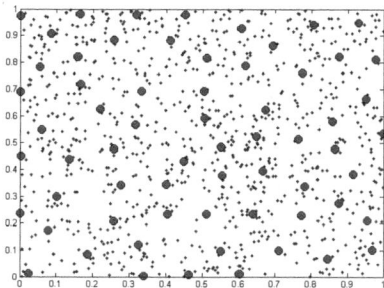

Fig. 1. Cluster-head positions after having applied the Matérn Hard core Poisson process where the intensity of the nodes is $\lambda = 1000$ and the distance $h = 0.1$ m. The side length of the square is 1 m.

The Fig. 1 shows the location of the cluster heads when the sensors are distributed via a Poisson point process with intensity 1000. Here, m(x) is the coordinate of a node. Initially, a node is randomly selected. Then, the Matérn hard-core process is applied and points are selected as cluster heads. The basic principle to use rC-MHP as the clustering mechanism is that inside an area/sphere there can only be a single cluster head. If the node falls in a zone where a CH is already present, it can't become a CH. In the real implementation, the Node_id will be the mark $(m(x))$ and h will be the LQI.

According to Baccelli et al. [7], the Matérn hard core process is a natural model for the access scheme of HiPERLAN (High Performance Radio LAN) type 1 and the MAC of HiPERLAN type 1 actually uses an advanced version of CSMA. [8] suggests that the MHP distribution gives regular points.

1.2 Description of Max-Min Cluster Formation Heuristic

The WSN can be modeled as a graph $G = (V, E)$, where two nodes are connected by an edge if they can communicate with each other. Let $x \in V$ be a node in the WSN. $N(x)$ is the set of neighbors of the node x and $W(x)$ is the weight of the node, which in our case is the degree of connectivity denoted $D(x)$. The clusterheads form a subset S of V which is a $d - dominating$ set over G. Let X be the image set of V by v ; v is a bijection of V over X. The reverse function is denoted v^{-1}: $\forall x \in V, \quad v^{-1}(v(x)) = x$.

Max-Min uses $2d + 1$ rounds, where d is the number of hops. A node can be d hops away from its CH. The algorithm includes $2d$ runs. The d first runs constitute the *Max phase*. The d last runs constitute the *Min phase*[2]. Each node updates two lists, *Winner* and *Sender*, of $2d + 1$ records. Winner is a list of

---

[2] The min is intended to avoid that a node which is too far from the max node gets to be CH even if it has lower mark. So, the idea is that a CH may be the minimum among the max marked, and it will get the lower mark nodes.

elements of X. Sender is a list of elements of V. Let us denote $W_k(x)$ and $S_k(x)$ the images of x for the functions W_k and S_k, defined by induction.

The basic idea of the $d-dominating$ set is: in the first phase, the Max phase, a node determines its dominating node (for a given criterion) among its d hop neighbors; then, in the Min phase, a node knows whether it is a dominating node for one of its neighbor nodes. If it is the case, this node belongs to the set S. For a given criterion, the only dominating set is built from this very simple process.

Initial Phase $k = 0$

$$\forall y \in V, W_0 = D(x), S(x) = x \tag{3}$$

FloodMax Phase: $k \in [|1; d|]$
Assuming that $\forall x \in V$, $W_{k-1}(x)$ and $S_{k-1}(x)$ are known in the previous step. Let $y_k(x)$ be a unique node in $N(x)$ defined by:

$$\forall y \in N(x)\backslash y_k(x), W_{k-1}(y_k(x)) > W_{k-1}(y) \tag{4}$$

W_k and S_k are calculated as follows:

$$\forall x \in V \quad W_k(x) = W_{k-1}(y_k(x)) \quad S_k(x) = y_k(x) \tag{5}$$

FloodMin Phase: $k \in [|d+1; 2d|]$,
Assuming that $\forall x \in V$, $W_{k-1}(x)$ and $S_{k-1}(x)$ are known in previous step. Let $y_k(x)$ be the unique node in $N(x)$ defined by:

$$\forall y \in N(x)\backslash y_k(x), W_{k-1}(y_k(x)) < W_{k-1}(y) \tag{6}$$

W_k and S_k are calculated as follows:

$$\forall x \in V \quad W_k(x) = W_{k-1}(y_k(x)) \quad S_k(x) = y_k(x) \tag{7}$$

The set of CHs is defined as follows:

$$S = \{x \in V, W_{2d}(x) = v(x)\} \tag{8}$$

2   Implementation

As per the definition of the Zigbee standard [9], the Link Quality Indication (LQI) measurement is a "characterization of the strength and/or quality of a received packet". Therefore, we used the LQI as a channel quality indicator between two nodes to create clusters. In case of the CC2420 trans-receiver, the LQI values range from 50 to 110. The minimum transmission power of the radio is -25 dBm which corresponds to a 5-6 m communication range. The maximum transmission power is 0 dBm which corresponds to 70 m (indoor). Both clustering algorithms use beacons for routing and the Base Station (BS) is their final destination. To filter out the bad links, we use a threshold LQI of 106 which corresponds to 98% Packet Reception Rate (PRR) [10]. All the nodes are in direct line of sight. We use the default CSMA/CA available in the TinyOS stack.

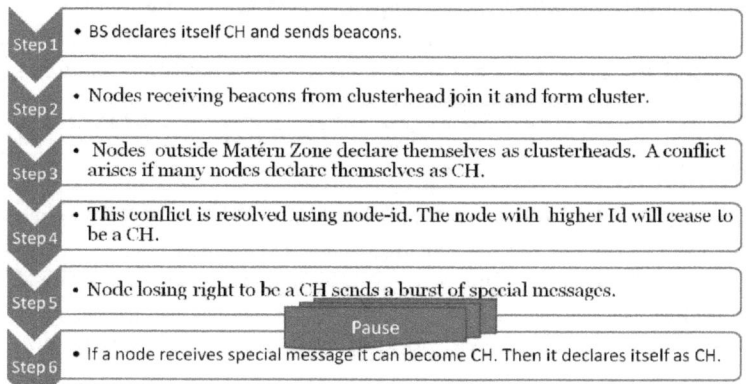

Fig. 2. rC-MHP

2.1 rC-MHP: A Rapid Clustering Inspired from MHP

rC-MHP as described in Fig. 2 is implemented in 6 steps:

Step 1: The BS declares itself as cluster head and sends beacons. When other nodes receive those beacons they check the LQI of these received beacons. If the LQI is ≥ 106, they join the BS as cluster nodes and form a cluster.

Step 2: Nodes receiving beacons from the BS with LQI < 106 declare themselves as CHs.

Step 3: A conflict may arise if more than one node in the same Matèrn zone declare themselves as CHs.

Step 4: This conflict is resolved using the node-id. Nodes with higher Id will cease to be a CH.

Step 5: A node ceasing to be a CH sends special messages: "I am no longer a CH".

Step 6: The other nodes receiving these special messages from their former CH participate again in the clustering process. A node waits/pauses before declaring itself to be CH to check (1) if it is still under influence of some other CH or (2) if all of its neighbors have had enough time to send the beacon. This makes clustering more efficient.

Once the clustering is over, the HybridLQI algorithm [11] is used as the inter-cluster routing algorithm. The HybridLQI improves the performance of the Mul-tihopLQI in asymmetric wireless link sensor networks. In HybridLQI, every node maintains the number of messages it sends to each of its neighbors and how many of them are being acknowledged. In this manner, the packet loss percentage (up-link channel quality) over the links is calculated. Beacons and the LQI are used to estimate the down-link channel. Therefore, without adding any extra cost to the network, the bidirectional channel quality is calculated. The Multiho-pLQI obtains the link quality from the LQI calculated from the received beacon (downlink) and assumes that the bidirectional links are symmetric.

Functioning: Initially, all nodes (except the BS) are unconnected (i.e., there is no path to the BS). All clustering will be done from beacon messages which are also being used for routing. BS (default CH) starts sending beacon messages. The reception of beacons is handled via three cases:

Case 1: If a node receives a beacon message from a CH with $LQI \geq 106$, three more conditions are possible:

- If the node is not connected, it will join that CH and set its status as connected node. Then, it starts sending beacon messages.
- If the node is connected but itself is not a CH, it will keep this CH in its cluster table.
- If it is a CH, it will check (via node id) if it can remain CH or not.

Case 2: If an unconnected node receives a beacon message from a connected node (either a cluster head or any other node which is connected), it checks if it can declare itself as clusterhead or not.

Case 3: The CHs use these beacons to discover a route towards the BS. The other nodes which are not CH will simply route to their clusterhead. Now, the CHs will send this packet to the BS. If two CHs are out of range they will communicate via a intermediate node of the other CH. A CH will never route its packets via a node which is in its cluster.

2.2 Max-Min Algorithm

Max-Min algorithm (for $d = 1$) is described in Fig. 3. It is implemented in 5 phases.

1. **Phase 1:** Each node broadcasts its Phase 1 message with $TTL = 1$ (Time To Live). This is done for the neighbor discovery. As in the case of rC-MHP, to be considered as a neighbor, the LQI between two nodes should be $>= 106$. In this way, the nodes determine their own weight (degree of connectivity, $D(x)$) and store this information in the neighbor table (a single entry per neighbor).

   - Max-Min's requirement to store information for all of its neighbors is inconvenient and needs memory.

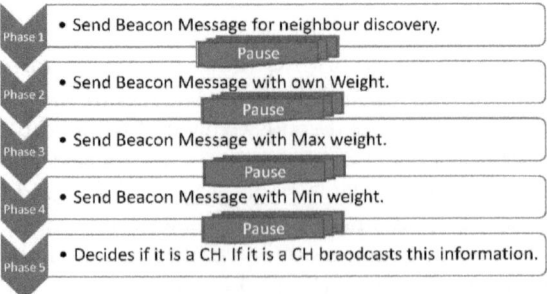

Fig. 3. Max-Min Algorithm

- If any node misses these phase packets, the selection of the CH may not be sound. Therefore, a node must send its Phase 1 messages repeatedly to ensure that all of its neighbors have received a phase message. The nodes which are in Phase 1 can only participate in Max-Min cluster formation heuristic.

2. **Phase 2:** Once a node finishes computing its weight, it needs to broadcast that weight again with $TTL = 1$. So, it sends the Phase 2 messages. The nodes which are in Phase 2 can only process these received phase packet (under condition $LQI \geq 106$). Based on these packets, at the end of the Phase 2, each node knows the Node_Id associated with the maximum weight.

3. **Phase 3:** The nodes send phase messages with their maximum weight as information. Accordingly, the neighbor tables are updated. This marks the end of the floodmax phase.

4. **Phase 4:** When the max weight is known, the nodes perform another set of calculations to know the minimum weight associated with the node and then broadcast it.

5. **Phase 5:** Finally, a node can decide if it can become a CH or not. A CH then sends a burst of beacon messages. On reception of these beacon messages, a non-CH node joins a cluster.

Why to pause after each phase: Initially, the nodes are switched ON at different time. So, to avoid a phase mismatch among the nodes, the nodes must wait before entering a new phase. During the pause, a node waits to receive phase messages from the other nodes, if it detects a new node then it retransmits its own phase packet.

After clustering, the HybridLQI is used for inter-cluster routing. If two CHs are out of the range they will communicate via a intermediate node of the other CH. The CH will never route its packets via a node which is in its cluster.

3 Analysis

The Table 1 lists the network deployment parameters for the rest of the paper. In the case of Max-Min, the term "Phase/Signal Message" refers to control messages that are used to form clusters. In case of rC-MHP, the term "Special Message" refers to control messages that are used to form/maintain clusters. A *BEACON_TIME_OUT* may be triggered if a node does not receive some beacon from its CH. A *BEACON_TIME_OUT* will lead to the eviction of the CH from the cluster/parent table of the node. However, the value of the *BEACON_TIME_OUT* should be sufficiently large to avoid a network instability. The cluster maintenance is also important as radio links are unreliable [1].

3.1 Effect of the Node Density on Max-Min

In Max-Min, the selection of CHs needs a lot of run time memory. Due to the limited available memory in the sensors, the size of the neighbor table was set

Table 1. Deployment Parameters

Packet Size	78 bytes
Frequency	6 packets per minute
Beacon Frequency	2 per minute
BEACON_TIME_OUT	4 x Number of nodes
Maximum Number of retries	5
Transmission Power	-25 dBm
Maximum Transmission Range Possible	5-6 meters
Channel	11
Threshold LQI	106
Maximum Distance for Threshold LQI	3 meters

to 12. The effect of the node density is shown via an example. We conducted two tests: first with 12 nodes whereas the second with 30 nodes. Both of these tests were conducted over an area of 5 m × 4 m. Basically, in the second set, we increased the node density of the network. In our implementation, Max-Min can only store up to 12 nodes. Further, by choosing 30, we are testing the case when the number of immediate neighbors are around 250% of the memory capacity.

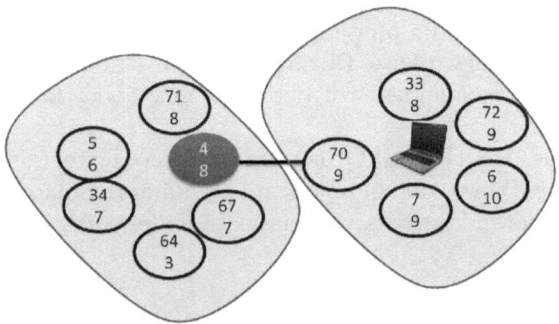

Fig. 4. Node id with its own weight. Clusters produced by Max-Min. There are 12 nodes including the BS.

Case A: 12 nodes including the BS: The Fig. 4 illustrates that Max-Min formed two clusters. In this experiment, the node 4 became a CH after completion of all the phases of Max-Min. So, the node 4 sent its beacon declaring that "I am a CH". Similarly, the BS (default CH) also sent its beacon declaring itself to be a CH. On the reception of these beacons, Non-CH nodes decided to join the respective cluster (based on first come first served basis).

Case B: 30 nodes including the BS: Fig. 5 illustrates that several clusters were produced by Max-Min when 30 nodes were deployed in the network. As the size of the neighbor was set to 12, the nodes were able to store up to 12 neighbors. We can see that most of the nodes have a weight equal to 12. What

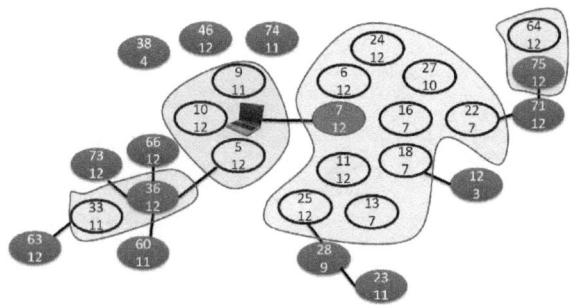

Fig. 5. Max-Min cluster formation, 30 nodes with node id and weight

happened is, the node A was in the neighbor table of the node B but the node B was not in the neighbor table of the node A. Due to this mismatch, several nodes became CH and more nodes became singleton clusters as fewer nodes were available to join many CHs.

Due to the limited memory capacity of the sensors, the scalability of Max-Min's is inhibited by its basic requirement: to know and store the weights of each of its neighbor. Hence, Max-Min is not scalable in practice.

3.2   Effect of the Node Density in rC-MHP

- rC-MHP does not need to store any clustering criteria for any of its neighbors. If a node is under the area of influence of a CH, it cannot declare itself as a CH. Therefore, the memory needed is independent of the node density.
- In rC-MHP, if the nodes have changed their state (CH ⇔ non-CH), they need to broadcast (TTL = 1) very few packets. Sensors don't need to know their degree of connectivity. So, a short burst of special message packets is enough in rC-MHP to diffuse changes. Underlying CSMA/CA ensures the channel availability.
- After the *BEACON_TIME_OUT*, a node will either join a CH or declare itself as a CH. Similarly, if the channel quality changed, a node may declare itself as a CH. Hence, rC-MHP will inherently do the clustering.

To validate the above hypothesis, we deployed rC-MHP in sparse (12 nodes) and dense (30 nodes) networks (5 m × 4 m). Fig. 6(a) Fig. 6(b) show the clustering for rC-MHP. Some nodes during the initialization phase or after the BEACON_TIME_OUT phase declare themselves as CHs. If the node detects that it is under other CH and looses the contention, it sends a burst of special messages stating "I am no longer a CH".

In case of sparse network (Fig. 6(a)), mostly, only a single CH (BS) could be observed. Sometimes, due to wireless channel problems or other factors such as BEACON_TIME_OUT, the nodes 17/63/64 declared themselves as CH(only one node at a time). The nodes in the Case 1 and 2 were deployed in similar conditions.

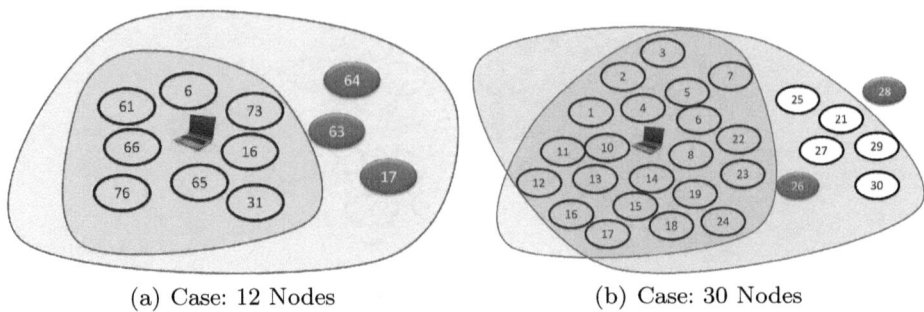

(a) Case: 12 Nodes   (b) Case: 30 Nodes

Fig. 6. rC-MHP for different numbers of nodes in the same area

Similarly, in the case of a dense network (Fig. 6(b)), for most of the time, only a single cluster could be observed. Once during the experiments, due to wireless channel problems or other factors such as BEACON_TIME_OUT, the node 28 declared itself as a CH and few other nodes joined it (only one node at a time). To summarize, the node density has no effect on rC-MHP, hence rC-MHP behaves like a scalable algorithm.

3.3   rC-MHP and Max-Min in Large Network

The Fig. 7(a) and Fig. 7(b) illustrate the positions of CHs when 63 sensors were deployed in an area of $17m \times 27m$. The transmission power was set to $-20dbm$ and the threshold LQI was 106. In case of rC-MHP, we can see that the distribution of the CHs is regular while in case of Max-Min the CHs distribution is not ordered. It is due to the fact that, in case of rC-MHP, only one node can remain CH in its area of influence. Since the sensors are regularly distributed, the result is uniform, while in Max-Min it depends on the last phase.

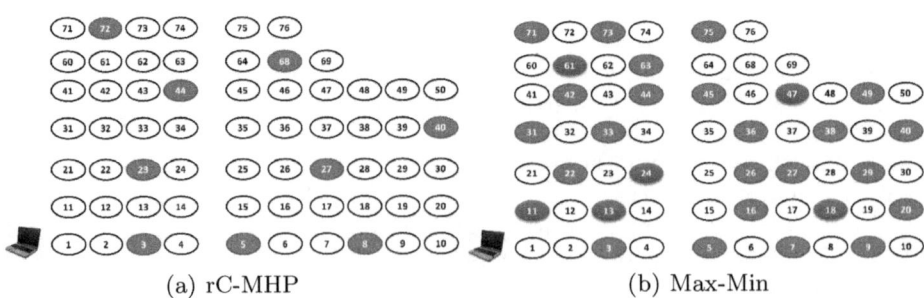

(a) rC-MHP   (b) Max-Min

Fig. 7. Location of Cluster Heads

3.4   rC-MHP Vs Max-Min - Performance Analysis

We compare rC-MHP and Max-Min based on the overhead to create and maintain clusters. Therefore, our performance criterion is "Number of Cluster/Phase

Table 2. Average number of clusters, data packets and hop counts per node

	Max-Min			rC-MHP			
	Signal Pkts per time out	Data Pkts per time out	Hop time Cnt.	Signal Pkts per time out	Signal Pkts with maintenance	Data Pkts with maintenance	Hop Cnt.
Sparse (12 nodes)	200	300	1.75	2.2	30	4000	1.2
Dense (30 nodes)	280	310	2.3	3	35	4000	1.1
Very Sparse (63 nodes)	130	250	4.67	3.5	40	8000	2.9

messages" per data packets. The Table 2 compares the performance of rC-MHP with Max-Min. In Max-Min, increasing node density resulted in a higher number of phases messages. For rC-MHP, the relative number of phase messages remained the same.

The sensor nodes were switched ON at different time. So, they entered different phases at different time. Therefore, Max-Min needed more time with high numbers of redundant phase messages to fully diffuse its clustering criteria. So, the number of phase messages increased along with number of neighbors. On average, after sending close to 300 data packets, at least one node in the network experienced a *BEACON_TIME_OUT*. So, the clustering procedure had to be repeated which increased the overhead. For the large network, nodes were spread over a larger area. Therefore the node density decreased, hence the relative number of cluster messages decreased.

From Table2, it can be easily seen that rC-MHP easily outperforms Max-Min. rC-MHP is independent of the node density. For sparse networks, only 30 cluster messages were needed to send 4000 (approx.) data packets. Even in the larger networks, only 40 cluster messages were used to send close to 8000 data packets. In rC-MHP, the cluster head maintenance is in-built. Very few messages were needed by using the underlying CSMA to convey the message "I am no longer a CH" or "I am a CH". Of course, there were several BEACON_TIME_OUT but we did not need to redo the clustering. So, it resulted into lesser cluster/signal packets. In case of Max-Min, the heuristic had to be rerun.

We mainly focused on the number of clusters in the network. As expected, the average hop count is also lower for rC-MHP than for Max-Min. So, rC-MHP consumes less energy. Though, we have not observed any problem for rC-MHP during the experiments with 65 sensors, we do intend to extend this work via simulations on large scale.

Complexity: Max-Min has a $O(n \times d)$ memory requirement where n is the number of neighbors while rC-MHP needs only 4 bytes for clustering. On a real implementation, Max-Min needed 4 Timers while rC-MHP needed one Timer.

4   Conclusion and Discussion

In this paper, we propose rC-MHP, a rapid clustering algorithm inspired from the Matérn Hardcore Process and implement it on Tmote Sky sensors. rC-MHP does not need to store any information of its immediate neighbors so it is very practical. It is very simple to code on sensors. It does not need much synchronization between the nodes, hence it can be concluded that rC-MHP is a natural algorithm which can be used with CSMA. We compare it with Max-Min cluster algorithm and empirically show that the clustering overhead for rC-MHP is negligible and it is feasible in real sensors. rC-MHP inherently does the cluster maintenance. For rC-MHP no memory is required for clustering (except for the clusterhead and the routing table) and the number of cluster messages does not increase with the node density.

References

1. Becker, M., Beylot, A.-L., Dhaou, R., Gupta, A., Kacimi, R., Marot, M.: Experimental Study: Link Quality and Deployment Issues in Wireless Sensor Networks. In: Fratta, L., Schulzrinne, H., Takahashi, Y., Spaniol, O. (eds.) NETWORKING 2009. LNCS, vol. 5550, pp. 14–25. Springer, Heidelberg (2009)
2. Gupta, A., Diallo, C., Marot, M., Becker, M.: Understanding topology challenges in the implementation of wireless sensor network for cold chain. In: IEEE Radio and Wireless Symposium, RWS 2010, New Orleans, USA (January 2010)
3. Woo, A., Tong, T., Culler, D.: Taming the underlying challenges of reliable multihop routing in sensor networks. In: SenSys (2003)
4. Stoyan, D., Kendall, W.S., Mecke, J.: Stochastic Geometry and Its Applications, 2nd edn. (September 1995)
5. Amis, A., Prakash, R., Vuong, T., Huynh, D.: Max-min d-cluster formation in wireless ad hoc networks. In: Nineteenth Annual Joint Conference of the IEEE Computer and Communications Societies (2000)
6. De Clauzade De Mazieux, A.D., Marot, M., Becker, M.: Correction, Generalisation and Validation of the "Max-Min d-Cluster Formation Heuristic". In: Akyildiz, I.F., Sivakumar, R., Ekici, E., Oliveira, J.C.d., McNair, J. (eds.) NETWORKING 2007. LNCS, vol. 4479, pp. 1149–1152. Springer, Heidelberg (2007)
7. Baccelli, F., Błaszczyszyn, B., Mühlethaler, P.: An aloha protocol for multihop mobile wireless networks. IEEE Transactions on Information Theory 52, 421–436 (2006)
8. Hoydis, J., Petrova, M., Mahonen, P.: Effects of topology on local throughput-capacity of ad hoc networks. In: IEEE 19th International Symposium on Personal, Indoor and Mobile Radio Communications, PIMRC 2008 (2008)
9. Ieee std. 802.15.4 - 2003: Wireless medium access control (mac) and physical layer (phy) specifications for low rate wireless personal area networks (lr-wpans)
10. Srinivasan, K., Levis, P.: Rssi is under appreciated. In: Third Workshop on Embedded Networked Sensors, EmNets (2006)
11. Gupta, A., Sharma, M., Marot, M., Becker, M.: Hybridlqi: Hybrid multihopplqi for improving asymmetric links in wireless sensor networks. In: The Sixth Advanced International Conference on Telecommunications, Barcelona, Spain, May 9-15 (2010)

AWPS – An Architecture for Pro-active Web Performance Management

Gabriele Kotsis and Martin Pinzger

Department of Telecooperation, Johannes Kepler University,
Altenberger Str. 69, 4040 Linz, Austria
Gabriele.Kotsis@jku.at, Mpinzger@gmail.com

Abstract. The growing demand for quality and performance has become a discriminating factor in the field of software applications. Specifically in the area of web applications, performance has become a key factor for success creating the need for new types of performance evaluation models and methods capable of representing the dynamic characteristics of web environments.

In this paper we will recall seminal work in this area and present AWPS, a tool for automatic web performance simulation and prediction. AWPS is capable of (a) automatically creating a web performance simulation and (b) conducting trend analysis of the system under test. The operation and usage of this tool is demonstrated in a case study of a two-tier architecture system.

Keywords: Web Performance, Simulation, Automation.

1 Introduction and Related Work

Today, performance represents the key to successful applications, especially in the field of web applications. In new software projects, performance aspects are often already part of the development process [6], yet there are still many projects where performance aspects have not been considered and the number of realizable improvements are limited.

The traditional approach to evaluate the performance of computer systems and networks is an "off-line" performance analysis cycle (see for example [11]). Starting with a characterisation of the system under study and a characterisation of the load, a performance model is built and performance results are obtained by applying performance evaluation techniques (including analytical, numerical, or simulation techniques). Alternatively, performance measurements of the real system can replace the modelling and evaluation step. In any case, an interpretation of results follows which can trigger either a refinement of the model (or new measurements) if the results do not provide the required insight, or which will lead to performance tuning activities (system side or load side) if performance deficiencies are detected.

It follows quite natural (and has been observed in the history of performance evaluation, see for example the report given in [8]) that changes in the computing environment (parallel and distributed computing, network computing,

K.A. Hummel et al. (Eds.): PERFORM 2010 (Haring Festschrift), LNCS 6821, pp. 215–226, 2011.
© IFIP International Federation for Information Processing 2011

mobile computing, pervasive computing, etc.) and new system features (security and reliability mechanisms, agent-based systems, intelligent networks, adaptive systems, etc.) pose new challenges to performance evaluation by raising the need for new analysis methodologies.

From a load modelling point of view, the difference in using computing resource has changed the type of model for workload characterization. While in the early days of computing (70s) the typical systems were used in batch or interactive mode, static workload models [4] could adequately represent the user behaviour. In the 80s, dynamic workload models were introduced, which were able to represent variabilities in user behavior ([1,5]). In the 90s, generative workload models [3,17] have been proposed as a suitable method for bridging the gap between the user's application oriented view of the load and the actual load (physical, resource oriented requests) submitted to the system. Hierarchical descriptions of workloads have proven their usability for representing workloads of parallel and distributed systems [7,18,2], or web traffic characterisation [13].

But the many aspects of emerging computing environments create a variety of performance influencing factors which cannot be adequately represented in a model. Applications in such environments are typically characterised by a complex, irregular, data-dependent execution behavior which is highly dynamic and has time varying resource demands. The performance delivered by the execution platform is hard to predict as it is usually constituted of heterogeneous components.

In addition, a demand is observed for immediate, embedded performance tuning actions replacing the traditional off-line approach of performance analysis. The human expert driven process of constructing a model, evaluating it, validating and interpretating the results and finally putting performance tuning activities into effect is no longer adequate if real-time responsiveness of the system is needed.

Therefore we follow an on-line, dynamic performance management approach as outlined in [12]. Several approaches have been proposed to support an automated performance evaluation and prediction, see for example [14,10,19], but only a few presenting fully integrated frameworks such as [21].

In this paper we demonstrate this general concept of performance management in web based information systems. The paper is separated into two main parts. First AWPS, a framework for automated web performance simulation and prediction, will be described, then a case study dealing with the two-tier architecture system will be presented.

Finally conclusions based on the case study will be drawn and from these, the next necessary steps will be derived.

2   AWPS Concept

With AWPS a system which is capable of automatically creating a web performance simulation and conducting a trend analysis of the SUT is presented [15]. The concept of AWPS is based on building a system which enables system administrators with little experience in the field of web performance simulation and prediction to analyze and monitor their system. The basic concept is to integrate the AWPS into the SUT in a transparent and non-intrusive way. From the three key functions of AWPS, data collection, simulation and prediction, only the datacollection part needs to be integrated into the SUT. The simulation and prediction component provide the user access to the system from an observer point of view, requiring no specific experiences in simulation or prediction.

2.1   Data Collection Component

The data collection component is made up of three elements which are the configuration component, the data recording component and the monitoring component. The configuration component provides access to static information about the SUT.

The task of the data recording component is to store and pre-process the collected information about the SUT, key features here are: capability to serve several components at the same time *(multi-threaded / multi-processor)*, to provide *high performance access to the stored data* and to *incrementally add data*.

The monitoring component provides the input interface to the SUT, for its realisation software monitoring [11] is used. Thereby three types based on their interaction with the environment are distinguished, active, passive and active & passive software monitoring. All three supported types are described in detail in [15], by default the passive software monitoring approach (e.g. network sniffing) is used, as it has no influence on the SUT.

2.2   Simulation Component

The simulation component represents a core component of the AWPS. The functions of this component are quite demanding concerning system resources and time. Therefore a key requirement for the so called *on-line simulation* is that the time required for the simulation, must not be higher than the real response time of the system. In the following subsections a description of the subcomponents is given.

Model Generation Component

The model generation component deals with the task of initially generating the simulation model. For this the component depends on the information provided by the data collection component. To accomplish this task, two solution types could be used in principle. These types are:

1. *Minimum complexity simulation model*: The idea is to start with the simplest possible simulation model, in case of queuing-network-models this would be a single class / single queue / single server model.
2. *Maximum complexity simulation model*: In a maximum complexity simulation model the system starts with a simulation model, which attempts to

include all possible combinations at startup. An example for this could be a queuing-network-model which is initialized with a multi class / multi queue / multi server model.

Currently, the AWPS uses the *maximum complexity simulation model* strategy, with the constraint that at least one monitoring point per server is provided.

As example for the simulation model generation process two examples with four monitoring points are provided. The first example is a system with a web server and an application server, the second example is a system with a web server, an application server and a database server. The basic monitoring points *Global In* and *Global Out* were placed at the Input and Output of the system, additonally one monitoring point was placed at the *Application Server Input.*

In the first example the fourth monitoring point is the *Application Server Out* inserted before the Global Out, by the AWPS the following formula are generated and used for the calculation of the request delay time.

$$TotalSystemTime = GlobalOut - GlobalIn \tag{1}$$

$$WebServerTimeA = AppServerIn - GlobalIn \tag{2}$$

$$AppServerTime = AppServerOut - AppServerIn \tag{3}$$

$$WebServerTimeB = GlobalOut - AppServerOut \tag{4}$$

$$WebServerTime = WebServerTimeA + WebServerTimeB \tag{5}$$

Based on this a simulation model as shown in Figure 1 is created. In this case it consists of two servers, two queues supporting multiple classes.

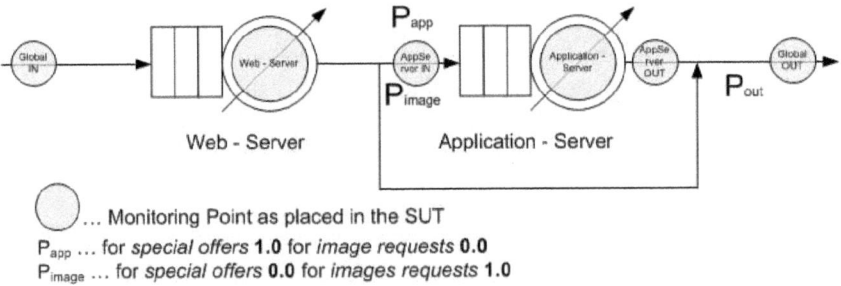

Fig. 1. Generated Simulation Model with 4 Monitoring Points

In the second example the fourth monitoring point is the *Database Server In* placed at the Database Server Input / at the Application Server Output. The following formula are generated and used for calculation by the AWPS.

$$TotalSystemTime = GlobalOut - GlobalIn \tag{6}$$

$$WebServerTime = AppServerIn - GlobalIn \tag{7}$$

$$AppServerTime = DBServerIn - AppServerIn \qquad (8)$$
$$DBServerTime = GlobalOut - DBServerIn \qquad (9)$$

In this case the setup would lead to a simulation model with three servers and three queues supporting multiple classes, as shown in Figure 2.

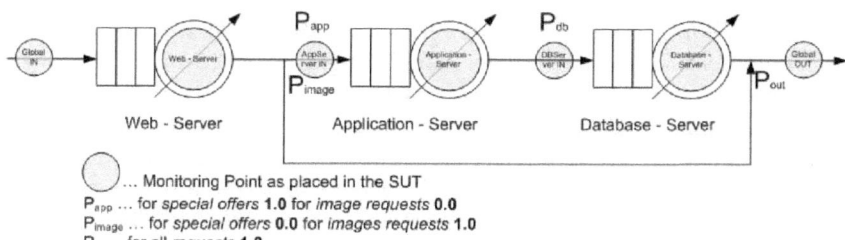

Fig. 2. Generated Simulation Model with 4 Monitoring Points (database server)

Model Comparison Component
To be able to determine whether to execute the model adjustment component or to keep the simulation model as it is, it is necessary to compare the results of the simulation model with the results of the SUT. For this the model comparison component uses on the one hand the monitoring information provided by the data collection component and on the other hand the results generated by the model simulation component which executes the simulation model.

The comparison process is structured in the following way. When a request is sent to the SUT, it is also inserted into the simulation model. Then the response should be created with the same time delay by the SUT and the simulation model. If there is a discrepancy (error) which is higher than the acceptable value, the simulation model needs to be adjusted.

Model Adjustment Component
The model adjustment component has several options to adjust the simulation model of the SUT. Depending on the performance of the simulation observed by the model comparison component, parameter changes (e.g. delay of a server) are applied or strategies are changed. The option to change a configuration value of the simulation model should be observed more closely. Four strategies for estimating the correct parameters for a link (server delay) in the simulation model should be mentioned: AVG strategy, median strategy, ARMA strategy and ARMA with 60 sec. grouping strategy [16].

Model Simulation Component
The model simulation component implements the execution environment for the simulation model. As already mentioned for the AWPS it is very important to have a simulation which is at least as fast as the SUT to work properly. Given

the demands it is the goal to create a model simulation component which is independent from a basic simulation environment, to be able to replace the simulation environment if the performance demands are not fulfilled. At the current state of the AWPS the JSIM [9] simulation framework is used as simulation environment.

2.3   Prediction Component

The prediction component is separated into four components, these are: the statistical component, the scenario generation & execution component, the longterm analysis component and reporting component.

The task of this *Statistical Component* is to provide a statistical analysis of the data provided by the data collection component. This analysis includes basic statistical information about the SUT, like for example min/max/mean/avg response time, min/max/mean/avg utilization, number of requests and number of errors.

The *Scenario Generation & Execution Component* uses the simulation model to execute generated scenarios and test for possible events which could occur in the SUT. Results provided by these experiments are reported back to the data collection component.

The basis for the *Longterm Analysis Component* is the information provided by the data collection component, the results from the scenario generation & execution component are of special importance. This information is processed and analyzed, to identify trends for the SUT concerning the perspective of availability and service level agreement.

The *Reporting Component* is intended as a base for the conversion of data which should be shared with the environment of the AWPS. To accomplish this, the reporting component has to provide a high flexibility and a suitable set of communication interfaces.

3   AWPS Environment Interaction

In this section the interaction of the AWPS with its environment is described. Especially information concerning the current development status of the AWPS and resulting limitations or simplifications are provided. In general there are two areas to deal with, the system setup where the components setup, which was used during the case study is described and the influence of the AWPS on the SUT.

3.1   System Setup

The AWPS is installed on a separate system which has been placed in the same network segment (collision domain) as the SUT, as a consequence the passive software monitoring strategy can be used. For the case study a multi class / single queue / multi server model is created, by the system, additional informations about the parameters can be found in [16].

The prediction component is integrated in a basic version so that, it is possible to evaluate the results and compare the predictions with the actually occurred events. The data recording component is fully integrated, for evaluation purposes a development feature which provides pre-recorded data is used.

3.2   Influence of the AWPS on the Productive System

The influence of the AWPS on the SUT is given by the overhead produced by the monitoring method and the additional network traffic coming from the effort to transfer the monitored information to the AWPS (data collection component). The AWPS setup used in the case study presented in section 4, does not influence the SUT since only passive software monitoring is used.

4   Case Study

The case study was done on a two tier web application, which provides as functionality a web page where you can search and book space flights. The SUT represents a fully operable environment which was developed by a company to illustrate their product. For proprietary reasons it cannot be referenced directly. The system uses a JSP front-end (41 JSP-Files) and a Web Service as back-end (30 Class-Files), in the following sections at first the system environment and then the detailed objectives of the case study are described. Then the used settings and the simulation model which is generated by the AWPS are presented, finally the results of the case study are discussed.

4.1   System Environment

The system environment for the two tier web application consists of a back-end system and a front-end system.

For the back-end system an Intel Core Duo with 2.20 GHz and 2 GB RAM with Microsoft Windows XP and Service Pack 2 was used. The front-end system was an Intel Centrino with 1.60 GHz and 1 GB RAM with Microsoft Windows XP and Service Pack 3. As web server for the front-end and as application server for the back-end the JBOSS in version 4.2.3 was used. The connection between the front-end system and the back-end system was established via a direct 100 Mbit Ethernet Network connection.

For the AWPS as platform an Intel Centrino with 1.60 GHz and 1 GB RAM with a Microsoft Windows XP with Service Pack 3, a PostgreSQL 8.3 database, Mathematica 5.2 and a WampServer 2.0 were used. The monitoring of the two tier web application was done by passive software monitoring as described in section 4.3.

4.2   Objectives of the Case Study

The objective of the case study is to provide an insight into the AWPS and to prove the simulation model generation process is feasible. Additionally the different integrated strategies used for simulation model adjustments are evaluated.

4.3   Case Study Settings

The case study settings can be separated in three parts on the *system load generation*, the *AWPS configuration* and the *monitoring configuration*.

At first the *system load generation* should be specified, the artificial system load for the case study was realized using the Microsoft Web Application Stress 1.1 Tool [20]. The tool was configured to send requests to the system requesting 12 different web pages, in the time frame of an hour 69621 request were sent.

Secondly the *monitoring configuration* should be provided, as mentioned in the system environment section, passive software monitoring was used. The monitoring was done on three points:

1. *Web Server In / Global In*, this monitoring point represents the first of the two minimal required monitoring points. In this and in the common case this monitoring point is realized by the capturing of the traffic (requests) sent to the web server.
2. *Application Server In*, based on this monitoring point the AWPS can determine that more than one server is used. Additionally the distribution of the response time between the front-end and the back-end can be determined.
3. *Web Server Out / Global Out*, this monitoring point represents the second of the two minimal required monitoring points. As the *Global In* monitoring point it captures the messages sent between the server and the client, in this case the response message (HTTP) is captured.

Finally the *AWPS configuration* is provided for the 12 executed test cases: The four model adjustment strategies (AVG strategy, Median strategy, ARMA strategy, ARMA with 60 sec. Grouping strategy) were evaluated with step size of 1000, 100 and 10. With the step size the number of request considered by the system before reevaluating the configuration of the AWPS is meant. This value also defines the number of request which are predicted by the AWPS. Additionally the system was configured to support multiple classes of requests, in concrete the requests for images (GIF) and the requests for sites which require a query for special offers (do?action=special) in the back-end were considered. It should be stated that the specification of the considered request types is optional, if no information is provided the system learns the request types automatically.

Predefined values of the AWPS configuration which were used during the evaluation are a *AVG Strategy Windows Size of 5000*, a *Median Strategy Position in percent of 50* and a *pMax and qMax of 5 for the ARMA and ARMA with 60 sec. grouping Strategy*.

4.4   Generated Simulation Model

The generated simulation model is based on the AWPS configuration and on the information provided by the monitoring points. The three monitoring points listed in section 4.3 are used to calculate a request delay time for the SUT automatically, the following formula illustrate this (no user interaction is required).

$$TotalSystemTime = GlobalOut - GlobalIn \tag{10}$$

$$WebServerTime = AppServerIn - GlobalIn \tag{11}$$
$$AppServerTime = GlobalOut - AppServerIn \tag{12}$$

This calculation is done for each request class, out of this the time consumption for the images (GIF) turns out to be zero for the Application Server. This is interpreted by the system and results in the conclusion that the images (GIF) request do not need the Application Server they only require the Web Server. For the special offers (do?action=special) request the system returns a time for the Web Sever and the Application Server which is interpreted accordingly.

Based on this request delay time structure a similar simulation model as shown in Figure 1 is created automatically. Only the indicated monitoring point *AppServer Out* is not included, as this monitoring point is not configured in the system setup (section 4.3).

4.5   Case Study Results

The data can be separated into three parts, the calculation time consumption, the comparison for significant differences between simulated and reference data and the comparison of the delta error between the integrated simulation model adjustment strategies.

The *calculation time consumption* consists of the total time required for the system to generate, adjust and execute the simulated model. The periods in which the AWPS was idle waiting for new requests is not included. Out of these a calculation time for the different integrated adjustment strategies for the 60 minutes case study can be provided, see Table 1. The results shown in the table show that with reducing the step size the calculation time increases, however for all methods for a step size of 1000 and 100 the time consumption is below 60 minutes so that means that the system is online capable. For a step size of 10 the system shows problems with the ARMA strategy. These limitations can be solved by additional hardware. Preliminary test results indicate that an Intel Dual Core with 2.4 Ghz and 2GB of RAM is already sufficient.

The *significant difference between simulation and reference data*, is evaluated with a double sided t-Test with a threshold of 5 percent between the results. In Table 2 these results are presented and indicate that the differences are not significant for the special offers (do?action=special) request class for a step size of 1000 for the AVG, ARMA and ARMA G. method. There is no significant

Table 1. Calculation time consumption. Values below 3600 sec. mean that the adjustment method is online-capable.

Strategy	Step Size 1000	Step Size 100	Step Size 10
AVG	244 s	511 s	2659 s
Median	489 s	541 s	3243 s
ARMA	339 s	945 s	**6706 s**
ARMA G.	288 s	827 s	2947 s

difference for a step size of 100 and 10 for the ARMA. However a significant difference in the image (GIF) request class is given. The reason for this is that the difference in the response time of the request class is different to the special offers (do?action=special) request class response time by a factor of 1000. The cacheing strategies of the simulated server differ in this special case too much from the cacheing strategies of the web server of the SUT. If only the image (GIF) request class or a *slow* image (GIF) request class is considered the difference between the SUT and the simulates system again is not significant, because the influence of the right chasing strategy on the server is reduced.

Table 2. Significant difference between simulation and reference data. The value in the brackets represents the double sided t-Test value.

Strategy	Step Size 1000		Step Size 100		Step Size 10	
	image	special	image	special	image	special
AVG	Yes.	**No (0.980230)**	Yes	Yes (0.935309)	Yes	Yes (0.789822)
Median	Yes.	Yes (0.758424)	Yes	Yes (0.335950)	Yes	Yes (0.002655)
ARMA	Yes.	**No (0.982876)**	Yes	**No (0.990521)**	Yes	**No (0.998247)**
ARMA G.	Yes.	**No (0.992849)**	Yes	Yes (0.289593)	Yes	Yes (0.000000)

Table 3. Mean values and Variance values in seconds referring to the delta error, for the special offers (do?action=special) request class

Strategy	Step Size 1000		Step Size 100		Step Size 10	
	mean	variance	median	variance	median	variance
AVG	0.002228	0.000002	0.005419	0.000013	0.023877	0.000514
Median	0.017624	0.000052	0.017543	0.000084	0.020431	0.001013
ARMA	0.005091	0.000012	0.006319	0.000022	0.023590	0.000568
ARMA G.	0.006824	0.000019	0.088866	0.055627	0.230810	0.060259

The significant difference in the delta error of the integrated simulation adjustment method, meaning the absolute value of the error between the simulated values and the measured values is analyzed with a two sided t-Test with a threshold of 5 percent. The analysis is done to determine if there is a significant difference between the integrated adjustment strategies. The results clearly indicate that the ARMA strategy provides the best results for all step sizes, the difference between ARMA and ARMA G. is minimal at a step size of 1000 but still significant. For a step size of 10 the t-Test indicates the AVG method is the best approximation, however the difference is still significant. In Table 3 the mean values and the variance values of the integrated simulation adjustment methods are presented to provide some insight concerning the range of the values, keeping in mind the values, reference to the delta error and the fact that reducing the step size also means that the smoothing effect by the average calculation is reduced.

5   Conclusion

As demonstrated by the case study, AWPS works as expected and provides representative results. The simulation model generation process works autonomously and is sufficiently fault-tolerant. The integrated strategies for the adjustment of the simulation model work accurately, especially when the autoregressive moving average (ARMA) model strategy is used. The AWPS provides suitable results not only for artificial systems shown in [16], it also provides representative results for a productive system, as demonstrated with this case study.

The known limitation of the AWPS is that at least two monitoring points are required and that for each server which should be considered in the simulation at least one monitoring point is required, two monitoring points per server ensure accurate and reliable results as presented in the case study but at the costs of slightly higher level of intrusion.

Nevertheless, improvements of the AWPS are necessary in two ways: the functionality should be enhanced (e.g. adaptive scenario generation), and additional case studies need to be done, including the analysis of a productive system under high (real) load.

References

1. Calzarossa, M., Italiani, M., Serazzi, G.: A Workload Model Representative of Static and Dynamic Characteristics. Acta Informatica 23, 255–266 (1986)
2. Calzarossa, M., Merlo, A.P., Tessera, D., Haring, G., Kotsis, G.: A hierarchical approach to workload characterization for parallel systems. In: HPCN Europe, pp. 102–109 (1995)
3. Dulz, Q., Hofman, S.: Grammar-based workload modeling of communication systems. In: Proc. of Intl. Conference on Modeling Techniques and Tools For Computer Performance Evaluation, Tunis (1991)
4. Ferrari, D.: Workload Characterization and Selection in Computer Performance Measurement. Computer 5(4), 18–24 (1972)
5. Haring, G.: On Stochastic Models of Interactive Workloads. In: Agrawala, A.K., Tripathi, S.K. (eds.) PERFORMANCE 1983, pp. 133–152. North-Holland (1983)
6. Haring, G., Ferscha, A.: Performance oriented development of parallel software with capse. In: Proceedings of the 2nd Workshop on Environments and Tools for Parallel Scientific Computing. SIAM (1994)
7. Haring, G., Kotsis, G.: Workload modeling for parallel processing systems. In: Dowd, P., Gelenbe, E. (eds.) Proc. of the 3rd Int. Workshop on Modeling, Analysis and Simulation of Computer and Telecommunication Systems, MASCOTS 1995, pp. 8–12. IEEE Computer Society Press (1995) ISBN 0-8186-6902-0 (invited Paper)
8. Haring, G., Lindemann, C., Reiser, M. (eds.): Performance Evaluation of Computer Systems and Communication Networks. Dagstuhl-Seminar-Report, No 189 (1997)
9. Hou, J.: J-sim official (January 2005), http://sites.google.com/site/jsimofficial/
10. Israr, T.A., Lau, D.H., Franks, G., Woodside, M.: Automatic generation of layered queuing software performance models from commonly available traces. In: WOSP 2005: Proceedings of the 5th International Workshop on Software and Performance, pp. 147–158. ACM, New York (2005)

11. Jain, R.: The Art of Computer Systems Performance Analysis, pp. 93–110. John Wiley and Sons, Inc. (1991)
12. Kotsis, G.: Performance Management in Dynamic Computing Environments. In: Calzarossa, M.C., Gelenbe, E. (eds.) MASCOTS 2003. LNCS, vol. 2965, pp. 254–264. Springer, Heidelberg (2004) (ISBN 3-540-21945-5)
13. Kotsis, G., Krithivasan, K., Raghavan, S.V.: Generative workload models of internet traffic. In: Proceedings of the ICICS Conference, pp. 152–156. IEEE, Singapore (1997)
14. Mos, A., Murphy, J.: A framework for performance monitoring, modelling and prediction of component oriented distributed systems. In: WOSP 2002: Proceedings of the 3rd International Workshop on Software and Performance, pp. 235–236. ACM, New York (2002)
15. Pinzger, M.: Automated web performance analysis. In: ASE, pp. 513–516. IEEE (2008)
16. Pinzger, M.: Strategies for automated performance simulation model adjustment, preliminary results. In: Workshop on Monitoring, Adaptation and Beyond (MONA+) at the ServiceWave 2008 Conference. Universität Duisburg-Essen, ICB-Research Report No. 34 (2009)
17. Raghavan, S.V., Vasuki Ammaiyar, D., Haring, G.: Generative networkload models for a single server environment. In: SIGMETRICS, pp. 118–127 (1994)
18. Raghavan, S.V., Vasuki Ammaiyar, D., Haring, G.: Hierarchical approach to building generative networkload models. Computer Networks and ISDN Systems 27(7), 1193–1206 (1995)
19. Sancho, P.P., Juiz, C., Puigjaner, R.: Automatic performance evaluation and feedback for mascot designs. In: WOSP 2005: Proceedings of the 5th International Workshop on Software and Performance, pp. 193–194. ACM, New York (2005)
20. Microsoft AppCenter Service Team. Microsoft web application stress 1.1 (1999), `http://webtool.rte.microsoft.com/`
21. Zhang, L., Liu, Z., Riabov, A.V., Schulman, M., Xia, C.H., Zhang, F.: A Comprehensive Toolset for Workload Characterization, Performance Modeling, and Online control. In: Kemper, P., Sanders, W.H. (eds.) TOOLS 2003. LNCS, vol. 2794, pp. 63–77. Springer, Heidelberg (2003)

Analysis of Web Logs: Challenges and Findings

Maria Carla Calzarossa and Luisa Massari

Dipartimento di Informatica e Sistemistica
Università di Pavia
I-27100 Pavia, Italy
{mcc,massari}@unipv.it
http://peg.unipv.it

Abstract. Web logs are an important source of information to describe and understand the traffic of the servers and its characteristics. The analysis of these logs is rather challenging because of the large volume of data and the complex relationships hidden in these data. Our investigation focuses on the analysis of the logs of two Web servers and identifies the main characteristics of their workload and the navigation profiles of crawlers and human users visiting the sites. The classification of these visitors has shown some interesting similarities and differences in term of traffic intensity and its temporal distribution. In general, crawlers tend to re-visit the sites rather often, even though they seldom send bursts of requests to reduce their impact on the servers resources. The other clients are also characterized by periodic patterns that can be effectively represented by few principal components.

1 Introduction

The Web has become a phenomenon of growing social, economic and cultural importance and an essential component of the modern society that attracts million of users and accesses daily. On the Web, users distribute and share information and knowledge, conduct businesses, communicate, socialize and develop relationships. To discover, locate and retrieve the huge amount of information published on the Web, crawling has emerged as a key enabling technology [17].

Many applications and services rely on crawling. For example, to facilitate navigation and provide users with up-to-date information, search engines periodically crawl Web sites to index, group and cache Web content. Other applications crawl the Web for different legitimate or malicious purposes: to maintain a site, discover Web services, collect email addresses and personal information, extract business intelligence, exploit vulnerabilities.

Crawling employs programs, known as Web crawlers or robots, that automatically access and download Web pages without continuous involvement of human users. These programs are expected to comply with the Robot Exclusion Protocol [11], a standard that allows Web site administrators to specify, in the `robots.txt` file, the rules of operation of the crawlers. Nevertheless, some of them ignore the file and the rules, thus leading to potential performance problems as well as to privacy and security concerns [24]. It is then important to

K.A. Hummel et al. (Eds.): PERFORM 2010 (Haring Festschrift), LNCS 6821, pp. 227–239, 2011.

identify the presence of both ethical and malicious crawlers as they might have a considerable impact on the infrastructures, thus hindering normal user accesses and causing damages and even economic losses.

Web access logs represent an important source of information to describe and understand Web server traffic and usage patterns as well as users behavior. Logs provide useful inputs to a large variety of engineering activities, ranging from the improvement of the site structure and organization, to the provisioning of personalized content, the development of recommendation systems, the selection of prefetching and caching policies, the formulation of content distribution and replication strategies. Moreover, the multiplicity of statistics, metrics and diagrams about the visitor traffic derived by the tools specialized in the analysis of the content of Web logs, can be used for commercial purposes, to develop, for example, customized marketing strategies or new business models or to attract advertisements.

In this paper we study Web servers access logs with the objective of modeling the access patterns of the visitors and identify typical navigation profiles as well as clients trying to compromise the servers by issuing malicious requests. The outcomes of this analysis could be used to develop proactive policies aimed at enhancing server availability, security and performance as well as to define the input of load generators used, for example, for benchmarking experiments.

Our study is experimental, that is, based on the analysis of the logs collected during more than one year on two Web servers. The choice of these servers is motivated by their characteristics, such as, potential users and traffic, that make them particularly suitable to assess our methodological approach. One server hosts an academic site mainly used by students and researchers of Computer Science [19], whereas the other hosts the European mirror of the SPEC (Standard Performance Evaluation Corporation) Web site [21] whose content is of interest for the entire community of IT specialists.

The paper is organized as follows. Section 2 briefly discusses the state of the art in the area of the analysis of Web workload. The main characteristics of the two Web servers considered in our study and the results of the preliminary exploratory analysis of their traffic are presented in Section 3. The methodological approach applied for the identification of the navigation profiles and its outcomes are described in Section 4. Finally, Section 5 concludes the paper by pointing out the major findings and challenges encountered in the analysis of Web logs.

2   Related Work

Logs have been used as the basis of many studies since the early days of the Web (see e.g., [2,3,8,14,15,20,25]). Most of these studies focused on the characteristics of the workload being processed by the servers and used the information extracted from their access logs to describe the properties of the workload in terms of various metrics, such as, document types and popularity, file size distribution, concentration of references, inter-reference time. In particular, Arlitt and Williamson presented in [2] ten invariants, i.e., characteristics common to all the sites that are likely to persist over time. A more recent paper [25] has actually shown that, even though the Web traffic has dramatically increased in ten years, the same invariants can accurately capture its properties.

Other studies focused on the analysis of Web logs with the objective of identifying the users behavior. Graph-based models were proposed to represent the navigation profiles of the customers of e-commerce sites [16].

As Web crawlers are responsible of a large fraction of the Web traffic, several authors addressed specifically their attention to the identification and characterization of this type of traffic (see, e.g., [1,4,13,22,23]). Some of these studies analyzed its overall characteristics, whereas others took into account some more specific aspects. For example, Dikaiakos et al. [4] characterized and compared the behavior of the crawlers of five popular search engines by analyzing access logs collected on various academic Web servers. The study introduced a set of metrics that provide a qualitative description of the behavior of these crawlers. Lee et al. [13] analyzed a very large number of transactions recorded by a commercial server over a 24 hours period to investigate the characteristics of various Web robots. Metrics associated with HTTP traffic features and resource types were then used for the classification of the robots.

In [23] Tai and Kumar introduced the concept of Web robot sessions and used some access features derived from the Web server logs for their identification and classification. Sessions considered in the framework of search engines were studied in [6] where a multidimensional approach was applied to Web search logs to derive a systematic classification of users as humans or robots.

On the contrary, the classification of Web robots presented in [5] took into account the influence exerted by the goals and the functions performed by robots on their navigational patterns. Mouse clicks streams were used in [18] to infer whether the traffic source is a human or a robot.

Our study complements these studies because of the perspective adopted for the analysis of Web logs. More specifically, starting from the access patterns of the individual visitors, we apply various types of statistical techniques to highlight differences, similarities and peculiarities in the behavior of Web crawlers and human users.

3   Data Sets

The data sets used in our investigation are represented by the logs collected on two Web servers, hosting an academic site at the University of Pavia in Italy and the European SPEC mirror site, respectively.

Both servers record the details of the HTTP transactions being processed according to the Extended Log File Format [7]. In particular, a transaction is described by the IP address of the client that issued the HTTP request, the timestamp of the transaction, the method and resource requested, the status code of the server response, the number of bytes transmitted by the server, the referrer of the previous site visited by the client, and the user agent, that is, the browser used by the client to issue the request.

As a first step, we performed an exploratory analysis of the Web logs to derive from these large volumes of information some preliminary insights in the characteristics of the workload of these Web servers. Note that the information stored in the two logs files accounts for about 50Mbytes and 970Mbytes, respectively.

Table 1 summarizes the main characteristics of the transactions processed by the two servers. As already pointed out, the sites considered in our study

Table 1. Main characteristics of the servers workload

	Academic server	SPEC server
Measurement interval	14 months	12 months
Total number of transactions	239,081	5,098,621
Total number of 2xx transactions	144,081	3,977,929
Total number of 4xx transactions	27,197	143,863
Total GBytes transmitted	8	129
Number of clients	7,940	19,135
Number of one-time clients	1,034	3,364

differ in terms of potential users, hence, their traffic intensity is quite different. In a period of approximately 14 months, since April 2009, the academic server processed some 560 HTTP transactions per day and transmitted 18.5Mbytes of data. The SPEC server, with its 14,000 transactions per day, was much busier. In 12 months, it processed more than five million transactions and transmitted 129Gbytes of data in total.

The transactions with status code 2xx refer to the requests of the client successfully received, understood and accepted by the server, whereas the 4xx status code refers to bad requests due to client errors. As can be seen, the large majority, i.e., 78%, of the transactions processed by the SPEC server was successful and bad transactions were very few: their fraction did not reach the 3%. It is also worth noting that 1,556 requests could not be processed because of temporary server errors. On the contrary, for the academic server about 60% of the requests were successful but a good fraction of requests, i.e., 11.4%, was bad. Most of these requests were to non-existing resources, e.g., various types of PHP scripts mainly developed to exploit server vulnerabilities.

Another clear indicator of the different behavior of the two servers is represented by their hourly traffic. As Fig. 1 shows, the traffic over the 24 hours of the academic server follows a typical diurnal pattern with its highest peak at noon, whereas for the SPEC mirror the traffic is basically flat with transactions evenly distributed and no significant difference between day and night.

As we will explain in more details later on, this is mainly due to the very strong presence of crawlers that are responsible for the majority of the traffic of this server.

In what follows, we focus on the analysis of the visitors identified by means of the IP addresses specified in the logs. Although these addresses do not univocally

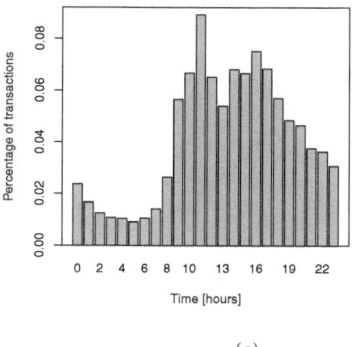
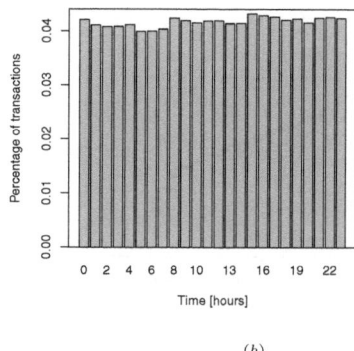

(a)   (b)

Fig. 1. Percentage of transactions processed by the academic server (a) and by the SPEC server (b) over the 24 hours

identify individuals because of the dynamic assignment of addresses and of their management within organizations, they seemed appropriate for the purposes of our work.

The behavior of the clients in terms of total number of requests and total number of bytes transmitted by the servers during our measurement interval are shown in Fig. 2 and in Fig. 3, respectively. Note that the y axes of the plots are in log scale.

The clients of the academic server issued on average 30 requests each, even though one of them issued as many as 11,953 requests. Moreover, three-quarter of the clients sent at most 20 requests, and 13% one request only, that is, they are the so-called "one-timers". The clients of the SPEC server exhibit a rather different behavior: one client is responsible of the 7.2% of the total traffic of this server and some 30 clients account for half of the traffic. Moreover, three-quarter of the clients issued 27 requests at most.

In terms of bytes, the number of bytes downloaded by three-quarter of the clients of both servers did not exceed 375Kbytes, nevertheless, few clients downloaded most of the bytes transmitted by the servers. For example, one client downloaded from the SPEC server as many as 9.2Gbytes. It is worth noting that about 5% of the clients of the academic server did not download any byte because their HTTP requests either used a HEAD method or specified an "If-modified-since" header and the corresponding pages were not transmitted as they were not modified by the server since their latest download. These requests represent about 4% of the workload of this server. In summary, on average clients of the academic server downloaded 1Mbytes of data each, compared to 6.8Mbytes of the SPEC clients.

Before studying the navigational profiles of the clients, we did some preprocessing of the log files to identify "well-known" crawlers and assess their impact on the overall traffic of the servers. More precisely, we recognized clients as being crawlers, either because they accessed the `robots.txt` file or because of some explicit information in the user agent field.

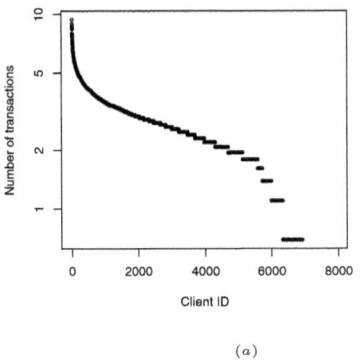
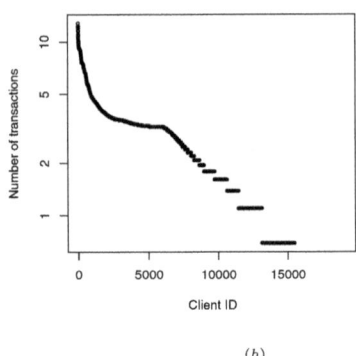

(a)   (b)

Fig. 2. Total number of transactions per client of the academic server (a) and the SPEC server (b)

With this pre-processing, we classified as crawlers about 16% of the clients of the SPEC server, namely, 3,108, and 12% of the clients of the academic server, namely, 974. It is interesting to outline that in terms of traffic, while crawlers account for about 15% of the traffic of the academic server, the situation is completely different on the other server, where crawlers are responsible for the vast majority of its traffic, namely, for about 4.8 million requests, out of approximately five million, and of 122Gbytes of data, out of 129Gbytes. The crawlers of three major search engines, i.e., Google, Microsoft and Yahoo, emerged as the top crawlers on both servers as they generated about 80% of their traffic. Table 2 presents the main characteristics of the traffic produced by these three top crawlers on the SPEC server.

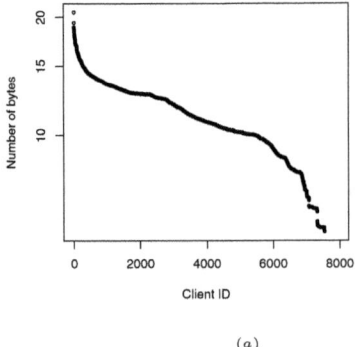
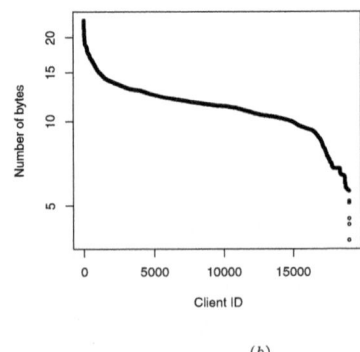

(a)   (b)

Fig. 3. Total number of bytes transmitted per client by the academic server (a) and the SPEC server (b)

Table 2. Main characteristics of the traffic of the three top crawlers of the SPEC server

	Google	Microsoft	Yahoo
Total number of transactions	1,429,954	2,147,582	238,202
Total number of 2xx transactions	1,156,072	1,434,838	227,640
Total number of 4xx transactions	36,952	34,742	1,150
Total GBytes transmitted	48	46	6
Number of clients	535	958	268

From now on, we investigate separately the behavior of clients identified as crawlers and of the remaining clients, that might include human users as well as crawlers that did not identify themselves mainly because of their malicious intentions.

4   Navigation Profiles

The methodological approach followed for the analysis and characterization of the navigation profiles of the visitors of the two Web servers is based on the selection of the parameters that describe their behavior and the application of various statistical techniques to uncover differences and similarities among profiles.

The parameters used to describe the navigation profiles of the individual clients were related to the traffic generated by the clients and their temporal distribution. More specifically, the inter-reference time, that is, the time elapsed between two consecutive requests of a given client, is a good metric to describe the profiles in terms of traffic intensity.

Figure 4 shows the details of the distributions of the inter-reference times measured on the academic server for all the requests of the crawlers and of the other clients.

The average inter-reference time of crawlers is much larger than for the other clients; the 90-th percentile of the distribution is about 310,000 seconds, that is, more than 86 hours, compared to 22 seconds for the other clients.

This investigation has shown that, whenever the inter-reference time was larger than 240 seconds for the crawlers and 120 seconds for the other clients, a navigation session was basically over, that is, the client will start a new session. A session is then defined as the sequence of requests issued by a client and characterized by inter-reference times smaller than the selected thresholds. As a consequence, the navigation profile of a client can be described in terms of number of sessions and their duration, number of requests per session and inter-session time, that is, the time between two consecutive sessions of a given client. Table 3 presents the average characteristics of the navigation profiles of the crawlers and of the other clients of the SPEC server in terms of these parameters.

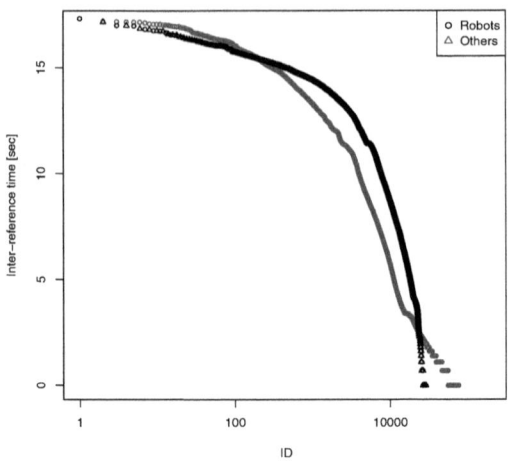

Fig. 4. Inter-reference times for the clients identified as crawlers and for the remaining clients of the academic server

Table 3. Main characteristics of the navigation profiles of the SPEC clients

	Crawlers	Others
Number of sessions per client	462.44	2.62
Number of transactions per session	29.43	13.68
Session duration [sec]	467.09	22.40
Inter-session time [sec]	18,697.00	468,279.00
Number of one-transaction sessions	869,678.00	14,953.00
Number of one-session clients	485.00	8,752.00

Crawlers sessions were bigger in terms of average duration and average number of transactions. Nevertheless, these sessions were characterized by a large variability across clients. The standard deviations of these parameters were an order of magnitude larger than the corresponding averages. From these results, it appears that crawlers that identify themselves tend to behave and do not send their requests in bursts to reduce their impact on the server resources.

In terms of re-visit patterns, crawlers re-visit the site very often: on average every 31 minutes and with requests distributed across many sessions. This is mainly related to the use of some sort of distributed crawling policies to speedup the process. In details, the majority of the crawlers (i.e., 88%) re-visited the site at least three times, whereas this was the case of very few of the other clients. It is also interesting to outline that after a session with a large number of transactions, crawlers were likely to re-visit the site very soon, that is, their inter-session times were small. For example, clients identified as Google crawlers sent as many as 6,470 requests each, distributed across as many as 870 sessions and spanning over a time period of more than four months. On the other hand, we have discovered that the number of sessions characterized by one transaction was not negligible for both crawlers and the other clients, namely, 69% and 44% of the total number of sessions, respectively, and about one third of the clients characterized by one session were the so-called "one-timers".

Another metric used to describe the navigation profiles was related to distribution of the requests across months, days of the month and hours of the day. In particular, for each client we counted the number of requests issued in each of these time periods. We then obtained a tuple of N parameters, N being equal to 69 for the academic server and 67 for the SPEC server, because its logs contained the measurements of 12 months instead of 14.

These parameters allows us to identify clients with similar patterns and discover periodic patterns, i.e., clients visiting the site regularly, for example, in the first day of the month at noon or in last day of May and August.

To make this large number of parameters more manageable, we applied various multivariate statistical techniques in combination [9,10]. The Principal Component Analysis was used to linearly transform these correlated parameters into a much smaller set of uncorrelated parameters, the principal components. The Correspondence Analysis was used to visually display the clients and the parameters used for their description. Finally, the application of hierarchical clustering techniques allowed us to discover groups of clients with homogeneous behavior.

The rest of this section is dedicated to present the classification of the hourly patterns of the crawlers and of the other clients of the academic server, each described in terms of number of requests issued in the various hours of the day, i.e., 24 parameters. Moreover, to take into account the distribution of the requests of each client across months and days of the month, we used two additional parameters, that is, the total number of months and the total number of days during which the client sent at least one request. Note that for this analysis we used the FactoMineR package [12].

The application of the Principal Component Analysis to both sets of clients described by these 26 parameters has shown that few principal components could summarize very well the variability in the original data. More specifically, the first two principal components computed for the other clients accounted for 55% of their variance, whereas in the case of crawlers the principal components could capture their behavior even better. The first two principal components accounted for approximately the 70% of their variance and four principal components covered 80% of the variance. The weights associated with the first principal component are about equal. This means that each of the parameters is equally represented in the linear composition, i.e., this component represents crawlers that do not differentiate their traffic among the various hours of the day. On the contrary, the second principal component represents the contrast between day traffic and night traffic.

For the other clients, the first principal component mainly describes the traffic sent during business hours, i.e., between 8am and 5pm, whereas the second

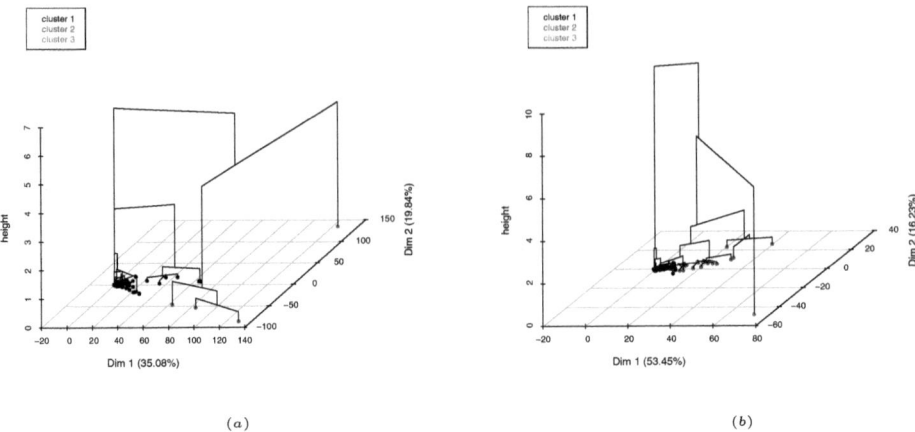

(a)   (b)

Fig. 5. Clusters obtained for the other clients (a) and for the crawlers (b) of the academic server

component represents the difference between day and night traffic and traffic sent across few months and days.

We then applied hierarchical clustering techniques to the data of both sets of clients represented in the principal components space. The partitions in three clusters obtained for both sets are shown in Fig. 5. Each plot represents the projection of the data in the space of the first two principal components. Thus, these diagrams highlight the structure of the data and their similarities and differences. Clients close to each other in this space were similar to each other in their original data.

From the figure, we can notice that the behavior of the other clients is quite homogeneous: two of the three clusters consist of one and three clients, respectively, while the remaining clients belong to one big tight group. It is worth noting that the client belonging to cluster 3 is characterized by requests concentrated in one month and in two days. A further analysis of this client has shown that it was probably a crawler that did not identify itself as being such: all its HTTP requests used the HEAD method and specified the root of the site (i.e., /) as resource. On the contrary, the three clients of cluster 2 are actually human users, whose behavior is fully explained by the first principal component: all their requests were issued during the working hours only.

For the crawlers, clustering identified one very persistent crawler (belonging to cluster 3) and two other groups of crawlers with requests distributed over the 24 hours, sometimes in the day and other times at night. It is interesting to outline that many of the Google crawlers were grouped in cluster 2.

All these conclusions are also supported by the diagrams of Fig. 6 obtained by applying the Correspondence Analysis. This geometric representations display the relationships between clients and parameters. and point out their mutual

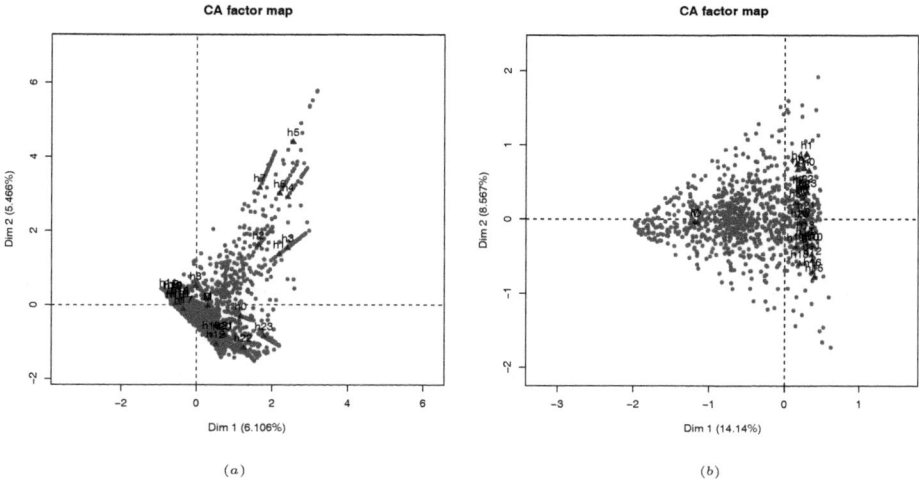

Fig. 6. Correspondence Analysis maps of the other clients (a) and of the crawlers (b) of the academic server

influence and ease to identify visually clustered observations. As can be seen, while the diagram obtained for the crawlers does not show any specific association between them and the parameters used for their description, there is a stronger association between some of the other clients and the parameters describing night traffic. Note that not to clutter the presentation, we did not plot the identifiers of the clients, represented by red dots.

5   Conclusions

The analysis of Web logs is very useful to discover interesting properties in the traffic of the servers and in the behavior of their clients. This analysis becomes rather challenging especially for very busy servers that receive many requests by many different clients. Our study focused on the characterization of the access patterns and navigation profiles of the clients of two Web servers. The traffic of one of the servers was heavily dominated by some very persistent crawlers. Nevertheless, on both servers we have noticed that, despite the intensity of their traffic, clients that identify themselves as being crawlers tend to behave and avoid sending bursts of requests. The application of various statistical techniques in combination has allowed us to highlight similarities and differences among the navigation profiles of the clients. The other clients, that is, human users and unidentified crawlers, are characterized by a rather homogeneous behavior. Nevertheless, while some clients return to the sites periodically to download the pages available on the servers, some others visit the site with the only intention of discovering vulnerabilities and compromising the servers.

  This work has a number of possible extensions and improvements. For example, it will be necessary to develop more reliable criteria to identify and classify clients that visit the sites for malicious purposes. Scrapers are an example of these clients in that they send legitimate requests with the objective of automatically creating copies of Web sites to be used for malicious purposes, such as, phishing. Moreover, starting from these results we plan to develop proactive policies aimed at improving the security of Web sites.

References

1. Almeida, V., Menascé, D., Riedi, R., Peligrinelli, F., Fonseca, R., Meira Jr., W.: Analyzing Web robots and their impact on caching. In: Proc. of the Sixth Web Caching and Content Delivery Workshop (2001)
2. Arlitt, M.F., Williamson, C.L.: Web server workload characterization: the search for invariants. In: Proc. of the ACM SIGMETRICS International Conference on Measurement and Modeling of Computer Systems, pp. 126–137 (1996)
3. Crovella, M., Bestavros, A.: Self-similarity in World Wide Web traffic: evidence and possible causes. IEEE/ACM Trans. on Networking 5(6), 835–846 (1997)
4. Dikaiakos, M.D., Stassopoulou, A., Papageorgiou, L.: An investigation of web crawler behavior: characterization and metrics. Computer Communications 28(8), 880–897 (2005)
5. Doran, D., Gokhale, S.: Discovering new trends in web robot traffic through functional classification. In: Proc. of the International Symposium on Network Computing and Applications, pp. 275–278. IEEE Computer Society (2008)
6. Duskin, O., Feitelson, D.G.: Distinguishing humans from robots in web search logs: preliminary results using query rates and intervals. In: Proc. of the Workshop on Web Search Click Data, pp. 15–19. ACM (2009)
7. Hallam-Baker, P.M., Behlendorf, B.: Extended Log File Format. W3C Working Draft WD-logfile-960323 (1996)
8. Iyengar, A.K., Squillante, M.S., Zhang, L.: Analysis and characterization of large-scale Web server access patterns and performance. World Wide Web 2(1-2), 85–100 (1999)
9. Johnson, R.A., Wichern, D.W.: Applied Multivariate Statistical Data Analysis, 6th edn. Pearson Prentice Hall (2007)
10. Jolliffe, I.T.: Principal Component Analysis, 2nd edn. Springer, Heidelberg (2002)
11. Koster, M.: A method for Web Robots control. Network Working Group - Internet Draft (1996)
12. Lê, S., Josse, J., Husson, F.: FactoMineR: An R Package for Multivariate Analysis.. Journal of Statistical Software 25(1), 1–18 (2008)
13. Lee, J., Cha, S., Lee, D., Lee, H.: Classification of web robots: An empirical study based on over one billion requests. Computers & Security 28(8), 795–802 (2009)
14. Mahanti, A., Williamson, C., Wu, L.: Workload characterization of a large systems conference Web server. In: Proc. of the Seventh Annual Communication Networks and Services Research Conference, pp. 55–64. IEEE Computer Society (2009)
15. Menascé, D.A., Almeida, V.A.F., Riedi, R., Ribeiro, F., Fonseca, R., Meira Jr., W.: A hierarchical and multiscale approach to analyze E-business workloads. Performance Evaluation 54(1), 33–57 (2003)
16. Menascé, D.A., Almeida, V.: Capacity Planning for Web Services: metrics, models, and methods. Prentice Hall (2001)
17. Olston, C., Najork, M.: Web Crawling. Journal of Foundations and Trends in Information Retrieval 4(3), 175–246 (2010)

18. Park, K., Pai, V.S., Lee, K.-W., Calo, S.: Securing web service by automatic robot detection. In: Proc. of USENIX 2006, pp. 23–23. USENIX Association (2006)
19. Performance Evaluation Group Web site – University of Pavia: http://peg.unipv.it
20. Pitkow, J.E.: Summary of WWW characterizations. World Wide Web 2(1-2), 3–13 (1999)
21. SPEC Web site – European mirror: http://spec.unipv.it
22. Stassopoulou, A., Dikaiakos, M.D.: Web robot detection: A probabilistic reasoning approach. Computer Networks 53(3), 265–278 (2009)
23. Tan, P.N., Kumar, V.: Discovery of Web robot sessions based on their navigational patterns. Data Mining and Knowledge Discovery 6(1), 9–35 (2002)
24. Thelwall, M., Stuart, D.: Web crawling ethics revisited: Cost, privacy, and denial of service. Journal of the American Society for Information Science and Technology 57(13), 1771–1779 (2006)
25. Williams, A., Arlitt, M., Williamson, C., Barker, K.: Web workload characterization: Ten years later. In: Tang, X., Xu, J., Chanson, S.T. (eds.) Web Content Delivery. Web Information Systems Engineering and Internet Technologies, vol. 2, pp. 3–21. Springer, US (2005)

A Matrix-Analytic Solution for Randomized Load Balancing Models with PH Service Times

Quan-Lin Li[1], John C.S. Lui[2], and Yang Wang[3,*]

[1] School of Economics and Management Sciences, Yanshan University
[2] Department of Computer Science & Engineering,
The Chinese University of Hong Kong
[3] Department of Computer Science and Technology, Peking University

Abstract. In this paper, we provide a matrix-analytic solution for randomized load balancing models (also known as *supermarket models*) with phase-type (PH) service times. Generalizing the service times to the phase-type distribution makes analysis of the supermarket models more difficult and challenging than that of the exponential service time case which has been extensively discussed in the literature. We describe the supermarket model as a system of differential vector equations, provide a doubly exponential solution to the fixed point of the system of differential vector equations, and analyze the exponential convergence of the current location of the supermarket model to its fixed point.

1 Introduction

In the past few years, a number of companies (e.g., Amazon, Google and Microsoft) are offering the *cloud computing* service to enterprises, and many content publishers and application service providers are increasingly using *Data Centers* to host their services. This emerging computing paradigm allows service providers and enterprises to concentrate on developing and providing the own services/goods without worrying about computing system maintenance or upgrade, and thereby significantly reduce their operating cost. For companies that offer cloud computing service in the data centers, they can take advantage of the variation of computing workloads from these customers and achieve the computational multiplexing gain. One of the important technical challenges that they have to address is how to utilize these computing resources in the data centers efficiently since many of these servers can be used. There is a growing interest to examine simple and robust load balancing strategies to efficiently utilize the computing resource of the server farms.

Randomized load balancing is a simple and effective mechanism to fairly utilize computing resources. It can deliver surprisingly good performance measures such as reducing collisions, waiting times and backlogs. In a supermarket model, each arriving job randomly picks a small subset of servers and examines their instantaneous workload, and the job is routed to the least loaded server. When

---

* Corresponding author.

K.A. Hummel et al. (Eds.): PERFORM 2010 (Haring Festschrift), LNCS 6821, pp. 240–253, 2011.

a job is committed to a server, jockeying is not allowed and each server uses the first-come-first-service (FCFS) discipline to process all jobs, e.g., see Mitzenmacher [7]. For the supermarket models, most of recent research applied density dependent jump Markov processes to deal with the simple case with Poisson arrival processes and exponential service times, and illustrated that there exists a fixed point which decreases doubly exponentially. Readers may refer to, such as, a simple supermarket model by [7,14]; simple variations by [8,9,11,4]; load information by [2,10]; fast Jackson network by Martin and Suhov [6,5,12]; and general service times by Bramson, Lu and Prabhakar [1]. When the arrival processes or the service times are more general, the available results of the supermarket models are few up to now. The purpose of this paper is to provide a novel approach for studying a supermarket model with PH service times, and show that the fixed point decreases doubly exponentially. Also, the PH approximation of order 2 gains new numerical insights on practical applications of the general service times to supermarket models.

The remainder of this paper is organized as follows. In the next section, we describe the supermarket model with the PH service times as a system of differential vector equations based on density dependent jump Markov processes. In Section 3, we set up a system of nonlinear equations satisfied by the fixed point, provide a doubly exponential solution to the system of nonlinear equations, and compute the expected sojourn time of a tagged arriving customer. In Section 4, we study exponential convergence of the current location of the supermarket model to its fixed point.

2   Supermarket Model

In this section, we describe a supermarket model with a Poisson arrival process and PH service times as a system of differential vector equations based on density dependent jump Markov processes.

Let us formally describe the supermarket model, which is abstracted as a multi-server multi-queue stochastic system. Customers arrive at a queueing system of $n > 1$ servers as a Poisson process with arrival rate $n\lambda$ for $\lambda > 0$. The service time of each customer is of phase type with irreducible representation (α, T) of order m. Each arriving customer chooses $d \geq 1$ servers independently and uniformly at random from the n servers, and waits for service at the server which currently contains the fewest number of customers. If there is a tie, servers with the fewest number of customers will be chosen randomly. All customers in every server will be served in the FCFS manner. Please see Figure 1 for an illustration of the supermarket model.

For the supermarket models, the PH distribution allows us to model more realistic systems and understand their performance implication under the randomized load balancing strategy. As indicated in [3], the process lifetime of many parallel jobs, in particular, jobs to the data centers, tends to be non-exponential. For the PH service time distribution, we use the following irreducible representation: (α, T) of order m, the row vector α is a probability vector whose jth entry is the probability that a service begins in phase j for $1 \leq j \leq m$; T is an

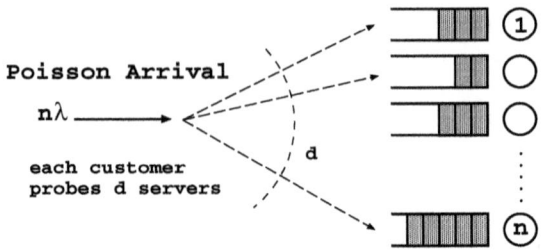

Fig. 1. The supermarket model: each customer can probe the loading of d servers

$m \times m$ matrix whose $(i,j)^{th}$ entry is denoted by $t_{i,j}$ with $t_{i,i} < 0$ for $1 \leq i \leq m$, and $t_{i,j} \geq 0$ for $1 \leq i, j \leq m$ and $i \neq j$. Let $T^0 = -Te \gneq 0$, where e is a column vector of ones with a suitable dimension in the context. The expected service time is given by $1/\mu = -\alpha T^{-1} e$. Unless we state otherwise, we assume that all the random variables defined above are independent, and that the system is operating in the stable region $\rho = \lambda/\mu < 1$.

We define $n_k^{(i)}(t)$ as the number of queues with at least k customers and the PH service time in phase i at time $t \geq 0$. Clearly, $0 \leq n_k^{(i)}(t) \leq n$ for $k \geq 1$ and $1 \leq i \leq m$. Let

$$X_n^{(0)}(t) = \frac{n}{n} = 1,$$

and $k \geq 1$

$$X_n^{(k,i)}(t) = \frac{n_k^{(i)}(t)}{n},$$

which is the fraction of queues with at least k customers and the service time in phase i at time $t \geq 0$. We write

$$X_n^{(k)}(t) = \left(X_n^{(k,1)}(t), X_n^{(k,2)}(t), \ldots, X_n^{(k,m)}(t) \right), \quad k \geq 1,$$

$$X_n(t) = \left(X_n^{(0)}(t), X_n^{(1)}(t), X_n^{(2)}(t), \ldots \right).$$

The state of the supermarket model may be described by the vector $X_n(t)$ for $t \geq 0$. Since the arrival process to the queueing system is Poisson and the service time of each server is of phase type, the stochastic process $\{X_n(t), t \geq 0\}$ is a Markov process whose state space is given by

$$\Omega_n = \{ \left(g_n^{(0)}, g_n^{(1)}, , g_n^{(2)} \ldots \right) : g_n^{(0)} = 1, g_n^{(k)} \geq g_n^{(k+1)} \geq 0,$$

$$\text{and } n g_n^{(k)} \text{ is a vector of nonnegative integers for } k \geq 1 \}.$$

Let

$$s_0(n,t) = E \left[X_n^{(0)}(t) \right]$$

and $k \geq 1$

$$s_k^{(i)}(n,t) = E \left[X_n^{(k,i)}(t) \right].$$

Clearly, $s_0(n, t) = 1$. We write

$$S_k(n, t) = \left(s_k^{(1)}(n, t), s_k^{(2)}(n, t), \ldots, s_k^{(m)}(n, t)\right), \quad k \geq 1.$$

As shown in Martin and Suhov [6] and Luczak and McDiarmid [4], the Markov process $\{X_n(t), t \geq 0\}$ is asymptotically deterministic as $n \to \infty$, this is due to the fact that from a block-structured point of view, the PH distribution can keep many excellent properties of the exponential distribution. Thus $\lim_{n \to \infty} E\left[X_n^{(0)}(t)\right]$ and $\lim_{n \to \infty} E\left[X_n^{(k,i)}\right]$ always exist by means of the law of large numbers. Based on this, we write

$$S_0(t) = \lim_{n \to \infty} s_0(n, t) = 1,$$

for $k \geq 1$

$$s_k^{(i)}(t) = \lim_{n \to \infty} s_k^{(i)}(n, t),$$

$$S_k(t) = \left(s_k^{(1)}(t), s_k^{(2)}(t), \ldots, s_k^{(m)}(t)\right)$$

and

$$S(t) = (S_0(t), S_1(t), S_2(t), \ldots).$$

Let $X(t) = \lim_{n \to \infty} X_n(t)$. Then it is easy to see from the Poisson arrivals and the PH service times that $\{X(t), t \geq 0\}$ is also a Markov process whose state space is given by

$$\Omega = \left\{\left(g^{(0)}, g^{(1)}, g^{(2)}, \ldots\right) : g^{(0)} = 1, g^{(k)} \geq g^{(k+1)} \geq 0 \text{ for } k \geq 1\right\}.$$

If the initial distribution of the Markov process $\{X_n(t), t \geq 0\}$ approaches the Dirac delta-measure concentrated at a point $g \in \Omega$, then its steady-state distribution is concentrated in the limit on the trajectory $S_g = \{S(t) : t \geq 0\}$. This indicates a law of large numbers for the time evolution of the fraction of queues of different lengths. Furthermore, the Markov process $\{X_n(t), t \geq 0\}$ converges weakly to the fraction vector $S(t) = (S_0(t), S_1(t), S_2(t), \ldots)$, or for a sufficiently small $\varepsilon > 0$,

$$\lim_{n \to \infty} P\{\|X_n(t) - S(t)\| \geq \varepsilon\} = 0,$$

where $\|a\|$ is the L_∞-norm of vector a.

In what follows we provide a system of differential vector equations in order to determine fraction vector $S(t)$. To that end, we introduce the *Hadamard Product* of two matrices $A = (a_{i,j})$ and $B = (b_{i,j})$ as follows:

$$A \odot B = (a_{i,j} b_{i,j}).$$

Specifically, for $k \geq 2$ we have

$$A^{\odot k} = \underbrace{A \odot A \odot \cdots \odot A}_{k \text{ matrix } A}.$$

To determine the fraction vector $S(t)$, we need to set up a system of differential vector equations satisfied by $S(t)$. To that end, we provide a heuristic description in terms of a concrete example, and indicate how to derive the differential vector equations.

In the supermarket model with n servers, we determine the expected change in the number of queues with at least k customers over a small time period of length dt. The probability vector that during the time period $[0,dt)$, any arriving customer joins a queue with $k-1$ customers is given by

$$n\left[\lambda S_{k-1}^{\odot d}(n,t) - \lambda S_k^{\odot d}(n,t)\right]dt.$$

Similarly, the probability vector that during the time period $[0,dt)$, a customer leaves a server queued by k customers is given by

$$n\left[S_k(n,t)T + S_{k+1}(n,t)T^0\alpha\right]dt.$$

Therefore we can obtain

$$\begin{aligned}dE\left[n_k(n,t)\right] =&\, n\left[\lambda S_{k-1}^{\odot d}(n,t) - \lambda S_k^{\odot d}(n,t)\right]dt\\&+ n\left[S_k(n,t)T + S_{k+1}(n,t)T^0\alpha\right]dt,\end{aligned}$$

which leads to

$$\frac{dS_k(n,t)}{dt} = \lambda S_{k-1}^{\odot d}(n,t) - \lambda S_k^{\odot d}(n,t) + S_k(n,t)T + S_{k+1}(n,t)T^0\alpha. \tag{1}$$

Taking $n \to \infty$ in both sides of Equation (1), we have

$$\frac{dS_k(t)}{dt} = \lambda S_{k-1}^{\odot d}(t) - \lambda S_k^{\odot d}(t) + S_k(t)T + S_{k+1}(t)T^0\alpha. \tag{2}$$

Using a similar analysis to Equation (2), we obtain a system of differential vector equations for the fraction vector $S(t) = (S_0(t), S_1(t), S_2(t), \ldots)$ as follows:

$$S_0(t) = 1,$$

$$\frac{d}{dt}S_0(t) = -\lambda S_0^d(t) + S_1(t)T^0, \tag{3}$$

$$\frac{d}{dt}S_1(t) = \lambda\alpha S_0^d(t) - \lambda S_1^{\odot d}(t) + S_1(t)T + S_2(t)T^0\alpha, \tag{4}$$

and for $k \geq 2$,

$$\frac{d}{dt}S_k(t) = \lambda S_{k-1}^{\odot d}(t) - \lambda S_k^{\odot d}(t) + S_k(t)T + S_{k+1}(t)T^0\alpha. \tag{5}$$

Note that the above derivations or the ordinary differential equations (ODE) descriptions are the approximation to the large scale stochastic system. One can be made the formal derivation using the *"mean field"* or the *"stochastic density jump Markov process"* techniques. While it is in form of ODE, we believe readers can easily understand the ODE of the above equations since the supplementary variable method are extensively used in the study of stochastic systems.

3   A Matrix-Analytic Solution

In this section, we provide a doubly exponential solution to the fixed point of the system of differential vector equations (3), (4) and (5).

A row vector $\pi = (\pi_0, \pi_1, \pi_2, \ldots)$ is called a fixed point of the fraction vector $S(t)$ if $\lim_{t \to +\infty} S(t) = \pi$. In this case, it is easy to see that

$$\lim_{t \to +\infty} \left[\frac{d}{dt} S(t) \right] = 0.$$

Therefore, as $t \to +\infty$ the system of differential vector equations (3), (4) and (5) can be simplified as

$$-\lambda \pi_0^d + \pi_1 T^0 = 0, \tag{6}$$

$$\lambda \alpha \pi_0^d - \lambda \pi_1^{\odot d} + \pi_1 T + \pi_2 T^0 \alpha = 0, \tag{7}$$

and for $k \geq 2$,

$$\lambda \pi_{k-1}^{\odot d} - \lambda \pi_k^{\odot d} + \pi_k T + \pi_{k+1} T^0 \alpha = 0. \tag{8}$$

In general, it is more difficult and challenging to express the fixed point of the supermarket models with more general arrival processes or service times. Fortunately, we can derive a closed-form expression for the fixed point $\pi = (\pi_0, \pi_1, \pi_2, \ldots)$ for the supermarket model with PH service times by means of a novel matrix-analytic approach.

Noting that $S_0(t) = 1$ for all $t \geq 0$, it is easy to see that $\pi_0 = 1$. It follows from Equation (6) that

$$\pi_1 T^0 = \lambda. \tag{9}$$

To solve Equation (9), we denote by ω the stationary probability vector of the irreducible Markov chain $T + T^0 \alpha$. Obviously, we have

$$\omega T^0 = \mu,$$

$$\frac{\lambda}{\mu} \omega T^0 = \lambda. \tag{10}$$

Thus, we obtain $\pi_1 = \frac{\lambda}{\mu} \omega = \rho \cdot \omega$. Based on the fact that $\pi_0 = 1$ and $\pi_1 = \rho \cdot \omega$, it follows from Equation (7) that

$$\lambda \alpha - \lambda \rho^d \cdot \omega^{\odot d} + \rho \cdot \omega T + \pi_2 T^0 \alpha = 0,$$

which leads to

$$\pi_2 T^0 = \lambda \rho^d \omega^{\odot d} e.$$

Let $\theta = \omega^{\odot d} e$. Then it is easy to see that $\theta \in (0, 1)$, and

$$\pi_2 T^0 = \lambda \theta \rho^d.$$

Using a similar analysis to Equation (10), we have

$$\pi_2 = \frac{\lambda\theta\rho^d}{\mu}\omega = \theta\rho^{d+1}\cdot\omega. \tag{11}$$

Based on $\pi_1 = \rho\cdot\omega$ and $\pi_2 = \theta\rho^{d+1}\cdot\omega$, it follows from Equation (8) that for $k = 2$,

$$\lambda\rho^d\cdot\omega^{\odot d} - \lambda\theta^d\rho^{d^2+d}\cdot\omega^{\odot d} + \theta\rho^{d+1}\cdot\omega T + \pi_3 T^0\alpha = 0,$$

which leads to

$$\pi_3 T^0 = \lambda\theta^{d+1}\rho^{d^2+d}.$$

Using a similar analysis on Equation (10), we have

$$\pi_3 = \frac{\lambda\theta^{d+1}\rho^{d^2+d}}{\mu}\omega = \theta^{d+1}\rho^{d^2+d+1}\cdot\omega. \tag{12}$$

Based on Equations (11) and (12), we may infer that there is a structured expression $\pi_k = \theta^{d^{k-2}+d^{k-3}+\cdots+d+1}\rho^{d^{k-1}+d^{k-2}+\cdots+d+1}\cdot\omega$ for $k \geq 1$. To that end, the following theorem states this important result.

Theorem 1. *The fixed point* $\pi = (\pi_0, \pi_1, \pi_2, \ldots)$ *is given by*

$$\pi_0 = 1, \qquad \pi_1 = \rho\cdot\omega$$

and for $k \geq 2$,

$$\pi_k = \theta^{d^{k-2}+d^{k-3}+\cdots+1}\rho^{d^{k-1}+d^{k-2}+\cdots+1}\cdot\omega, \tag{13}$$

or

$$\pi_k = \theta^{\frac{d^{k-1}-1}{d-1}}\rho^{\frac{d^k-1}{d-1}}\cdot\omega = \rho^{d^{k-1}}(\theta\rho)^{\frac{d^{k-1}-1}{d-1}}\cdot\omega.$$

Proof: By induction, one can easily derive the above result.

It is clear that Equation (13) is correct for the cases with $l = 2, 3$ according to Equations (11) and (12). Now, we assume that Equation (13) is correct for the cases with $l = k$. Then it follows from Equation (8) that for $l = k+1$, we have

$$\lambda\theta^{d^{k-2}+d^{k-3}+\cdots+d}\rho^{d^{k-1}+d^{k-2}+\cdots+d}\cdot\omega^{\odot d} - \lambda\theta^{d^{k-1}+d^{k-2}+\cdots+d}\rho^{d^k+d^{k-1}+\cdots+d}\cdot\omega^{\odot d}$$

$$+ \theta^{d^{k-2}+d^{k-3}+\cdots+1}\rho^{d^{k-1}+d^{k-2}+\cdots+1}\cdot\omega T + \pi_{k+1}T^0\alpha = 0,$$

which leads to

$$\pi_{k+1}T^0 = \lambda\theta^{d^{k-1}+d^{k-2}+\cdots+d+1}\rho^{d^k+d^{k-1}+\cdots+d}.$$

By a similar analysis to (10), we have

$$\pi_{k+1} = \frac{\lambda\theta^{d^{k-1}+d^{k-2}+\cdots+d+1}\rho^{d^k+d^{k-1}+\cdots+d}}{\mu}\omega$$

$$= \theta^{d^{k-1}+d^{k-2}+\cdots+d+1}\rho^{d^k+d^{k-1}+\cdots+d+1}\cdot\omega.$$

This completes the proof. ∎

Now, we compute the expected sojourn time T_d that a tagged arriving customer spends in the supermarket model. For the PH service times, a tagged arriving customer is the kth customer in the corresponding queue with probability vector $\pi_{k-1}^{\odot d} - \pi_k^{\odot d}$. When $k \geq 1$, the head customer in the queue has been served, and so its service time is residual and is denoted as X_R. Let X be of phase type with irreducible representation (α, T). Then X_R is of phase type with irreducible representation (ω, T). Clearly, we have

$$E[X] = \alpha (-T)^{-1} e, \quad E[X_R] = \omega (-T)^{-1} e.$$

Thus it is easy to see that the expected sojourn time of the tagged arriving customer is given by

$$E[T_d] = \left(\pi_0^d - \pi_1^{\odot d} e\right) E[X] + \sum_{k=1}^{\infty} \left(\pi_k^{\odot d} - \pi_{k+1}^{\odot d}\right) e \left\{E[X_R] + kE[X]\right\}$$

$$= \rho^d \theta (\omega - \alpha) (-T)^{-1} e + \alpha (-T)^{-1} e \left(1 + \sum_{k=1}^{\infty} \theta^{\frac{d^k-1}{d-1}} \rho^{\frac{d^{k+1}-d}{d-1}}\right). \quad (14)$$

When the arrival process and the service time distribution are Poisson and exponential, respectively, it is clear that $\alpha = \omega = \theta = 1$ and $\alpha (-T)^{-1} e = 1/\mu$, thus we have

$$E[T_d] = \frac{1}{\mu} \sum_{k=0}^{\infty} \rho^{\frac{d^{k+1}-d}{d-1}},$$

which is the same as Corollary 3.8 in Mitzenmacher [7].

In what follows we consider an interesting problem: When using the PH approximation, how many moments of the service time distribution are needed to obtain a better accuracy for computing the fixed point or the expected sojourn time. It is well-known from the theory of probability distributions that the first three moments are basic for analyzing such an accuracy. Also, we can construct a PH distribution of order 2 by using the first three moments. Telek and Heindl [13] provided a fitting procedure for matching a PH distribution of order 2 in terms of the first three moments exactly. It is necessary to list the fitting procedure as follows:

For a nonnegative random variable X, let $m_n = E[X^n]$, $n \geq 1$. We take a PH distribution of order 2 with the canonical representation (α, T), where $\alpha = (\eta, 1 - \eta)$ and

$$T = \begin{pmatrix} -\xi_1 & \xi_1 \\ 0 & -\xi_2 \end{pmatrix},$$

$0 \leq \eta \leq 1$ and $0 < \xi_1 \leq \xi_2$. Note that the three unknown parameters η, ξ_1 and ξ_2 can be obtained from the first three moments m_1, m_2 and m_3.

In Table 1, $c_X^2 = m_2/m_1^2 - 1$ which is the squared coefficient of variation. If the moments do not satisfy these conditions in Table 1, then we may analyze the following four cases:

Table 1. Specific Bounds of the First Three Moments

Moment	Condition	Bounds
m_1		$0 < m_1 < \infty$
m_2		$1.5 m_1^2 \leq m_2$
m_3	$0.5 \leq c_X^2 \leq 1$	$3 m_1^3 \left(3 c_X^2 - 1 + \sqrt{2} \left(1 - c_X^2 \right)^{\frac{3}{2}} \right) \leq m_3 \leq 6 m_1^3 c_X^2$
	$1 < c_X^2$	$\frac{3}{2} m_1^3 \left(1 + c_X^2 \right)^2 < m_3 < \infty$

(a.1) if $m_2 < 1.5 m_1^2$, then we take $m_2 = 1.5 m_1^2$;

(a.2) if $0.5 \leq c_X^2 \leq 1$ and $m_3 < 3 m_1^3 \left(3 c_X^2 - 1 + \sqrt{2} \left(1 - c_X^2 \right)^{\frac{3}{2}} \right)$, then we take

$m_3 = 3 m_1^3 \left(3 c_X^2 - 1 + \sqrt{2} \left(1 - c_X^2 \right)^{\frac{3}{2}} \right)$;

(a.3) if $0.5 \leq c_X^2 \leq 1$ and $m_3 > 6 m_1^3 c_X^2$, then we take $m_3 = 6 m_1^3 c_X^2$; and

(a.4) if $1 < c_X^2$ and $m_3 \leq \frac{3}{2} m_1^3 \left(1 + c_X^2 \right)^2$, then we take $m_3 = \frac{3}{2} m_1^3 \left(1 + c_X^2 \right)^2$.

Let $c = 3 m_2^2 - 2 m_1 m_3$, $d = 2 m_1^2 - m_2$, $b = 3 m_1 m_2 - m_3$ and $a = b^2 - 6cd$.
If the first three moments satisfy their specific bounds shown in Table 1 or the
exceptive four cases, then the three unknown parameters η, ξ_1 and ξ_2 can be
computed in the following three cases.

(1) If $c > 0$, then

$$\eta = \frac{-b + 6 m_1 d + \sqrt{a}}{b + \sqrt{a}}, \quad \xi_1 = \frac{b - \sqrt{a}}{c}, \quad \xi_2 = \frac{b + \sqrt{a}}{c}.$$

(2) If $c < 0$, then

$$\eta = \frac{b - 6 m_1 d + \sqrt{a}}{-b + \sqrt{a}}, \quad \xi_1 = \frac{b + \sqrt{a}}{c}, \quad \xi_2 = \frac{b - \sqrt{a}}{c}.$$

(3) If $c = 0$, then

$$\eta = 0, \ \xi_1 > 0, \ \xi_2 = \frac{1}{m_1}.$$

From the above discussion, we can construct a PH distribution of order 2 to
approximate an arbitrarily given general distribution under the same first three
moments. In fact, such an approximation achieves a better accuracy in computations of the fixed point and the expected sojourn time.

For the PH distribution of order 2, we have

$$T + T^0 \alpha = \begin{pmatrix} -\xi_1 & \xi_1 \\ 0 & -\xi_2 \end{pmatrix} + \begin{pmatrix} 0 \\ \xi_2 \end{pmatrix} \begin{pmatrix} \eta & 1 - \eta \end{pmatrix} = \begin{pmatrix} -\xi_1 & \xi_1 \\ \xi_2 \eta & -\xi_2 \eta \end{pmatrix},$$

which leads to

$$\omega = \left(\frac{\xi_2 \eta}{\xi_1 + \xi_2 \eta}, \frac{\xi_1}{\xi_1 + \xi_2 \eta} \right)$$

and

$$\theta = \frac{\xi_1^d + \xi_2^d \eta^d}{\left(\xi_1 + \xi_2 \eta \right)^d}.$$

Note that the PH distributions are dense in the set of all nonnegative random variables, we can numerically provide necessary understanding for the role played by the general service times in performance analysis of the supermarket model by means of the PH approximation of order 2. At the same time, all computations involved in the PH approximation of order 2 are very simple to implement.

In the remainder of this section, we provide two examples to illustrate that our approach is effective and efficient in analyzing the doubly exponential solution of supermarket models with non-exponential service requirements.

Example one (PH Distribution) Let $\lambda = 1, d = 5, m = 3, \alpha(1) = (1/3, 1/3, 1/3)$ and $\alpha(2) = (1/12, 7/12, 1/3)$,

$$T = \begin{pmatrix} -10 & 2 & 4 \\ 3 & -7 & 4 \\ 0 & 2 & -5 \end{pmatrix}.$$

Table 2 shows how the doubly exponential solution (π_1 to π_4) depends on the vectors $\alpha(1)$ and $\alpha(2)$, respectively.

Table 2. The doubly exponential solution depends on the vector α

	$\alpha = (\frac{1}{3}, \frac{1}{3}, \frac{1}{3})$	$\alpha = (\frac{1}{12}, \frac{7}{12}, \frac{1}{3})$
π_1	(0.0741, 0.1358 , 0.2346)	(0.0602, 0.1728, 0.2531)
π_2	(5.619e-05, 1.030e-05, 1.779e-04)	(7.182e-05, 2.063e-04, 3.020e-04)
π_3	(1.411e-20, 2.587e-20, 4.469e-20)	(1.739e-19, 4.993e-19, 7.311e-19)
π_4	(1.410e-98, 2.586e-98, 4.466e-98)	(1.444e-92, 4.148e-92, 6.074e-92)

Example two (Expected Sojourn Time) We consider an m-order PH distribution with irreducible representation (α, T), where $m = 2$, $\alpha = (1/2, 1/2)$ and

$$T = \begin{pmatrix} -4 & 3 \\ 2 & -7 \end{pmatrix}.$$

Let $\lambda = 1$, $\mu = 2.7500$ and $d = 2$. Figure 2 provides a comparison for the expected sojourn times between the exponential service time and the PH service time when they have a same expected service time. This figure shows that the PH service time makes the lower expected sojourn time.

4   Exponential Convergence to the Fixed Point

In this section, we study exponential convergence of the current location $S(t)$ of the supermarket model to its fixed point π.

We provide some notation for comparison of two vectors. Let $a = (a_1, a_2, a_3, \ldots)$ and $b = (b_1, b_2, b_3, \ldots)$. We write $a \prec b$ if $a_k < b_k$ for some $k \geq 1$ and $a_l \leq b_l$ for $l \neq k, l \geq 1$; and $a \preceq b$ if $a_k \leq b_k$ for all $k \geq 1$.

The following proposition analyzes how the current location $S(t)$ of the supermarket model can be affected by the initial point $S(0)$ for $t > 0$.

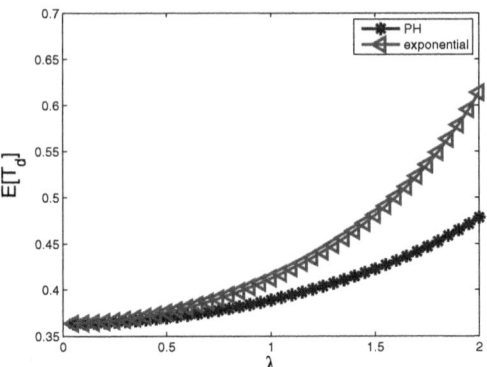

Fig. 2. The expected sojourn times correspond to the exponential and PH distributions

Proposition 1. *If $S(0) \preceq \widetilde{S}(0)$, then $S(t) \preceq \widetilde{S}(t)$.*

It follows from the system of differential vector equations (3), (4) and (5) that for $d \geq 2$

$$\frac{\mathrm{d}}{\mathrm{d}t} S(t) = S^{\odot d}(t) Q_1 + S(t) Q_2, \tag{15}$$

where

$$Q_1 = \begin{pmatrix} -\lambda I & \lambda I & & \\ & -\lambda I & \lambda I & \\ & & -\lambda I & \lambda I \\ & & & \ddots & \ddots \end{pmatrix}, \quad Q_2 = \begin{pmatrix} 0 & & & \\ T^0\alpha & T & & \\ & T^0\alpha & T & \\ & & \ddots & \ddots \end{pmatrix}.$$

It follows from (15) that

$$\frac{\mathrm{d}}{\mathrm{d}t} S^{\odot(1-d)}(t) = eQ_1 + S^{\odot(1-d)}(t) Q_2.$$

Let $W(t) = S^{\odot(1-d)}(t)$. Then

$$\frac{\mathrm{d}}{\mathrm{d}t} W(t) = eQ_1 + W(t) Q_2.$$

Hence we obtain

$$W(t) = \left[W(0) + e \int_0^t \exp\{Q_2 x\} \, \mathrm{d}x \right] \exp\{Q_1 t\}.$$

Since $W(t) = S^{\odot(1-d)}(t)$, we have

$$W(0) = S^{\odot(1-d)}(0)$$

and

$$S(t) = \left[S^{\odot(1-d)}(t) \right]^{\odot \frac{1}{1-d}} = [W(t)]^{\odot \frac{1}{1-d}}.$$

Therefore, the solution to the system of differential vector equations (3), (4) and (5) is given by

$$S(t) = \left\{ \left[S^{\odot(1-d)}(0) + e \int_0^t \exp\{Q_2 x\}\, dx \right] \exp\{Q_1 t\} \right\}^{\odot \frac{1}{1-d}}. \tag{16}$$

Note that $\int_0^t \exp\{Q_2 x\}dx > 0$ and $\exp\{Q_1 t\} > 0$, it is easy to see from (16) that if $S(0) \preceq \widetilde{S}(0)$, then $S(t) \preceq \widetilde{S}(t)$ for $t > 0$. This completes the proof. ∎

Based on Proposition 1, the following theorem shows that the fixed point π is an upper bound of the current location $S(t)$ for all $t \geq 0$.

Theorem 2. *For the supermarket model, if there exists some k such that $S_k(0) = 0$, then the sequence $\{S_k(t)\}$ has an upper bound sequence which decreases doubly exponentially for all $t \geq 0$, that is, $S(t) \preceq \pi$ for all $t \geq 0$.*

Proof: Let $\widetilde{S}_k(0) = \pi_k$ for $k \geq 1$. Then for each $k \geq 1$, $\widetilde{S}_k(t) = \widetilde{S}_k(0) = \pi_k$ for all $t \geq 0$, since $\widetilde{S}(0) = \left(\widetilde{S}_1(0), \widetilde{S}_2(0), \widetilde{S}_2(0), \ldots \right)$ is a fixed point in the supermarket model. If $S_k(0) = 0$ for some k, then $S_k(0) \prec \widetilde{S}_k(0)$ and $S_j(0) \preceq \widetilde{S}_j(0)$ for $j \neq k, j \geq 1$, thus $S(0) \prec \widetilde{S}(0)$. It is easy to see from Proposition 1 that $S_k(t) \preceq \widetilde{S}_k(t) = \pi_k$ for all $k \geq 1$ and $t \geq 0$. Thus we obtain that for all $k \geq 1$ and $t \geq 0$

$$S_k(t) \leq \theta^{\frac{d^{k-1}-1}{d-1}} \rho^{\frac{d^k-1}{d-1}} \cdot \omega.$$

This completes the proof. ∎

To show the exponential convergence, we define a Lyapunov function $\Phi(t)$ as

$$\Phi(t) = \sum_{k=1}^{\infty} w_k [\pi_k - S_k(t)] e$$

in terms of the fact that $S_k(t) \preceq \pi_k$ for $k \geq 1$ and $\pi_0 = S_0(t) = 1$, where $\{w_k\}$ is a positive scalar sequence with $w_{k+1} \geq w_k \geq w_1 = 1$ for $k \geq 2$.

The following theorem illustrates that the distance between the fixed point and the current location quickly comes close to zero with exponential convergence.

Theorem 3. *For $t \geq 0$, $\Phi(t) \leq c_0 e^{-\delta t}$, where c_0 and δ are two positive constants. In this case, the potential function $\Phi(t)$ is exponentially convergent.*

Proof: Note that

$$\Phi(t) = \sum_{k=1}^{\infty} w_k [\pi_k - S_k(t)] e,$$

we have

$$\frac{d}{dt} \Phi(t) = -\sum_{k=1}^{\infty} w_k \frac{d}{dt} S_k(t) e.$$

It follows from Equations (3) to (5) that

$$\frac{d}{dt}\Phi(t) = -w_1[\lambda S_0^d(t)\alpha - \lambda S_1^{\odot d}(t) + S_1(t)T + S_2(t)T^0\alpha]e$$
$$-\sum_{k=1}^{\infty} w_k[\lambda S_{k-1}^{\odot d}(t) - \lambda S_k^{\odot d}(t) + S_k(t)T + S_{k+1}(t)T^0\alpha]e.$$

By means of $S_0(t) = 1$ and $Te = -T^0$, we can obtain

$$\frac{d}{dt}\Phi(t) = -w_1[\lambda - \lambda S_1^{\odot d}(t)e - S_1(t)T^0 + S_2(t)T^0]$$
$$-\sum_{k=2}^{\infty} w_k[\lambda S_{k-1}^{\odot d}(t)e - \lambda S_k^{\odot d}(t)e - S_k(t)T^0 + S_{k+1}(t)T^0]. \quad (17)$$

We take some nonnegative constants $c_k(t)$ and $d_k(t)$ for $k \geq 1$ such that

$$\lambda = f_1(t)S_1(t)T^0,$$

for $k \geq 1$

$$\lambda S_k^{\odot d}(t)e = c_k(t)[\pi_k - S_k(t)]e$$

and

$$S_k(t)T^0 = d_k(t)[\pi_k - S_k(t)]e.$$

Then it follows from (17) that

$$\frac{d}{dt}\Phi(t) = -\{[(w_2 - w_1)]c_1(t) + w_1[f_1(t) - 1]d_1(t)\} \cdot [\pi_1 - S_1(t)]e$$
$$-\sum_{k=2}^{\infty}[(w_{k+1} - w_k)c_k(t) + (w_{k-1} - w_k)d_k(t)] \cdot [\pi_k - S_k(t)]e.$$

For a constant $\delta > 0$, we take

$$w_1 = 1,$$
$$[(w_2 - w_1)]c_1(t) + w_1[f_1(t) - 1]d_1(t) \geq \delta w_1$$

and

$$(w_{k+1} - w_k)c_k(t) + (w_{k-1} - w_k)d_k(t) \geq \delta w_k.$$

In this case, it is easy to see that

$$w_2 \geq 1 + \frac{\delta + 1 - f_1(t)}{c_1(t)}$$

and for $k \geq 2$

$$w_{k+1} \geq w_k + \frac{\delta w_k}{c_k(t)} + \frac{d_k(t)}{c_k(t)}(w_k - w_{k-1}).$$

Thus we have

$$\frac{d}{dt}\Phi(t) \leq -\delta\sum_{k=0}^{\infty} w_k[\pi_k - S_k(t)]e = -\delta\Phi(t),$$

which can leads to

$$\Phi(t) \leq c_0 e^{-\delta t}.$$

This completes the proof.   ∎

Acknowledgements. The work of Quan-Lin Li was supported by the National Science Foundation of China under grant No. 10871114, John C.S. Lui was supported by the RGC grant, and Yang Wang was supported by the National Science Foundation of China under grant No. 61001075.

References

1. Bramson, M., Lu, Y., Prabhakar, B.: Randomized load balancing with general service time distributions. In: Proceedings of the ACM SIGMETRICS International Conference on Measurement and Modeling of Computer Systems, pp. 275–286 (2010)
2. Dahlin, M.: Interpreting stale load information. IEEE Transactions on Parallel and Distributed Systems 11, 1033–1047 (1999)
3. Harchol-Balter, M., Downey, A.B.: Exploiting process lifetime distributions for dynamic load balancing. ACM Transactions on Computer Systems 15, 253–285 (1997)
4. Luczak, M., McDiarmid, C.: On the maximum queue length in the supermarket model. The Annals of Probability 34, 493–527 (2006)
5. Martin, J.B.: Point processes in fast Jackson networks. The Annals of Applied Probability 11, 650–663 (2001)
6. Martin, J.B., Suhov, Y.M.: Fast Jackson networks. The Annals of Applied Probability 9, 854–870 (1999)
7. Mitzenmacher, M.D.: The power of two choices in randomized load balancing. PhD thesis, University of California at Berkeley, Department of Computer Science, Berkeley, CA (1996)
8. Mitzenmacher, M.D.: Analyses of load stealing models using differential equations. In: Proceedings of the Tenth ACM Symposium on Parallel Algorithms and Architectures, pp. 212–221 (1998)
9. Mitzenmacher, M.D.: On the analysis of randomized load balancing schemes. Theory of Computing Systems 32, 361–386 (1999)
10. Mitzenmacher, M.D.: How useful is old information? IEEE Transactions on Parallel and Distributed Systems 11, 6–20 (2000)
11. Mitzenmacher, M.D., Richa, A., Sitaraman, R.: The power of two random choices: a survey of techniques and results. In: Pardalos, P., Rajasekaran, S., Rolim, J. (eds.) Handbook of Randomized Computing, vol. 1, pp. 255–312 (2001)
12. Suhov, Y.M., Vvedenskaya, N.D.: Fast Jackson Networks with Dynamic Routing. Problems of Information Transmission 38, 136–153 (2002)
13. Telek, M., Heindl, A.: Matching moments for acyclic discrete and continuous phase-type distributions of second order. International Journal of Simulation: Systems, Science & Technology 3, 47–57 (2002)
14. Vvedenskaya, N.D., Dobrushin, R.L., Karpelevich, F.I.: Queueing system with selection of the shortest of two queues: An asymptotic approach. Problems of Information Transmissions 32, 20–34 (1996)

Author Index